Topology and Maps

MATHEMATICAL CONCEPTS AND METHODS
IN SCIENCE AND ENGINEERING

Series Editor: **Angelo Miele**
Mechanical Engineering and Mathematical Sciences
Rice University, Houston, Texas

Volume 1   **INTRODUCTION TO VECTORS AND TENSORS**
Volume 1: Linear and Multilinear Algebra
Ray M. Bowen and C.-C. Wang

Volume 2   **INTRODUCTION TO VECTORS AND TENSORS**
Volume 2: Vector and Tensor Analysis
Ray M. Bowen and C.-C. Wang

Volume 3   **MULTICRITERIA DECISION MAKING**
AND DIFFERENTIAL GAMES
Edited by George Leitmann

Volume 4   **ANALYTICAL DYNAMICS OF DISCRETE SYSTEMS**
Reinhardt M. Rosenberg

Volume 5   **TOPOLOGY AND MAPS**
Taqdir Husain

Volume 6   **REAL AND FUNCTIONAL ANALYSIS**
A. Mukherjea and K. Pothoven

Volume 7   **PRINCIPLES OF OPTIMAL CONTROL THEORY**
R. V. Gamkrelidze

Volume 8   **INTRODUCTION TO THE LAPLACE TRANSFORM**
Peter K. F. Kuhfittig

Volume 9   **MATHEMATICAL LOGIC**
An Introduction to Model Theory
A. H. Lightstone

A Continuation Order Plan is available for this series. A continuation order will bring
delivery of each new volume immediately upon publication. Volumes are billed only upon
actual shipment. For further information please contact the publisher.

Topology and Maps

TAQDIR HUSAIN

McMaster University
Hamilton, Ontario, Canada

PLENUM PRESS · NEW YORK AND LONDON

Library of Congress Cataloging in Publication Data

Husain, Taqdir.
    Topology and maps.

    (Mathematical concepts and methods in science and engineering)
    Includes bibliographical references and index.
    1. Topology. 2. Topological spaces. 3. Mappings (Mathematics) I. Title.
QA611.H82   514'.3   77-10110

ISBN-13: 978-1-4615-8800-9   e-ISBN-13: 978-1-4615-8798-9
DOI: 10.1007/978-1-4615-8798-9

© 1977 Plenum Press, New York
Softcover reprint of the hardcover 1st edition 1977

A Division of Plenum Publishing Corporation
227 West 17th Street, New York, N.Y. 10011

TO
MY PARENTS

Preface

This work is suitable for undergraduate students as well as advanced students and research workers. It consists of ten chapters, the first six of which are meant for beginners and are therefore suitable for undergraduate students; Chapters VII–X are suitable for advanced students and research workers interested in functional analysis.

This book has two special features: First, it contains generalizations of continuous maps on topological spaces, e.g., almost continuous maps, nearly continuous maps, maps with closed graph, graphically continuous maps, w-continuous maps, and σ-continuous maps, etc. and some of their properties. The treatment of these notions appears here, in Chapter VII, for the first time in book form.

The second feature consists in some not-so-easily-available nuptial delights that grew out of the marriage of topology and functional analysis; they are topics mainly courted by functional analysts and seldom given in topology books. Specifically, one knows that the set $C(X)$ of all real- or complex-valued continuous functions on a completely regular space X forms a locally convex topological algebra, *a fortiori* a topological vector space, in the compact–open topology. A number of theorems are known: For example, $C(X)$ is a Banach space iff X is compact, or $C(X)$ is complete iff X is a k_r-space, and so on. Chapters VIII and X include this material, which, to the regret of many interested readers has not previously been available in book form (a recent publication (Weir [106]) does, however, contain some material of our Chapter X).

Since the treatment of the above-mentioned material requires a basic understanding of topological spaces, it is imperative to first deal with the basic facts in topology.

Chapter I is a conglomeration of basic notions and definitions from set theory, algebra, and category theory. Not all proofs of all theorems mentioned in this chapter have been provided, since they are available in standard books, e.g., [1], [11], and [38]. Moreover, it is not a comprehensive treatise and does not undertake to prove everything. This would have required another Bourbakian anthology, which was far from this author's intentions.

Chapter II deals with the fundamental notions of topology, e.g., elementary properties of open and closed sets, limit points, bases of topologies, and the various methods of defining topologies; for instance, metric, filter, and net are discussed.

Chapter III includes some material on separation axioms and characterization of spaces satisfying some separation axioms, especially Hausdorff, regular, completely regular, and normal spaces. A characterization of the latter spaces does show the existence of continuous functions which separate closed sets.

In Chapter IV, a few methods for constructing new topological spaces from the old are given. For example, subspaces, topological sum, topological product, quotient space, and inductive limits are introduced.

In Chapter V, a treatment of uniform spaces, in particular that of metric spaces, topological groups, and topological vector spaces, is given. Problems of completions, metrizability, and the existence of fixed points and some applications are given. Also, proximity spaces are defined.

In Chapter VI, we discuss some basic properties of compact, locally compact, k-, Baire, and pseudocompact (etc.) spaces, and the compactifications.

After discussing the properties of topological spaces, Chapter VII introduces maps. We define various kinds of generalizations of continuous maps, e.g., almost continuous, nearly continuous, graphically continuous, w-continuous, and approximately continuous maps. Some applications of almost continuity are given. For example, some closed graph and open mapping theorems on some topological vector spaces are proved.

Chapter VIII deals with function spaces—spaces of all maps, spaces of all continuous maps, etc.—with various topologies, e.g., point–open, compact–open, and $\mathfrak{S}$-topologies. Several criteria for equicontinuity of a family of maps are given. In particular, Ascoli's theorem is proved.

In Chapter IX, we discuss extensions of continuous maps from subspaces to the whole spaces, e.g., Tietze's theorem, the Hahn–Banach extension theorem, and a general theorem due to Bishop are proved.

The last chapter, Chapter X, includes the Stone–Weierstrass theorem and the characterizations of $C(X)$ in terms of X. For example, the criteria for

$C(X)$ to be a Banach space, Fréchet space, complete locally convex space, barreled, bornological, quasibarreled, and countably barreled space are given.

The last chapter is followed by a bibliography and index. The numbers in [...] refer to the bibliography. Each of Chapters II–X is followed by a section of Examples and Exercises. For brevity in the text, we sometimes refer to "Exercise x", meaning "Exercises and Examples x."

The author regrets that not all available relevant material on maps could be included in this book for fear of making this book encyclopedic in nature, which was not the intention. A subjective judgment had to be made as to what to include and what to omit.

Finally, the author gratefully acknowledges the assistance of S. M. Khaleelullah, D. Rosa, and I. Tweddle for reading the manuscript—especially S. M. Khaleelullah and I. Tweddle, who did a good job of going through the entire manuscript. Indeed, with great satisfaction, I thank the Canadian National Research Council for providing some financial support, and record my appreciation to several secretaries of the Department of Mathematics, McMaster University, for typing the manuscript, and to the editorial staff of Plenum Publishing for the careful production of this book.

Taqdir Husain

Contents

Notation . . . .   **xv**

Chapter I. Preliminaries **1**

  1. Fundamental Notions of Set Theory   1
  2. Relations and Mappings   4
  3. Partial and Linear Orderings; Cartesian Products   8
  4. Lattices   10
  5. Algebraic Structures   14
  6. Categories and Functors   17

Chapter II. Topological Spaces **21**

  7. Open and Closed Sets   21
  8. Topologies and Neighborhoods   25
  9. Limit Points   26
  10. Bases and Subbases   30
  11. First and Second Countable Spaces   32
  12. Metric Spaces   33
  13. Nets   36
  14. Filters   43
  15. Topologies Defined by Other Topologies   46
    Examples and Exercises   49

Chapter III. Continuity and Separation Axioms **53**

  16. Continuous and Open Mappings   53
  17. Topologies Defined by Mappings   60
  18. Separation Axioms   60
  19. Continuous Functions on Normal Spaces   68
    Examples and Exercises   74

Chapter IV.   Methods for Constructing New Topological Spaces from Old 77

20. Subspaces 77
21. Topological Sums 80
22. Topological Products 83
23. Quotient Topology and Quotient Spaces 90
24. Projective and Inductive Limits 95
     Examples and Exercises 99

Chapter V.   Uniform Spaces 103

25. Uniformities and Topologies 103
26. Uniformity and Separation Axioms 108
27. Uniformizable Spaces 111
28. Uniform Continuity and Uniform Spaces 113
29. Completeness in Uniform Spaces 117
30. Completeness, Compactness, and Completions 120
31. Topological Groups and Topological Vector Spaces . . . 126
32. Metrizability 131
33. Fixed Points 136
34. Proximity Spaces 141
     Examples and Exercises 144

Chapter VI.   Compact Spaces and Various Other Types of Spaces 149

35. Compact Spaces 149
36. Countable Compactness and Sequential Compactness . . . 157
37. Compactness in Metric Spaces 159
38. Locally Compact Spaces 163
39. MB-Spaces 168
40. k-Spaces and k_r-Spaces 171
41. Baire Spaces 175
42. Pseudocompact Spaces 177
43. Paracompact Spaces 180
44. Compactifications 186
     Examples and Exercises 190

Chapter VII.   Generalizations of Continuous Maps 195

45. Almost Continuous Maps 195
46. Closed Graphs 199
47. Almost Continuity and Closed Graphs 203
48. Graphically Continuous Maps 207
49. Nearly Continuous and w-Continuous Maps 210

50. Semicontinuous Maps 216
51. Approximately Continuous Functions 220
52. Applications of Almost Continuity 222
    Examples and Exercises 232

Chapter VIII. Function Spaces **237**

53. The Set of All Maps 237
54. Compact–Open Topology and the Topology of Joint Continuity . . 240
55. Subsets of F^E with Induced Topologies 242
56. The Uniformities on F^E 244
57. ℭ-Uniformities and ℭ-Topologies 248
58. Equicontinuous Maps 250
59. Equicontinuity and Metric Spaces 253
60. Sequential Convergence in Function Spaces 261
    Examples and Exercises 263

Chapter IX. Extensions of Mappings **267**

61. Extensions of Maps on Completely Regular and Metric Spaces . . . 267
62. The Hahn–Banach Extension Theorem 274
63. A General Extension Theorem 283
    Examples and Exercises 286

Chapter X. $C(X)$ Spaces **289**

64. Stone–Weierstrass Theorem 289
65. Embeddings of X into $C(X)$ 296
66. $C(X)$ Spaces for Compact Spaces X 301
67. Separability in $C(X)$ 305
68. $C(X)$ Spaces for Completely Regular Spaces X 308
69. Characterization of Banach and Fréchet Spaces $C(X)$ 310
70. Characterization of Locally Convex Spaces $C(X)$ 314
    Epilogue 322
    Examples and Exercises 322

BIBLIOGRAPHY 327
INDEX 333

Notation

Chapter I

$x \in A$ x is an element of the set A

$x \notin A$ x is not an element of the set A

$A \subset B$ A is a subset of B

$\{x\}$ singleton, or the subset consisting of x alone

$A \cap B = \{x \in A : x \in B\}$, the intersection of A and B

$A \cup B = \{x : x \in A \text{ or } x \in B\}$, the union of A and B

$\varnothing$ empty or void or null set

$A \backslash B = \{x \in A : x \notin B\}$, the difference of A and B

$A \bigtriangleup B = (A \backslash B) \cup (B \backslash A)$, the symmetric difference of A and B

$A^c = \{x \in X : x \notin A\} = X \backslash A = (\sim A)$, complement of A

$A \cap B = \varnothing$ A and B are disjoint sets

$\displaystyle\bigcup_{\alpha \in \Gamma} X_\alpha$ the union of a family of sets X_α ($\alpha \in \Gamma$)

$\displaystyle\bigcap_{\alpha \in \Gamma} X_\alpha$ the intersection of a family of sets X_α ($\alpha \in \Gamma$)

$\overline{\lim} \, A_n$ limit superior of sets $\{A_n\}_{n \geq 1}$

$\underline{\lim} \, A_n$ limit inferior of sets $\{A_n\}_{n \geq 1}$

$A_n \subset A_{n+1}$ ($n \geq 1$)   monotonic increasing sequence of sets $\{A_n\}$

$A_{n+1} \subset A_n$ ($n \geq 1$)   monotonic decreasing sequence of sets $\{A_n\}$

$A \times B = \{(x, y) : x \in A, y \in B\}$, product of A and B

$\bigtriangleup = \{(x, y) \in A \times B : x = y\}$, diagonal

$y \, R \, y$ means x is R-related to y

$f\colon X \to Y$ a mapping of X into Y, or f maps X into Y

$G_f = \{(x, y) \in X \times Y \colon y = f(x)\}$, the graph of $f\colon X \to Y$

$f \mid A$ restriction of $f\colon X \to Y$ on a subset A of X

$f(A) = \{f(x)\colon x \in A\}$

$f^{-1}(B) = \{x \in X \colon f(x) \in B\}$

$1\colon 1$ "one-to-one" or injective

$x \leq y$ reads "x is less than or equal to y"

$x \geq y$ reads "x is greater than or equal to y"

$p_\alpha\colon \prod\limits_{\alpha \in \Gamma} X_\alpha \to X_\alpha$ the αth projection, i.e., for each $x \in \prod\limits_{\alpha \in \Gamma} X_\alpha$, $p_\alpha(x)$
         $= x_\alpha \in X_\alpha$

$x \vee y$ join of x and y $\Big\}$ lattice operations

$x \wedge y$ meet of x and y

R real line or real numbers

C complex numbers

N the set of all positive integers

Q the set of all rational numbers

$f \circ g(x) = f(g(x))$; the composition of f with g

Chapter II

$(X, \mathcal{T}) = X_u$ a topological space with topology $\mathcal{T} = u = \{U\}$

$\mathrm{Cl}\, A = \mathrm{Cl}_{\mathcal{T}}\, A = \bar{A}$ the closure of a subset A in the topology $\mathcal{T}$

$\mathrm{Int}\, A = \mathrm{Int}_{\mathcal{T}}\, A = A^\circ$ the interior of a subset A in the topology

A' the set of all limit points of A (the derived set)

$\mathcal{U}_x = \mathcal{N}_x$ the system of neighborhoods at x

$\partial A = \bar{A} \cap \overline{(X \setminus A)}$, boundary of A

$\mathcal{P}(X)$ power set, i.e., the family of all subsets of X

$\{x_\alpha\colon \alpha \in \Gamma\}$ a net in a set X, if $x_\alpha \in X$

$\{x_n\}$ a sequence of elements x_n

$\mathcal{F} = \{U\}$ a filter

$\bigcap\limits_{\alpha} \mathcal{T}_\alpha = \wedge \mathcal{T}_\alpha = \{W\colon W \in \mathcal{T}_\alpha$ for each $\alpha\}$, intersection of topologies, i.e., the greatest lower bound of the topologies $\mathcal{T}_\alpha$

$\bigcup\limits_{\alpha} \mathcal{T}_\alpha = \{W\colon W \in \mathcal{T}_\alpha$ for some $\alpha\}$, base for the least upper bound of the topologies $\mathcal{T}_\alpha$

Chapter III

$$\mathscr{E} = \{U\} \quad \text{topology}$$

E_u topological space with topology u

A_i axioms $i = 0, 1, 2, 3, 4, 5$
T_i spaces $i = 0, 1, 2, 3, 4, 5$ see pages 61, 64

T_2 spaces $=$ Hausdorff spaces

$T_1 + T_3$ spaces   regular spaces

$T_1 + T_4$ spaces   completely regular spaces

$T_1 + T_5$ spaces   normal spaces

$$\| x \|_2 = \left(\sum_{i=1}^{\infty} | x_i |^2 \right)^{1/2}, \quad l_2\text{-norm of the sequence } x = \{x_n\}$$

$$\| x \|_p = \left(\sum_{i=1}^{\infty} | x_i |^p \right)^{1/p}, \quad l_p\text{-norm of the sequence } x = \{x_n\},$$
$$1 \le p < \infty$$

Ω first uncountable cardinal

$\aleph_0$ (aleph naught) first countable ordinal

Chapter IV

$$\sum_{\alpha} E_{\alpha} \quad \text{topological sum}$$

$d(x, y)$ metric

(E, d) metric space

$$\sum_{n=1}^{\infty} \frac{1}{2^n} \frac{d_n(x_n, y_n)}{1 + d_n(x_n, y_n)} \quad \text{Fréchet metric}$$

$E/\sim$ the set E with an equivalence relation $\sim$

$E^{\sim} = \{x^{\sim} : x^{\sim} \text{ equivalence class}\}$, quotient set

$(E_{\alpha}, \Gamma, f_{\alpha\beta})$ inverse system

$\varprojlim E_{\alpha}$ inverse limit

$(E_{\alpha}, \Gamma, f_{\beta\alpha})$ direct system

$\varinjlim E_{\alpha}$ direct or inductive limit

$[0, 1] = I$ closed unit interval

$(0, 1)$ open unit interval

Chapter V

$$\mathscr{U} = \{U\} \quad \text{uniformity}$$

$(E, \mathscr{U})\ (E, \{U\})$ uniform space

$U[A] = \{y\colon (x, y) \in U \text{ for some } x \in A\}$, where $U \subset E \times F$

$U[x] = U[\{x\}] = \{y\colon (x, y) \in U\}$

$U^{-1}[A] = \{y\colon (x, y) \in U^{-1} \text{ for some } x \in A\}$

$B_r(x) = \{y\colon d(x, y) < r\}$, open ball of radius r and center x

$\bar{B}_r(x) = \{y\colon d(x, y) \leq r\}$, closed ball of radius r and center x

$(\hat{E}, \hat{\mathscr{U}})$ completion of the uniform space $(E, \mathscr{U})$

$p(x)$ seminorm of x

$\| x \|$ norm of x

(E, δ) proximity space

Chapter VI

$k(u, \mathscr{C})$ k-extension of the topology u

k-space   see page 171

k_r-space   see page 175

$\tilde{E}$ compactification of E

βE the Stone–Čech compactification

Chapter VII

$G_{g \circ f}$ the graph of the composition map $g \circ f$

$m(A)$ Lebesgue measure of the subset A

$$\varrho_x(S) = \lim_{\delta \downarrow 0} \frac{m(S \cap I_\delta)}{m(I_\delta)}, \text{ metric density of } S$$

$C(E)$ the set of all continuous real- or complex-valued functions on the topological space E

Chapter VIII

$$T(K, U) = \{f\colon E \to F\colon f(x) \in U \text{ for all } x \in K\}$$

$$F^E = \{f\colon E \to F\} = \prod_{\alpha \in E} \{F_\alpha\colon F_\alpha = F\}$$

$\mathscr{C}_p$ the topology of pointwise convergence

$\mathscr{C}_c$ the compact–open topology

$[V] = \{(f, g): f, g \in F^E: (f(x), g(x)) \in V\}$ for a member V of the uniformity on F

$[V](f) = \{g \in F^E: (g, f) \in [V]\}$

$\mathfrak{S}$-topology   see page 248

$CB(E, F)$ the set of all continuous bounded maps from a topological space E to a metric space F

$d(f, g) = \sum\limits_{i=1}^{\infty} \dfrac{1}{2^i} \dfrac{d_i^+(f, g)}{1 + d_i^+(f, g)}$, Fréchet metric on $C(E, F)$

$\mathscr{C}_\beta$ strong topology on E'

$E' = E_u'$ the dual of a topological vector space E_u

$A^{\circ} = \{f \in E': |f(x)| \le 1 \text{ for all } x \in A\}$, the polar set of A

$A^{\circ\circ} = (A^{\circ})^{\circ}$, the bipolar set of A

Chapter IX

Com category of compact Hausdorff spaces

$\delta(A)$ diameter of A

σ_n simplex

LCV   category of locally convex Hausdorff spaces

$C(X)$ the set of all real- or complex-valued continuous functions on a topological space X

$M(X)$ the set of all Radon measures

$[C(X)]' = M(X)$, the dual of $C(X)$

$A^{\perp}$ the annihilator set of A

$Z(f) = \{x: f(x) = 0\}$, zero set of $f: E \to \mathbb{R}$ or $\mathbb{C}$

$l_p\ (1 \le p < \infty)$ the set of all real or complex sequences $x = \{x_n\}$ such that $\| x \|_p^p = \sum\limits_{n=1}^{\infty} |x_n|^p < \infty$

l_{∞} set of all bounded real or complex sequences

$I^n = \{(x_1, \dots, x_n): |x_i| < 1, i = 1, \dots, n\}$, n-dimensional unit hypercube

T^* transpose or adjoint of an operator T

Chapter X

$C(X, R) = C(X)$, the set of all real continuous functions on a topological space X

$C(X, C) = C(X)$, the set of all complex continuous functions on a topological space X

$\|f\|_\infty = \sup\limits_{x \in E} |f(x)|$ if $f: E \to R$ is bounded

$C_\infty(X)$ the set of all continuous functions on X which vanish at infinity

$BC(X)$ the set of all bounded continuous real- or complex-valued functions on X

$f \vee g(x) = \max\,(f(x), g(x))$, $x \in E$ $(f, g: E \to R)$

$f \wedge g(x) = \min\,(f(x), g(x))$, $x \in E$ $(f, g: E \to R)$

$C_R(X)$ the subset of all real parts of continuous complex-valued functions

$\mathfrak{I}_{x_0} = \{f \in C(X): f(x_0) = 0\}$

R^N the set of all real sequences or the countable product of the real line R by itself

Supp ℓ support of a functional ℓ on $C(X)$

Supp M support of a set M of functionals on $C(X)$

I

Preliminaries

In this chapter, we collect the basic notions from set theory, lattice theory, modern algebra, and category theory. The reader interested in detailed information about these notions may consult, e.g., Bourbaki [13], Birkhoff [9], and Freyd [38].

§1. Fundamental Notions of Set Theory

A collection of "distinguished" objects is called a *set*. The objects themselves are called *elements* or *members* of the set. The symbol "$a \in A$" means that a is an element of the set A; $a \notin A$ means that a is not an element of A. If A and B are two sets, then "$A \subset B$" means that each element of A is also in B and A is called a *subset* of B. Clearly, every set is a subset of itself. Two sets A and B are said to be *equal* (in symbols $A = B$) if and only if $A \subset B$ and $B \subset A$. If A and B are not equal, we write $A \neq B$. If $A \subset B$ and $A \neq B$, then A is said to be a proper *subset* of B. A set consisting of a single element x is called a *singleton* and is sometimes denoted by $\{x\}$.

A useful method of constructing a set is propositional, viz., if P is a proposition then

$$\{x: P \text{ for } x \text{ holds}\}$$

denotes the set consisting of all elements x for which proposition P holds or is true. Very often the phrase "P for x" is denoted by "$P(x)$."

If A and B are any two sets, the set $\{x: x \in A \text{ and } x \in B\}$ is called the *intersection* of A and B, and is denoted by $A \cap B$. The set $\{x: x \in A$

1

or $x \in B$} is called the *union* of A and B, and is denoted by $A \cup B$. It is quite clear from the definitions that $A \cap B$ is a subset of A and B, while each of A and B is a subset of $A \cup B$. The set defined by $\{x : x \neq x\}$ is called a *void, empty,* or *null* set and is denoted by $\emptyset$.

The following proposition gives some characterizations of subsets.

Proposition 1.   Let A and B be two sets. The following statements are equivalent:

(a) $A \subset B$;

(b) $A \cap B = A$;

(c) $A \cup B = B$.

Proof.   (a) implies (b) [in symbols (a) $\Rightarrow$ (b)]: We must show that $A \cap B \subset A$ and $A \subset A \cap B$. Since $A \cap B$ is a subset of A, $A \cap B \subset A$. But since $A \subset B$, each $a \in A$ is also in B and hence $A \subset A \cap B$. Thus (b) follows.

(b) $\Rightarrow$ (c):   By definition, $B \subset A \cup B$. To show $A \cup B \subset B$, each $x \in A \cup B$ implies $x \in A$ or $x \in B$. If $x \in A$, then (b) implies $x \in A \cap B$ and hence $x \in B$. Therefore $A \cup B \subset B$. Hence (c) follows.

(c) $\Rightarrow$ (a):   If (a) is not true, then there exists $x \in A$ and $x \notin B$. But this means that $x \in A \cup B$ and $x \notin B$. This shows that $A \cup B \neq B$, which violates (c).

If $A \cap B = \emptyset$ for any two sets A and B, then A and B are said to be *disjoint*. A is said to *intersect* B if $A \cap B \neq \emptyset$, i.e., there is at least one common element in both A and B.

If A and B are two sets, $\{x : x \in A$ and $x \notin B\}$ denotes the *difference* of A and B and is denoted by $A \backslash B$.

For any two sets A and B, $A \triangle B = (A \backslash B) \cup (B \backslash A)$ is called the *symmetric difference*.

If $A \subset B$, then $B \backslash A$ is called the *complement of A relative to B*. For any set A, $\{x : x \notin A\}$ is called the *absolute complement* or simply *complement* of A and is denoted by A^c or $\sim A$.

If A and B are subsets of a set X then each of the statements in Proposition 1 is equivalent to each of the following:

(a′) $X \backslash B \subset X \backslash A$;

(b′) $A \cap (X \backslash B) = \emptyset$;

(c′) $(X \backslash A) \cup B = X$.

If the elements of a set are sets or subsets of a given set, then we shall call it a *family* or *class* instead of a set in order to avoid the well-known Russell paradox in set theory (Abian [1]). A family $\mathscr{A}$ of sets is said to be a *disjoint family* if no two members of $\mathscr{A}$ intersect. Sometimes such a family is also called a *family of mutually pairwise disjoint sets*.

Let Γ denote an index set. For each $\alpha \in \Gamma$, let X_α be a set associated to α. Then $\bigcup_{\alpha \in \Gamma} X_\alpha$ denotes the union of X_α ($\alpha \in \Gamma$), i.e.,

$$\bigcup_{\alpha \in \Gamma} X_\alpha = \{x \colon x \in X_\alpha \text{ for some } \alpha \in \Gamma\}.$$

Moreover, $\bigcap_{\alpha \in \Gamma} X_\alpha$ denotes the intersection of X_α ($\alpha \in \Gamma$), i.e.,

$$\bigcap_{\alpha \in \Gamma} X_\alpha = \{x \colon x \in X_\alpha \text{ for all } \alpha \in \Gamma\}.$$

The basic results in this area are compiled below; an understanding of these will facilitate later work.

Proposition 2. Let A, B, C be sets and let D_α ($\alpha \in \Gamma$) be a family of subsets of a set D. Then:

(a) $B \setminus (B \setminus A) = A \cap B$;

(b) $A \triangle B = (A \cup B) \setminus (A \cap B)$ or $(A \triangle B) \cup (A \cap B) = A \cup B$;

(c) $A^{cc} = A$;

(d) $A \cup B = B \cup A$ and $A \cap B = B \cap A$ (commutative laws);

(e) $A \cup (B \cup C) = (A \cup B) \cup C$ and $A \cap (B \cap C) = (A \cap B) \cap C$ (associative laws);

(f) $A \cap (B \cup C) = (A \cap B) \cup (A \cap C)$ and $A \cup (B \cap C)$ $= (A \cup B) \cap (A \cup C)$ (distributive laws);

(g) $D \setminus \bigcap_{\alpha \in \Gamma} D_\alpha = \bigcup_{\alpha \in \Gamma} (D \setminus D_\alpha)$ and $D \setminus (\bigcup_{\alpha \in \Gamma} D_\alpha) = \bigcap_{\alpha \in \Gamma} (D \setminus D_\alpha)$ (De Morgan's laws).

Proof. (a) $x \in B \setminus (B \setminus A)$ if and only if $x \in B$ and $x \notin B \setminus A$. But $x \notin B \setminus A$ if and only if $x \notin B$ or $x \in A$. Since $x \in B$ and $x \notin B$ are contradictory, we conclude that $x \in B \setminus (B \setminus A)$ if and only if $x \in B$ and $x \in A$ or equivalently $x \in A \cap B$. This proves (a).

(b) Clearly, $x \in A \triangle B = (A \setminus B) \cup (B \setminus A)$ if and only if $x \in A \setminus B$ or $x \in B \setminus A$. But $x \in A \setminus B$ if and only if $x \in A$ and $x \notin B$, and $x \in B \setminus A$ if and only if $x \in B$ and $x \notin A$. Thus, combining the two cases we see that $x \in A \triangle B$ if and only if $x \in A \cup B$ and $x \notin A \cap B$. i.e., $x \in (A \cup B) \setminus (A \cap B)$.

Proofs of statements (c) to (e) are immediate.

(f)   Clearly $x \in A \cap (B \cup C)$ if and only if $x \in A$ and $x \in B \cup C$. But $x \in B \cup C$ if and only if $x \in B$ or $x \in C$. Therefore $x \in A \cap (B \cup C)$ if and only if "$x \in A$ and $x \in B$" or "$x \in A$ and $x \in C$." Since "$x \in A$ and $x \in B$" is equivalent to $x \in A \cap B$ and "$x \in A$ and $x \in C$" is equivalent to $x \in A \cap C$, we conclude that $x \in A \cap (B \cup C)$ if and only if "$x \in A \cap B$ or $x \in A \cap C$." But the latter is equivalent to $x \in (A \cap B) \cup (A \cap C)$. The other part follows similarly.

(g)   Clearly $x \in D \setminus \bigcap_{\alpha \in \Gamma} D_\alpha$ if and only if $x \in D$ and $x \notin D_\alpha$ for all $\alpha \in \Gamma$, which is equivalent to $x \in D \setminus D_\alpha$ for all $\alpha \in \Gamma$, which implies that $x \in \bigcap_{\alpha \in \Gamma}(D \setminus D_\alpha)$. The other part follows similarly.

Let $\{A_n\}$ be a sequence of sets. The set of all elements $x \in \bigcup_{n \geq 1}^{\infty} A_n$ which belong to a countably infinite number of sets A_n, i.e., $x \in A_{n_k}$ for all $k \geq 1$, where $\{n_k\}$ is a subsequence of positive integers, is called the *limit superior* of the sequence $\{A_n\}$ and is denoted by $\overline{\lim}\, A_n$. The set of all elements which belong to A_n for all $n \geq n_0$ for some n_0 is called the *limit inferior* of the sequence $\{A_n\}$ and is denoted by $\underline{\lim}\, A_n$.

It is clear that $\underline{\lim}\, A_n \subset \overline{\lim}\, A_n$. Whenever $\underline{\lim}\, A_n = \overline{\lim}\, A_n$, we say that *the limit set* of $\{A_n\}$ exists and we shall denote $\overline{\lim}\, A_n$ or $\underline{\lim}\, A_n$ by $\lim A_n$.

If the sequence $\{A_n\}$ is *monotonic*, i.e., $A_n \subset A_{n+1}$ for all $n \geq 1$, or $A_n \supset A_{n+1}$ for all $n \geq 1$ then $\lim_{n \to \infty} A_n$ exists (it may be equal to $\varnothing$).

The limits of any family $\{A_\alpha\}$, $\alpha \in \Gamma$, of sets will be defined later on.

Proposition 3.   Let $\{A_n\}$ be a sequence of sets. Then:

(a) $\overline{\lim}\, A_n = \bigcap_{m \geq 1} \left(\bigcup_{n \geq m} A_n \right)$;

(b) $\underline{\lim}\, A_n = \bigcup_{m \geq 1} \left(\bigcap_{n \geq m} A_n \right)$.

Proof.   Simple verification.

§2. Relations and Mappings

The notion of relation requires the concept of an "ordered pair." Each ordered pair has a first coordinate and a second coordinate. If x is the first coordinate and y the second coordinate, then the ordered pair is denoted by (x, y). Moreover, two ordered pairs (x_1, y_1) and (x_2, y_2) are equal, i.e., $(x_1, y_1) = (x_2, y_2)$, if and only if $x_1 = x_2$, $y_1 = y_2$.

A set of ordered pairs is called a *relation*. If R denotes a relation and $(x, y) \in R$, then sometimes we shall write $x\ R\ y$ (i.e., x is R-related to y) in place of $(x, y) \in R$.

A simple example of a relation is given by the following: Let A and B be two sets. The set of all ordered pairs (x, y) such that $x \in A$ and $y \in B$ is a relation. This relation is denoted by $A \times B$, which is known as the *Cartesian product* of A and B.

If R is a relation, the set $\{x: (x, y) \in R\}$ is called the *domain* of R and the set $\{y: (x, y) \in R\}$ is called the *range* of R.

The *diagonal* $\{(x, x): x \in A\}$ of a set $A \times A$ is called the *identity relation* on A.

If R is a relation, the set $\{(y, x): (x, y) \in R\}$ is called the *inverse* of R and is denoted by R^{-1}.

If R and S are relations, the set $\{(x, y):$ for some $z, (x, z) \in R$ and $(z, y) \in S\}$ is called the *composition* of R and S, and is denoted by $R \circ S$.

Let R be a relation such that $X = $ (domain R) $\cup$ (range R). R is said to be *reflexive* if $x\ R\ x$ for each $x \in X$. R is said to be *symmetric* if $y\ R\ x$ whenever $x\ R\ y$, or equivalently $R = R^{-1}$. R is said to be *transitive* if whenever $x\ R\ y$ and $y\ R\ z$ then $x\ R\ z$. A reflexive, symmetric, and transitive relation is called an *equivalence relation*. A subset A of X is called an *equivalence class* if there is an $x \in A$ such that $A = \{y: x\ R\ y\}$. An important result in this area is that each equivalence relation R such that $X = $ (domain R) $\cup$ (range R) partitions X into a disjoint family of equivalence classes and that each member of an equivalence class is R-related with every other member of the same class.

A special case of relations is of paramount significance not only in this book but in the entire mathematical literature.

A relation in which no two distinct members have the same first coordinate is called a *mapping, function, correspondence, map,* or *operator.*

It must be emphasized that this is not the usual definition of a mapping. The usual definition of a mapping is the following: Let X and Y be any two sets. If to each element $x \in X$ one can associate one and only one element of Y by some law, then this association is called a *mapping of X into Y*. Let g denote a mapping of X into Y in the usual sense. The subset $G_f = \{(x, y): y = f(x), x \in X\}$ of the Cartesian product $X \times Y$ is called the *graph* of f and $f(x)$ is called the *value* of f at $x \in X$, or the *image* of x in Y under f. Thus the above definition of a mapping via relations is what a graph of a mapping is in the usual sense. In other words, the notions of graph and mapping can be taken to be synonymous. Thus the two mappings are the same if and only if they have the same graphs.

A mapping f of a set X into Y is said to be *onto* or surjective if each element $y \in Y$ is an image of some $x \in X$. In other words, the range of f coincides with Y.

A mapping f of X into Y is said to be $1:1$ (*one-to-one*) or injective if and only if $f(x_1) \neq f(x_2)$ whenever $x_1 \neq x_2$, $x_1, x_2 \in X$, or if and only if $f(x_1) = f(x_2)$ implies $x_1 = x_2$. A mapping which is both surjective and injective is called *bijective*.

If f is a mapping of a set X into Y and if A is a subset of X, then the set $\{(x, y): y = f(x), x \in A\}$ is called the *restriction* of f on A and is denoted by $f \mid A$. A mapping f of a set X into Y is said to be an *extension* of a mapping g of a subset $A \subset X$ into Y if $g = f \mid A$, i.e., g is the restriction of f on A.

Since a mapping is a relation, we can define inverse and composition of mappings because the latter operations have already been defined for relations.

Let f be a mapping of a set X into a set Y. For a subset A of X, $f(A)$ denotes the set of all $y \in Y$ such that $y = f(x)$ for some $x \in A$; symbolically, $f(A) = \{f(x): x \in A\}$ and is called the *image* of A under f. If B is a subset of Y, then $f^{-1}(B) = \{x \in X: f(x) \in B\}$ is called the *inverse* (or *counter*) image of B.

A reasonable facility with the following results will be of great help to the reader in the sequel.

Proposition 4. Let f be a mapping of a set X into a set Y. Let A, A_α ($\alpha \in \Gamma$), and B be subsets of X. Then:

(a) $f(A \cup B) = f(A) \cup f(B)$;

(b) $f(A \cap B) \subset f(A) \cap f(B)$. But $f(A \cap B) = f(A) \cap f(B)$ if f is $1:1$;

(c) $f(\bigcup_{\alpha \in \Gamma} A_\alpha) = \bigcup_{\alpha \in \Gamma} f(A_\alpha)$.

Proof. Clearly (a) is a particular case of (c). (b) Since $A \cap B$ is a subset of A and B, the first part is obvious. To prove the second part, it is sufficient to show that $f(A) \cap f(B) \subset f(A \cap B)$ when f is $1:1$. Let $x \in f(A) \cap f(B)$. Then $x = f(a) = f(b)$ for some $a \in A$ and some $b \in B$. Since f is $1:1$, $a = b \in A \cap B$. Hence $f(a) \in f(A \cap B)$.

(c) Clearly $x \in f(\bigcup_{\alpha \in \Gamma} A_\alpha)$ if and only if $x = f(a)$ for some $a \in \bigcup_{\alpha \in \Gamma} A_\alpha$. But $a \in \bigcup_{\alpha \in \Gamma} A_\alpha$ means that for some $\alpha \in \Gamma$, $a \in A_\alpha$ and hence $f(a) \in \bigcup_{\alpha \in \Gamma} f(A_\alpha)$ and thus $x \in f(\bigcup_{\alpha \in \Gamma} A_\alpha)$ if and only if $x = f(a) \in \bigcup_{\alpha \in \Gamma} f(A_\alpha)$.

Remark. In general, $f(A \cap B)$ need not equal $f(A) \cap f(B)$. For example, let $f(x) = 1$ for all real numbers x. Let $A = \{x \text{ real}: x < 0\}$, $B = \{x \text{ real}: x > 0\}$. Then $A \cap B = \varnothing$ and so $f(A \cap B) = \varnothing$. But $f(A) \cap f(B) = \{1\} \neq f(A \cap B)$.

Proposition 5. Let A, B, and A_α $(\alpha \in \Gamma)$ be subsets of a set Y, where Γ is a nonempty set of indices. Let f be a mapping of a set X into a set Y. Then:

(a) $f^{-1}(A \setminus B) = f^{-1}(A) \setminus f^{-1}(B)$;

(b) $f^{-1}(\bigcup_{\alpha \in \Gamma} A_\alpha) = \bigcup_{\alpha \in \Gamma} f^{-1}(A_\alpha)$ and $f^{-1}(A \cup B) = f^{-1}(A) \cup f^{-1}(B)$;

(c) $f^{-1}(\bigcap_{\alpha \in \Gamma} A_\alpha) = \bigcap_{\alpha \in \Gamma} f^{-1}(A_\alpha)$ and $f^{-1}(A \cap B) = f^{-1}(A) \cap f^{-1}(B)$;

(d) If $A \cap B = \varnothing$ then $f^{-1}(A) \cap f^{-1}(B) = \varnothing$.

Proof. (a) $x \in f^{-1}(A \setminus B)$ if and only if $f(x) \in A$ and $f(x) \notin B$. But then $f(x) \in A$ if and only if $x \in f^{-1}(A)$ and $f(x) \notin B$ if and only if $x \notin f^{-1}(B)$. This proves that $x \in f^{-1}(A \setminus B)$ if and only if $x \in f^{-1}(A)$ and $x \notin f^{-1}(B)$. This proves (a).

(b) Clearly $x \in f^{-1}(\bigcup_{\alpha \in \Gamma} A_\alpha)$ if and only if $f(x) \in \bigcup_{\alpha \in \Gamma} A_\alpha$, if and only if $f(x) \in A_\alpha$ for some $\alpha \in \Gamma$, which holds if and only if $x \in f^{-1}(A_\alpha)$ for some $\alpha \in \Gamma$. In other words, $x \in f^{-1}(\bigcup_{\alpha \in \Gamma} A_\alpha)$ if and only if $x \in \bigcup_{\alpha \in \Gamma} f^{-1}(A_\alpha)$.

(c) Similar to (b).

(d) The proof is quite clear.

Two sets A and B are said to be *equipotent* if there exists a 1 : 1 mapping of A onto B. It is clear that the relation of being equipotent on a set of sets is an equivalence relation. Thus the class of sets is partitioned into a disjoint family of equivalence classes such that each set belongs uniquely to an equivalence class. The equivalence class to which a given set belongs is called the *cardinal number* or *power* of that set. It is clear that any two equipotent sets have the same cardinal number.

It is often desirable to know whether or not two given sets are equipotent. For this, a theorem of Schroeder–Bernstein provides a criterion: If there is a 1 : 1 mapping of a set A onto a subset of a set B and if there is a 1 : 1 mapping of B onto a subset of A, then A and B are equipotent or have the same cardinal number. An elegant proof of this theorem is given by Birkhoff and MacLane [11].

A set which is equipotent with the set of any finite number of positive integers is called a *finite set*. A set which is equipotent with the set of all positive integers is called *countable*. A set which is not countable is called *uncountable.* The set of all real numbers is uncountable, whereas the set of all rational numbers is countable.

Proposition 6.   (a) A subset B of a countable set A is either countable or finite.

(b) If a set contains an uncountable subset, then it must be uncountable.

(c) The image of a countable set under any mapping is either countable or finite.

Proof.   (a) Since A is countable, let a_1, a_2, a_3, ... be its enumeration. Now there is a first element in this enumeration which belongs to B. Call it b_1 and so on. Since each member of B is in A, we obtain $B = \{b_1, b_2, \ldots\}$. If after n steps all the members of B are accounted for, B is finite, i.e., countable.

(b) Let $A \subset B$ and assume A is uncountable. If B is not uncountable then B is countable or finite. Hence by (a), A is countable or finite, which is contrary to the assumption.

(c) Let $A = \{a_n: n \geq 1\}$ be a countable set and let f be a mapping of A. Since $f(A) = \{f(a_n): n \geq 1\}$, $\{f(a_n)\}$ has at least one element, i.e., $\{f(a_n)\}$ is either a finite subset of the image space or it follows that $f(A)$ is countable.

§3. Partial and Linear Orderings; Cartesian Products

The reader is familiar with the concept of ordering in the set of real numbers. For instance, one knows what it means to say that 2 is less than 3. A similar concept can be defined in a set.

Let X be a set. A relation, denoted by $\leq$ (or $<$), which is transitive is a *partial order* or *ordering* on X. (Observe that if the partial ordering is denoted by $<$ then it is not assumed to be reflexive.) If X is partially ordered by $\leq$ then $x \leq y$ reads that x is less than or equal to y. If X is partially ordered by $<$ then $x < y$ reads that x is less than y. If we wish to emphasize the partial ordering $\leq$ or $<$ of a set X we shall denote it by $(X, \leq)$ or $(X, <)$. A partial ordering $\leq$ on a partially ordered set $(X, \leq)$ is always reflexive, since $x \leq x$ trivially holds.

Sometimes the partial ordering is denoted by $\geq$ or $>$. In these cases, $x \geq y$ and $x > y$ read that "x is greater than or equal to y" and "x is greater than y," respectively.

Let A be a subset of a partially ordered set $(X, \leq)$. An element $x \in X$ is said to be an *upper bound* (or *lower bound*) of A if for each $a \in A$, $a \leq x$ (or $x \leq a$). An upper bound (or lower bound) x of A is said to be the *supremum* or (*infimum*) of A if x is the least upper bound (or greatest lower bound), i.e., for any upper (or lower) bound x' of A, $x \leq x'$ (or $x' \leq x$).

A partially ordered set $(X, \leq)$ is said to be *order-complete* if each nonempty subset of X which has an upper bound has a supremun. This is equivalent to the statement that each nonempty subset of a partially ordered set $(X, \leq)$ which has a lower bound has an infimum. (The proof is similar to that of the statement for the set of real numbers.)

A partially ordered set $(X, \leq)$ is said to be *linearly* (or *totally*) *ordered* if the following conditions hold:

(i) For any two distinct elements $x, y \in X$, one of the following relations is true: $x < y$ or $y < x$.

(ii) If $x \leq y$ and $y \leq x$ then $x = y$.

A partially ordered set satisfying (ii) is sometimes called a *partly ordered set* (e.g., Birkhoff [9]).

A linearly ordered set is also said to be a *chain*. The set of real numbers is a chain. Moreover, the set of all real numbers is order-complete. A linearly ordered set in which every nonempty subset has a least element (i.e., x is the *least* element of a subset A if $x < a$ for all $a \in A$) is called a *well-ordered* set. The set of positive integers is a well-ordered set. On the other hand, the set of all real numbers is not well-ordered under the usual ordering. But this does not mean that the set of real numbers is not well-ordered under any other ordering. As a matter of fact, it may be pointed out that for every set (in particular for the set of real numbers) there exists an ordering under which the given set is well-ordered (well-ordering principle). However, no such ordering for real numbers has yet been found.

A fundamental result in connection with the ordering is Zorn's lemma. First we define the following: An element x of a partially ordered set $(X, \leq)$ is *maximal* if there exists no $y \in X$ such that $x < y$.

Zorn's Lemma. A partially ordered set has a maximal element if each linearly ordered subset (or chain) has an upper bound.

Zorn's lemma is equivalent to various other statements, e.g., to the well-ordering principle (every set can be well-ordered), to Zermelo's axiom (if $\mathscr{A}$ is a disjoint family of sets, then it is possible to construct a set X such that $X \cap A$ consists of a single element of A for each A in $\mathscr{A}$), or to the axiom of choice [if A_α ($\alpha \in \Gamma$) is a nonempty family of nonempty sets then there exists a mapping f on Γ such that $f(\alpha) \in A_\alpha$ for each $\alpha \in \Gamma$]. See Fraenkel [37] for a proof.

Speaking of the axiom of choice, it is relevant to point out that this axiom actually is invoked when one wishes to establish the nonemptiness of the Cartesian product of an infinite family of nonempty sets. Let A_α $(\alpha \in \Gamma)$ be a family of nonempty sets. The set of all mappings x on Γ such that $x(\alpha) \in A_\alpha$ for each $\alpha \in A$ is called the *Cartesian product* of A_α $(\alpha \in \Gamma)$ and is denoted by $\prod_{\alpha \in \Gamma} A_\alpha$. The mapping p_α which assigns to each x its value $x(\alpha)$ at α is called the αth *projection* of $\prod_{\alpha \in \Gamma} A_\alpha$ into A_α.

If $(A_\alpha, \alpha \in \Gamma)$ is a nonempty family of nonempty sets, i.e., for each $\alpha \in \Gamma$, A_α is a nonempty set, then to show that $\prod_{\alpha \in \Gamma} A_\alpha$ is nonempty is equivalent to showing that there exists a mapping x such that $x(\alpha) \in A_\alpha$ for each $\alpha \in \Gamma$, which is equivalent to the axiom of choice as stated above.

If in a Cartesian product $\prod_{\alpha \in \Gamma} A_\alpha$, $A_\alpha = A$ (a fixed set) for each $\alpha \in \Gamma$, then $\prod_{\alpha \in \Gamma} A_\alpha$ is precisely the set of all mappings of Γ into A and is denoted by A^Γ.

§4. Lattices

A partially ordered set $(X, \leq)$ is said to be a *lattice* if for each pair x, y of X, the supremum and infimum of x and y exist uniquely and are members of X. The supremum of x and y is denoted by $x \vee y$, called the *join*, and the infimum of x, y is denoted by $x \wedge y$, called the *meet*; $\vee$ and $\wedge$ are called *lattice operations*.

The following results are easy to establish:

Proposition 7. In a lattice $(X, \leq)$, let $x, y, z \in X$. Then:

(i) $(x \wedge x) = x$ and $(x \vee x) = x$;

(ii) $(x \wedge y) = (y \wedge x)$ and $(x \vee y) = (y \vee x)$;

(iii) $x \wedge (y \wedge z) = (x \wedge y) \wedge z$ and $x \vee (y \vee z) = (x \vee y) \vee z$;

(iv) $(x \vee y) \wedge x = x$ and $(x \wedge y) \vee x = x$;

(v) $x \leq y \Leftrightarrow x \wedge y = y \wedge x = x$;

(vi) $x \wedge y = x \Leftrightarrow x \vee y = y$;

(vii) $x \wedge (y \vee z) = (x \wedge y) \vee (x \wedge z)$
and $x \vee (y \wedge z) = (x \vee y) \wedge (x \vee z)$.

Proof. (i)–(iv) are easy to verify.

(v) Let $x \leq y$ and put $z = x \wedge y$. Then $z \leq x$ and $z \leq y$. But $x \leq y$ implies $x \wedge y = x$ and hence $z = x$. Conversely, if $x \wedge y = x$ then $x \leq y$ is trivially true.

(vi) $x \wedge y = x$ implies $x \vee y = (x \wedge y) \vee y = (x \vee y) \wedge y = y$ by (v), since $y \leq x \vee y$. Similarly, the reverse implication follows.

(vii) $(x \vee y) \wedge (x \vee z) = [(x \vee y) \wedge x] \vee [(x \vee y) \wedge z]$
$$= x \vee [(x \vee y) \wedge z], \text{ by (v)}$$
$$= x \vee [(x \wedge z) \vee (y \wedge z)]$$
$$= [x \vee (x \wedge z)] \vee (y \wedge z), \text{ by (iii)}$$
$$= x \vee (y \wedge z).$$

One can prove the other implication similarly.

A lattice satisfying either of the conditions in (vii) of Proposition 7 is called *distributive*.

The statement (v) characterizes lattices in the following sense. In any set X in which $\vee$ and $\wedge$ are defined, one can introduce a partial ordering by means of (v) and thus X becomes a lattice with $\vee$ and $\wedge$ as its lattice operations.

The following examples of lattices are useful in understanding the concept:

1.   Let X be the set of all positive integers with the partial ordering $x \leq y$ meaning that x divides y. It is clear that X is a lattice under the operations: $x \wedge y =$ greatest common division of x and y, $x \vee y =$ least common multiple of x and y.

2.   Let X be the set of all real numbers with $x \leq y$ meaning that x is less than or equal to y. Let $x \wedge y =$ infimum (or minimum) of x and y, and $x \vee y =$ supremum (or maximum) of x and y. Then it is easy to check that X is a lattice.

3.   Let X be any nonempty set and let $\mathscr{P}(X)$ denote the family of all subsets of X. Let $A \leq B$ for $A, B \in \mathscr{P}(X)$ mean that A is a subset of B or A is contained in B. Let $A \wedge B = A \cap B$ (intersection of A and B) and $A \vee B = A \cup B$ (set union of A and B). Then $\mathscr{P}(X)$ is a lattice. It is easy to see that if X consists of more than one element, then $\mathscr{P}(X)$ is not a chain.

4.   Let X be a nonempty set and let R^X denote the set of all real-valued mappings on X. For $f, g \in R^X$, let $f \leq g$ mean that $f(x) \leq g(x)$ for all $x \in X$, where $\leq$ is the usual ordering of the real numbers. Let $h = f \wedge g, k = f \vee g$, where $h(x) = \min(f(x), g(x))$ and $k(x) = \max(f(x), g(x))$, for each $x \in X$. Then R^X is a lattice.

A lattice X is said to be *complete* if each subset of X has a supremum and infimum. The lattice of all subsets of a fixed set is complete.

A subset Y of a lattice X is said to be a *sublattice* if Y is a lattice under the lattice operations induced from X.

The set of all continuous real-valued functions on a closed interval $[a, b]$, $a \leq b$, of the real line is a sublattice of the lattice of all real-valued functions on $[a, b]$.

Let $(X, \geq)$ be a lattice. A subset Y of X is said to be an *ideal* if $y \geq x$ and $y \in Y$ then $x \in Y$ and if $y, z \in Y$ then $y \vee z \in Y$. Similarly A subset Y of X is said to be a *dual ideal* if $x \geq y$ and $y \in Y$ then $x \in Y$ and if $y, z \in Y$ then $y \wedge z \in Y$. A proper ideal $M \subset X$ is said to be *maximal* if for any ideal $L \supset M$, $L \neq M \Rightarrow L = X$.

Proposition 8. Let $(X, \leq)$ be a lattice and let A be a subset of X and B an ideal of X such that $A \cap B = \varnothing$. Let $\mathscr{L}$ be the family of all ideals containing B and disjoint from A. Then there exists a maximal ideal M in $\mathscr{L}$ containing B and disjoint from A.

Proof. It is clear that $\mathscr{L}$ is a partially ordered set with the order $\subset$ and is nonempty because $B \in \mathscr{L}$. Let $\{L_\alpha\}$ be a linearly ordered family of ideals in $\mathscr{L}$. Let $L = \bigcup_\alpha L_\alpha$. Then $B \subset L$ and $A \cap L = \varnothing$ because $L_\alpha \cap A = \varnothing$ each α. To show that L is an ideal, let $y \in L$ and $y \geq x$ for some x. Clearly $y \in L_\alpha$ for some α. Since L_α is an ideal, $y \geq x$ implies $x \in L_\alpha \subset L$. Further, if $y, z \in L$ then $y \in L_\alpha$ and $z \in L_\beta$ for some L_α, L_β in $\{L_\alpha\}$. Since $\{L_\alpha\}$ is linearly ordered, either $L_\alpha \subset L_\beta$ or $L_\beta \subset L_\alpha$. Therefore either $y, z \in L_\beta$ or L_α and hence $y \vee z \in L_\beta$ or L_α because L_β, L_α are ideals. Therefore L is an ideal. By Zorn's lemma, there exists a maximal ideal M in $\mathscr{L}$ containing B and disjoint from A.

Proposition 9. Let $(X, \geq)$ be a lattice and A, B be two disjoint ideals of X. If $\mathscr{L}$ is the family of all ideals containing A and disjoint from B, let M be a maximal ideal in $\mathscr{L}$ (assured by the above proof). Then there exists a maximal ideal M' containing B and disjoint from M.

Proof. This is similar to the above.

Proposition 10. Let M be an ideal of a lattice $(X, \geq)$ and let $x_0 \in X$. The smallest ideal containing M and x_0 is

$$K = \{x \colon x \leq x_0 \vee y \text{ for some } y \in M\}.$$

Proof. It is clear that K is an ideal. Since $x_0 \leq x_0 \vee y$, $x_0 \in K$. Now let H be any ideal containing M and x_0. Let $x \in K$. Then $x \leq x_0 \vee y$ for some $y \in M$. Since $x_0 \vee y \in H$ (because H is an ideal containing x_0 and M), it follows that $x \in H$ (because H is an ideal). Hence $K \subset H$. This proves that K is the smallest ideal containing M and x_0.

Proposition 11. Referring to Proposition 8, let M be the maximal ideal containing A and disjoint from B and let $x_0 \notin M \cup B$. Then $x_0 \vee x \in B$ for some $x \in M$.

Proof. Let $y \in B$. Since B is a dual ideal, for any z, $z \geq y$ implies $z \in B$. If $x_0 \vee x \notin B$ for any $x \in M$, let H be the smallest ideal containing M and x_0. Then $H = \{x: x \leq x_0 \vee y$ for some $y \in M\}$ by Proposition 10 and $H \cap B = \varnothing$. However, H cannot be disjoint from B since M is the largest ideal containing A and disjoint from B. Hence $H \subset M$. But $M \subset H$ and so $H = M$. That means $x_0 \in M$, a contradiction in view of the hypothesis and hence $x_0 \vee x \in B$ for some $x \in M$.

Proposition 12. Let M and M' be the same maximal ideals as in Proposition 9 in a distributive lattice. If $x_0 \notin M \cup M'$, then there exist $x \in M$ and $y \in M'$ such that $x_0 \vee x \in M'$ and $x_0 \wedge y \in M$. Hence $(x_0 \vee x) \wedge y \in M \cap M'$.

Proof. If there exist no $x \in M$ and $y \in M'$ such taht $x_0 \vee x \in M'$ and $x_0 \wedge y \in M$, then for all $x \in M$ and all $y \in M'$, $x_0 \vee x \notin M'$ and $x_0 \wedge y \notin M$. Let H be the smallest ideal containing M and x_0. Then $H = \{x: x \leq x_0 \vee y$, for some $y \in M\}$ by Proposition 10. Since $x_0 \vee x \notin M'$ for all $x \in M$, H is disjoint from M' and hence disjoint from B because $B \subset M'$ and H contains M. Therefore $H = M$ because M is the maximal ideal containing A and disjoint from B. That means $x_0 \in M$, a contradiction. Hence there exists $x \in M$ such that $x_0 \vee x \in M'$. Similar arguments lead to $y \in M'$ such that $x_0 \wedge y \in M$. Also

$$(x_0 \vee x) \wedge y = (x_0 \wedge y) \vee (x \wedge y) \in M \cap M'.$$

Theorem 1. Let $(X, \geq)$ be a distributive lattice and let A, B be two disjoint subsets of X such that A is an ideal, B a dual ideal in X. Then there exist disjoint maximal ideals M and M' such that $A \subset M$, $B \subset M'$, and $M \cup M' = X$.

Proof. Let M and M' be the maximal ideal and dual ideal containing A and B, respectively. Then it is sufficient to show that for any $x_0 \in X$, either $x_0 \in M$ or $x_0 \in M'$. Since $M \cap M' = \varnothing$, clearly x_0 cannot belong to both. Also clearly $M \cup M' \subset X$. If $x_0 \notin M \cup M'$, then by Proposition 12, there exist $x \in M$ and $y \in M'$ such that $(x_0 \vee x) \wedge y \in M \cap M' = \varnothing$, which is absurd. Hence $M \cup M' = X$.

§5. Algebraic Structures

A set G with a law of "composition" $a \cdot b$ for $a, b \in G$ is said to be a *semigroup* if

(i) $a \cdot b \in G$, for each pair $a, b \in G$;

(ii) $a \cdot (b \cdot c) = (a \cdot b) \cdot c = a \cdot b \cdot c$ for $a, b, c \in G$ (associative law);

A set G is said to be a *group* if it satisfies (i), (ii) and (iii), (iv) below:

(iii) G contains an element e called the *identity*, i.e., $e \cdot x = x \cdot e = x$ for each $x \in G$;

(iv) Each $x \in G$ has an *inverse* $x^{-1} \in G$, i.e., $x \cdot x^{-1} = x^{-1} \cdot x = e$.

If $a \cdot b = b \cdot a$ for all $a, b \in G$, G is said to be *commutative* or *abelian*.

We shall suppress "$\cdot$" in $a \cdot b$ and will denote it by "ab or $a + b$" depending upon our wish to emphasize the multiplicative ($a \times b$) or additive ($a + b$) aspect.

The identity of an additive group will be denoted by 0 and that of a multiplicative group by e. Clearly every group is a semigroup. But the converse is not true. For example the set $\{x \text{ real}: 0 < x < 1\}$ is a multiplicative semigroup but not a group.

Indeed, the sets of integers, rational numbers, real numbers, and complex numbers are additive groups, whereas the sets of all nonzero rational and nonzero real numbers form multiplicative groups.

A subset H of a group G is said to be a *subgroup* of G if H is a group under the law of composition induced from G. If H is a subgroup of G then $H = H^2 = H \cdot H$. A subgroup H of a group is *normal*, *distinguished*, or *invariant* if $aH = Ha$ for all $a \in G$ or $H = aHa^{-1}$.

Since, in a commutative group G, $ab = ba$ for all $a, b \in G$, each subgroup of a commutative group is *invariant*.

Let H be a subgroup of a group G. For each $a \in G$, $aH = \{ah: h \in H\}$ (respectively Ha) is called a left (or right) *coset* of G. If H is an invariant subgroup of G, one can define a group operation on the set G/H of all cosets in the following way:

$$(aH)(bH) = a(Hb)H = abH^2 = abH.$$

Thus G/H is closed under multiplication. Moreover,

$$(aH)H = aH^2 = aH \quad \text{and} \quad H(aH) = (aH)H = aH^2 = aH.$$

This shows that H is an identity of G/H. Furthermore, it is easily seen that the inverse of aH is $a^{-1}H$. Since the associative law is trivially verified, it follows that G/H is a group and is called a *quotient* or *factor group* of G.

Let $G_\alpha (\alpha \in A)$ be a family of groups. The set of all mappings $x \colon A \to \bigcup_\alpha G_\alpha$ such that $x(\alpha) \in G_\alpha$ for each $\alpha \in A$ is defined to be the *direct product* of the G_α's and denoted by $\prod_{\alpha \in A} G_\alpha$. It is easy to see that $\prod_{\alpha \in A} G_\alpha$ is a group, provided one defines addition as follows: For $x, y \in \prod_{\alpha \in A} G_\alpha$,

$$x + y(\alpha) = x(\alpha) + y(\alpha) = x_\alpha + y_\alpha \in G_\alpha$$

for each $\alpha \in A$. The identity e of $\prod_{\alpha \in A} G_\alpha$ is the mapping for which $e(\alpha) = e_\alpha \in G_\alpha$, where e_α is the identity of G_α.

The subset of $\prod_{\alpha \in A} G_\alpha$ consisting of all mappings x such that $x(\alpha) = e_\alpha$ for all $\alpha \in A$ except a finite subset of A is called the *direct sum* of G_α's.

Let G and H be two groups. A mapping of G into H is said to be a (group) *homomorphism* if for any $a, b \in G$, $f(ab) = f(a)f(b)$. It is easy to see that the identity of G goes into the identity of H and the inverses into inverses under a homomorphism. More specifically, $f(e) = e'$ and $f(x^{-1}) = [f(x)]^{-1}$, where e is the identity of G and e' the identity of H. If f is a homomorphism of G into H, the set $\{x \in G \colon f(x) = e'\} = f^{-1}(e')$ is called the *kernel* of f and $f^{-1}(e')$ is an invariant subgroup of G.

A set R with two laws of composition, viz., multiplication and addition is called a *ring* if

(a) R is an additive abelian group;

(b)   for $x, y \in R$, $xy \in R$;

(c)   the multiplication is associative, i.e., $x(yz) = (xy)z = xyz$ for $x, y, z \in R$;

(d)   the distributive laws hold, i.e.,

$$x(y + z) = xy + xz,$$
$$(y + z)x = yx + zx.$$

As one defines subgroups of a group, one can define *subrings* of a ring. A subset S of a ring R is said to be a *left* (or *right*) *ideal* if for $a, b \in S$, $a - b \in S$ and $rs \in S$ (or $sr \in S$) for all $s \in S$ and all $r \in R$ or equivalently, $RS \subset S$, where $RS = \{rs \colon r \in R, s \in S\}$ (or $SR \subset R$). An ideal which is both a left and a right ideal is called a *two-sided ideal*. Two-sided ideals in rings play the same role as invariant subgroups in a group. As for groups, one can define *quotient* rings, and the direct products of rings. A mapping f of a ring R into a ring S is called a *(ring)* *homomorphism* if

$$f(a + b) = f(a) + f(b),$$
$$f(ab) = f(a)f(b)$$

for all $a, b \in R$. Ker $f = f^{-1}(0)$ is a two-sided ideal of R. Every proper ideal in a commutative ring with identity is contained in a proper maximal ideal by Zorn's lemma.

A ring in which nonzero elements (i.e., elements different from the additive identity 0) form a commutative multiplicative group is called a *field*. The sets of rational, real, and complex numbers form fields. There are no *nontrivial* (i.e., other than $\{0\}$ and F) ideals in a field F.

Let E be an abelian group and K a field. Suppose for each $\lambda \in K$ and $x \in E$, λx is defined and $\lambda x \in E$ such that for all $\lambda, \mu \in K$ and $x, y \in E$,

   (i) $\lambda(x + y) = \lambda x + \lambda y$;
   (ii) $(\lambda + \mu)x = \lambda x + \mu x$;
   (iii) $\lambda(\mu x) = (\lambda\mu)x$;
   (iv) $1x = x$, where 1 is the identity element of K.

Then E is called a *vector* or *linear* space over the field K.

If the field K is that of all real or complex numbers, the vector space E is called a real or complex vector space.

Linear vector subspaces and quotient spaces of a vector space are defined in the same way as subgroups and ideals for groups and rings.

A mapping f of a vector space E into a vector space F (both over the same field K) is called *linear* if

$$f(\lambda x + \mu y) = \lambda f(x) + \mu f(y)$$

for all $\lambda, \mu \in K$ and $x, y \in E$. A (linear) mapping of a vector space over a field K into K is called a *(linear) functional*. The set of all linear functionals of a linear space E is called the *algebraic dual* of E and is denoted by E^*.

Finally, a linear space A over a field K is called an *algebra* if the following conditions hold: For $x, y \in A$, the product xy is defined and is in A. The product is associative and distributive. Moreover,

$$\lambda(xy) = (\lambda x)y = x(\lambda y)$$

for all $x, y \in A$ and $\lambda \in K$. In other words, A is a linear space as well as a ring with mixed distributive laws.

A subset B of an algebra A which is also an algebra is called a subalgebra of A. A subalgebra B of an algebra A is called a left, right, or two-sided ideal if $AB \subset B$, $BA \subset B$, or $AB \cup BA \subset B$ respectively. If B is a two-sided ideal of an algebra A, then the collection A/B of cosets forms an algebra.

A mapping f of an algebra A into an algebra B over the same field K is called an *algebra homomorphism* if f is linear and a ring homomorphism, i.e.,

$$f(\lambda x + \mu y) = \lambda f(x) + \mu f(y) \qquad \text{and} \qquad f(xy) = f(x)f(y)$$

for all $x, y \in A$ and $\lambda, \mu \in K$.

§6. Categories and Functors

Roughly speaking, a family of "objects" and "morphisms" with some defining laws is called a category.

Formally, a *category* is a family consisting of a class $\mathcal{O}$ of elements called *objects*, together with a class $\mathcal{M}$ of elements called *morphisms* such that the following conditions hold:

   (i)   Each element of $\mathcal{M}$ is a mapping from A into B, where $A, B \in \mathcal{O}$;

   (ii)   for each triple of objects A, B, C in $\mathcal{O}$, whenever $f \in \text{Hom}(A, B)$ (the set of all morphisms of A into B) and $g \in \text{Hom}(B, C)$, then the composition $g \circ f \in \text{Hom}(A, C)$;

   (iii)   for each $A \in \mathcal{O}$, the *identity mapping* I_A of A onto itself is in $\mathcal{M}$;

   (iv)   if $f \in \text{Hom}(A, B)$, $g \in \text{Hom}(B, C)$, $h \in \text{Hom}(C, D)$, then

$$(h \circ g) \circ f = h \circ (g \circ f);$$

   (v)   for each $f \in \text{Hom}(A, B)$, $I_B \circ f = f \circ I_A$.

An element f is said to be an identity of the category if it is an identity mapping of A for some $A \in \mathcal{O}$.

It should be remarked that the composition $g \circ f\colon A \to C$ of $f\colon A \to B$ and $g\colon B \to C$ is defined if the range of f is contained in the domain of g when g is not defined on the whole of B.

It is clear that the elements in $\mathcal{O}$ are in one-to-one correspondence with the identities of the category.

Example 1. The category of sets: In this category $\mathcal{O}$ consists of all sets and $\mathcal{M}$ consists of all maps from one set to another, i.e., for $A, B \in \mathcal{O}$, $\text{Hom}(A, B)$ is the set of all maps from A to B.

Example 2.   The category of groups: $\mathcal{O}$ consists of all groups and $\mathcal{M}$ consists of all homomorphisms from one group to another.

Example 3.   The category of abelian groups: $\mathcal{O}$ consists of all abelian groups and $\mathcal{M}$ consists of all group homomorphisms as in Example 2 above.

The category with which we shall specifically deal in the sequel is that of topological spaces and continuous mappings. We shall come across these concepts later on.

Example 4.   Each partly ordered set X can be regarded as a category by taking the elements of X as the objects of the category, the pairs (x, y): $y \to x$ with $y \leq x$ as morphisms, and $(x, y) \circ (y, z) = (x, z)$, $z \leq y \leq x$ as the law of composition. A category in which all maps are identity maps is called a *discrete* category. A category in which there is only one identity is called a *monoid*.

The notion of a functor (see below) between categories corresponds to that of a function between spaces.

Roughly speaking a functor is a mapping of a category into another category such that the basic structure of the first category is transformed into the basic structure of the other, i.e., the ranges, domains, identities, and composites in one category are transformed into their corresponding opposites in the other.

Formally, let $\mathbb{C}_1$ and $\mathbb{C}_2$ be two categories. A relation F of $\mathbb{C}_1$ into $\mathbb{C}_2$ is called a *functor* if for each object A in $\mathbb{C}_1$, $F(A)$ is an object of $\mathbb{C}_2$ and for each morphism $f \colon A \to B$ in $\mathbb{C}_1$, $F(f) \colon F(A) \to F(B)$ is a morphism in $\mathbb{C}_2$ such that:

(f_1):   For each identity map I_A in $\mathbb{C}_1$, $f(I_A)$ is the identity map of $F(A)$ in $\mathbb{C}_2$;

(f_2):   If $g \circ f$ is defined in $\mathbb{C}_1$, then $F(g \circ f)$ is defined in $\mathbb{C}_2$ and

$$F(g \circ f) = F(g) \circ F(f).$$

The above functor is usually called a *covariant* functor, in contrast with contravariant functors defined below.

A mapping F of a category $\mathbb{C}_1$ into another category $\mathbb{C}_2$ is called a *contravariant* functor if for each object A in $\mathbb{C}_1$, $F(A)$ is an object in $\mathbb{C}_2$, and for each morphism $f \colon A \to B$ in $\mathbb{C}_1$ $F(f) \colon f(B) \to f(A)$ is a morphism in $\mathbb{C}_2$ such that:

(cf_1): $F(I_A) = I_{F(A)}, \qquad A \in \mathbb{C}_1$;

(cf_2):   If $f \circ g$ is defined in $\mathbb{C}_1$, then $F(f \circ g) = F(g) \circ F(f)$.

Very often conditions (f_1) and (f_2) are given in compact form by the following diagram:

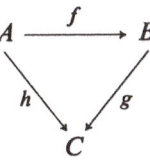

If the diagram commutes [i.e., $g \circ f = h$ or $g(f(a)) = h(a)$ for all $a \in A$], then

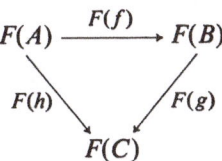

commutes.

Similarly one can express conditions (cf_1) and (cf_2) in diagrammatic form by reversing the arrows.

Let C be a category. An object I of C is said to be an *injective* if for any injective morphism $f\colon A \to B$ in the category C, and each morphism $h\colon A \to I$ there exists a morphism $\tilde{h}\colon B \to I$ such that the diagram

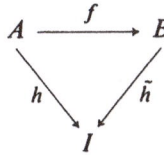

commutes.

II

Topological Spaces

In this chapter we study the elementary facts about topological spaces. Fundamental concepts of open and closed sets as well as those of interior, closure, and boundary are introduced. Bases and subbases of the family of open sets are defined. The chapter ends with a discussion of nets and their relations to topologies.

§7. Open and Closed Sets

Let X be a set and $\mathscr{E}$ a family of subsets of X. $\mathscr{E}$ is said to define a *topology* on X, or $\mathscr{E}$ is a topology on X, if the following axioms hold:

  (i) X and the empty set $\varnothing$ are in $\mathscr{E}$;
  (ii) the intersection of any finite subfamily of $\mathscr{E}$ is also in $\mathscr{E}$;
  (iii) the union of an arbitary family of members in $\mathscr{E}$ is also in $\mathscr{E}$.

A set X together with a family $\mathscr{E}$ of subsets of X satisfying (i)–(iii) is called a *topological space* and will sometime be denoted by $(X, \mathscr{E})$. [A set X together with a family $\mathscr{E}$ of subsets of X satisfying (i)–(ii) (or (i) and (iii)) is called an *infratopological* (respectively *supratopological*) space. Clearly, a set X together with a family of subsets is a topological space if and only if it is an infratopological space and a supratopological space.]

Each member U of $\mathscr{E}$ in a topological space $(X, \mathscr{E})$, is called an *open* set, or a $\mathscr{E}$-*open* set if one wishes to emphasize the topology $\mathscr{E}$ on X. This connotation is useful if one discusses more than one topology on a set X at the same time.

21

Observe that the family of open sets in a topological space is closed under the operations of taking finite intersections and arbitrary unions. In general, the intersection of arbitrary or a countable family of open sets need not be open.

[A topological space $(X, \mathscr{C})$ is said to be a *countably ultratopological space* or *P-space* if the intersection of any countable collection of open sets in $\mathscr{C}$ is also in $\mathscr{C}$. We will not discuss such general spaces.]

Let X be any set; the family consisting of X and $\varnothing$ defines a topology on X and is called the *indiscrete* topology. If the family $\mathscr{C}$ consists of all subsets of X, then $\mathscr{C}$ is called the *discrete* topology.

A subset A of a topological space $(X, \mathscr{C})$ is said to be *closed* (or $\mathscr{C}$-*closed*) if $X \backslash A$ is open (or $\mathscr{C}$-open).

Proposition 1. (a) Let $(X, \mathscr{C})$ be a topological space. Then X and $\varnothing$ are closed subsets;

(b) if $\{C_i : 1 \leq i \leq n\}$ is a finite family of closed subsets of a topological space, then $\bigcup_{i=1}^{n} C_i$ is also closed;

(c) the intersection of an arbitrary family of closed subsets is also a closed subset.

Proof. This is immediate in view of the De Morgan's laws (Proposition 2, §1).

Proposition 2. Let X be a set and let $\mathscr{C}$ be a family of subsets of X satisfying

(a) $X, \varnothing$ are in $\mathscr{C}$;

(b) the union of any finite number of sets in $\mathscr{C}$ is also in $\mathscr{C}$;

(c) the intersection of any arbitrary family of subsets in $\mathscr{C}$ is also in $\mathscr{C}$. Then there exists a topology $\mathscr{C}$ on X such that $\mathscr{C}$ is precisely the family of $\mathscr{C}$-closed sets.

Proof. Let $\mathscr{C}$ denote the family of all subsets $X \backslash C$, $C \in \mathscr{C}$. Then it is easy to verify that $\mathscr{C}$ defines a topology on X [i.e., $\mathscr{C}$ satisfies axiom (i)–(iii) of Definition 1]. Moreover, $\mathscr{C}$ is precisely the family of $\mathscr{C}$-closed sets, because of Proposition 2, §1.

Let X be a set with two topologies $\mathscr{C}_1$ and $\mathscr{C}_2$. $\mathscr{C}_1$ is said to be *finer* (or *stronger*) than $\mathscr{C}_2$ or equivalently $\mathscr{C}_2$ to be *coarser* (or *weaker*) than $\mathscr{C}_1$ if each $\mathscr{C}_2$-open set is $\mathscr{C}_1$-open. If $\mathscr{C}_1$ is finer than $\mathscr{C}_2$, we shall denote this relation by $\mathscr{C}_1 \supset \mathscr{C}_2$ or $\mathscr{C}_2 \subset \mathscr{C}_1$. (Sometimes $\mathscr{C}_1 \geq \mathscr{C}_2$ or $\mathscr{C}_2 \leq \mathscr{C}_1$ are used in place of $\mathscr{C}_1 \supset \mathscr{C}_2$.) By Proposition 2, it follows that $\mathscr{C}_1 \supset \mathscr{C}_2$ if and only if each $\mathscr{C}_2$-closed subset is also $\mathscr{C}_1$-closed. Whenever $\mathscr{C}_1 \subset \mathscr{C}_2$ and $\mathscr{C}_2 \subset \mathscr{C}_1$, $\mathscr{C}_1$ and $\mathscr{C}_2$ are said to be *equal* or *equivalent* and we write $\mathscr{C}_1 = \mathscr{C}_2$. Thus equal topologies have the same open and closed sets. Any topology on a set is coarser than the discrete topology and finer than the indiscrete topology. The problem of finding conditions under which two topologies are equal is deep and plays an important role in some branches of mathematics. We shall deal with this problem at a suitable place in the sequel.

Let A be a subset of a topological space $(X, \mathscr{C})$. The union of all open sets contained in A is called the *interior* of A and denoted by $A°$ or Int $A =$ Int$_\mathscr{C}$ A. The intersection of all closed subsets containing A is called the *closure* of A and denoted by $\bar{A}$ or Cl A or Cl$_\mathscr{C}$ A if we wish to emphasize the topology $\mathscr{C}$ with respect to which the closure is taken.

Proposition 3. Let $(X, \mathscr{C})$ be a topological space and A a subset of X. Then

(a) $A°$ is the largest open set contained in A;

(b) $\bar{A}$ is the smallest closed set containing A.

Proof. (a) By definition, $A°$ is open. Now let P be an open set such that $P \subset A$. Since P is open and $P \subset A$, $P \subset A°$ by definition.

(b) By definition, $\bar{A} = \cap \{C: C$ closed and $A \subset C\}$. If B is a closed set containing A, clearly $\bar{A} \subset B$.

Proposition 4.

(a) A set A is open if and only if $A = A°$;

(b) a set A is closed if and only if $A = \bar{A}$.

Proof. (a) Suppose A is open. Clearly by definition $A° \subset A$. Since A is open and $A \subset A$, it follows that $A \subset A°$. Hence $A = A°$. The "if" part is obvious because $A°$ is open.

(b) This follows similarly to (a).

Apart from two ways of describing topologies, viz., either by open sets or by closed sets, we can describe topologies by interior or closure operations as follows:

Theorem 1. (Kuratowski [57]) Let X be a topological space. Then the closure operation Cl $A = \bar{A}$ defined on the power set $\mathscr{P}(X)$ satisfies the following Kuratowski axioms:

   (a) $\text{Cl}\,\varnothing = \varnothing$ ($\varnothing$ is the empty set);
   (b) $A \subset \text{Cl}\,A$ for each $A \in \mathscr{P}(X)$ (or $A \subset \bar{A}$);
   (c) $\text{Cl}\,(\text{Cl}\,A) = \text{Cl}\,A$ (or $\bar{\bar{A}} = \bar{A}$);
   (d) $\text{Cl}\,(A \cup B) = \text{Cl}\,A \cup \text{Cl}\,B$ for $A, B \in \mathscr{P}(X)$ (or $\overline{A \cup B} = \bar{A} \cup \bar{B}$).

Conversely, if there exists an operation φ on $\mathscr{P}(X)$ satisfying (a)–(d), then there exists a unique topology $\mathscr{E}$ on X such that $\varphi(A) = \text{Cl}\,A$ for each $A \in \mathscr{P}(X)$.

   Proof.   It is easy to verify that the closure operator Cl satisfies (a)–(d). For the converse, let $\mathscr{E} = \{(X \backslash A): \varphi(A) = A, A \in \mathscr{P}(X)\}$. Then we show that $\mathscr{E}$ defines a topology. Clearly $\varphi(\varnothing) = \varnothing$ implies $X \backslash \varnothing = X \in \mathscr{E}$. Similarly $\varnothing \in \mathscr{E}$, because by (b) $X \subset \varphi(X)$ and $\varphi(X) \in \mathscr{P}(X)$ imply $X = \varphi(X)$. Let $A, B \in \mathscr{P}(X)$ such that $\varphi(A) = A$, $\varphi(B) = B$. Then

$$(X \backslash A) \cap (X \backslash B) = X \backslash (A \cup B) \quad \text{and} \quad \varphi(A \cup B) = \varphi(A) \cup \varphi(B) = A \cup B$$

imply that

$$(X \backslash A) \cap (X \backslash B) \in \mathscr{E}.$$

Hence the intersection of any finite number of subsets in $\mathscr{E}$ is also in $\mathscr{E}$. Furthermore, if $\{X \backslash A_\alpha\}$ is an arbitrary family in $\mathscr{E}$ then $\varphi(A_\alpha) = A_\alpha$ for each α. By De Morgan's law, $\bigcup_\alpha (X \backslash A_\alpha) = X \backslash \bigcap_\alpha A_\alpha$. But

$$\cap\, \varphi(A_\alpha) = \cap\, A_\alpha \subset \varphi(\textstyle\bigcap_\alpha A_\alpha) \subset \cap\, \varphi(A_\alpha)$$

implies $\varphi(\bigcap_\alpha A_\alpha) = \bigcap_\alpha A_\alpha$. Hence $(X \backslash \bigcap_\alpha A_\alpha) \in \mathscr{E}$. This proves that $\mathscr{E}$ is a topology. Now since $\mathscr{E}$ consists of all sets which are complements of those sets which satisfy $\varphi(A) = A$, it follows that A is $\mathscr{E}$-closed if and only if $\varphi(A) = A$. Now since $A \subset \text{Cl}_{\mathscr{E}}\,A$,

$$\varphi(A) \subset \varphi(\text{Cl}_{\mathscr{E}}\,A) = \text{Cl}_{\mathscr{E}}\,(A)$$

because $\text{Cl}_{\mathscr{E}}\,(A)$ is $\mathscr{E}$-closed. On the other hand, since $\varphi(\varphi(A)) = \varphi(A)$ by (c), $\varphi(A)$ is closed, and since $A \subset \varphi(A)$ and $\text{Cl}_{\mathscr{E}}\,(A)$ is the smallest closed set containing A, $\text{Cl}_{\mathscr{E}}\,(A) = \varphi(A)$. Since two topologies are equal if and only if they have the same open sets or closed sets, it follows that $\mathscr{E}$ is the unique topology for which $\text{Cl}_{\mathscr{E}}\,A = \varphi(A)$ for $A \in \mathscr{P}(X)$.

   In terms of the interior operator, we have the following:

Theorem 2. Let X be a topological space and let Int be the interior operation defined on $\mathscr{P}(X)$. Then:

(a) $\text{Int}\,\varnothing = \varnothing$;

(b) $\text{Int}\,A \subset A$ for each $A \in \mathscr{P}(X)$;

(c) $\text{Int}\,(\text{Int}\,A) = \text{Int}\,A$;

(d) $\text{Int}\,(A \cap B) = \text{Int}\,A \cap \text{Int}\,B$.

Conversely, if ψ is a mapping of $\mathscr{P}(X)$ into itself satisfying (a)–(d), then there exists a unique topology $\mathscr{T}$ such that $\text{Int}\,A = \psi(A)$, where $\text{Int}\,A$ is the interior of A taken with respect to $\mathscr{T}$.

Proof. Similar to that of Theorem 1.

§8. Topologies and Neighborhoods

Let $(X, \mathscr{T})$ be a topological space and let $x \in X$. A subset P of X is said to be a *neighborhood* (or $\mathscr{T}$-neighborhood if one wishes to emphasize the topology $\mathscr{T}$) of x if there exists an open set $U \subset X$ such that $x \in U \subset P$.

Observe that a neighborhood is not necessarily an open or closed set. However, open and closed sets in $(X, \mathscr{T})$ can be characterized by neighborhoods, as is the case in the following:

Proposition 5.

(a) A subset P in a topological space $(X, \mathscr{T})$ is open if and only if for each $x \in P$, P contains a neighborhood of x;

(b) a subset Q in a topological space $(X, \mathscr{T})$ is closed if and only if for each $x \in X \backslash Q$, $X \backslash Q$ contains a neighborhood of x.

Proof. (a) If P is open then clearly P is a neighborhood of each of its points. Conversely, if for $x \in P$ there exists $U_x \in \mathscr{T}$ such that $x \in U_x \subset P$, then $P = \bigcup_{x \in P} U_x$. Since $\bigcup_{x \in P} U_x$ is open by being an arbitrary union of open sets, so is P.

(b) Q is closed if and only if $X \backslash Q$ is open. Hence (b) follows from (a).

In the previous sections we have come across several methods of defining topologies. In the following theorem we give a method for defining topologies via neighborhoods.

Theorem 3. Let $(X, \mathscr{T})$ be a topological space and let $\mathscr{N}_x$ denote the system of all neighborhoods of x for each $x \in X$. Then:

(a)   For each U in $\mathcal{N}_x$, $x \in U$;

(b)   if U is in $\mathcal{N}_x$ and W is any set of X such that $U \subset W$, then W is in $\mathcal{N}_x$;

(c)   the intersection of each finite subfamily in $\mathcal{N}_x$ is also in $\mathcal{N}_x$;

(d)   for each U in $\mathcal{N}_x$ there exists a V in $\mathcal{N}_x$ such that $V \subset U$ and $U \in \mathcal{N}_y$ for each $y \in V$.

Conversely, given a set X and, for each $x \in X$, a nonempty family $\mathcal{U}_x$ of subsets of X satisfying (a)–(d), there exists a unique topology $\mathscr{C}$ on X such that $\mathcal{U}_x$ is precisely the system of all $\mathscr{C}$-neighborhoods of x for each $x \in X$.

Proof.   Let $\mathcal{N}_x$ be the system of all neighborhoods of some $x \in X$. Then (a) and (b) follow by definition. Let $\{U_i : (1 \leq i \leq n)\}$ be a finite subfamily in $\mathcal{N}_x$. Then for each i, there exists an open set P_i in $\mathscr{C}$ such that $x \in P_i \subset U_i$. Thus

$$x \in \bigcap_{i=1}^{n} P_i \subset \bigcap_{i=1}^{n} U_i.$$

Since $\bigcap_{i=1}^{n} P_i$ is open, by definition it follows that $\bigcap_{i=1}^{n} U_i \in \mathcal{N}_x$. Hence (c) is satisfied. To prove (d), let $U \in \mathcal{N}_x$. Then there exists an open set P such that $x \in P \subset U$. By Proposition 5, P is a neighborhood of each of its points. Hence $P \in \mathcal{N}_y$ for each $y \in P$.

Conversely, given a set X and for each $x \in X$, a nonempty family $\mathcal{U}_x$ satisfying (a)–(d). Let $\mathscr{C}$ denote the collection of sets comprising the empty set and all subsets U of X much that $U \in \mathcal{U}_x$ whenever $x \in U$. To verify the axioms, by (b) X is in $\mathcal{U}_x$ for each $x \in X$; hence X is in $\mathscr{C}$. Let $\{U_i : 1 \leq i \leq n\}$ be a finite subfamily of $\mathscr{C}$ such that $U_i \in \mathcal{U}_x$ where $x \in U_i$ for each $i \leq n$. Since $\bigcap_{i=1}^{n} U_i$ is in $\mathcal{U}_x$ by (c), it follows that $\bigcap_{i=1}^{n} U_i$ is in $\mathscr{C}$. Finally, let $\{U_\alpha\}$ be an arbitrary family of subsets in $\mathscr{C}$. Then for each α, $U_\alpha \in \mathcal{U}_x$ whenever $x \in U_\alpha$. Clearly $\bigcup_\alpha U_\alpha \supset U_\alpha$ for each α. Therefore again by (b), $\bigcup_\alpha U_\alpha$ is in $\mathscr{C}$. This proves that the family $\mathscr{C}$ defines a topology on X. By (d), it is clear that $\mathcal{U}_x$ is precisely the system of all $\mathscr{C}$-neighborhoods of x for each $x \in X$.

§9. Limit Points

Let $(X, \mathscr{C})$ be a topological space and A a subset of X. An element $x \in A$ is said to be a *limit* (or *cluster* or *accumulation*) *point* of A if each neighborhood of x contains an element of A other than x. The following simple results follow from definitions.

Proposition 6. (a) Each neighborhood of a point x has a nonempty intersection with A (i.e., $U \cap A \neq \varnothing$ for any neighborhood U of x) if and only if $x \in A$ or x is a limit point of A;

    (b) a set A is closed if and only if every limit point of A is in A;

    (c) a set A is open if and only if no limit point of $X \setminus A$ is in A;

    (d) the union of A and the set of all limit points of A is a closed set;

    (e) for each set A, $\bar{A}$ is the union of A and the set of all limit points of A.

Proof. (a) Suppose $x \in A$ or suppose x is a limit point of A. Then in the first case clearly, and in the second case by definition of limit point, each neighborhood of x has a nonempty intersection with A. Conversely, suppose each neighborhood of x has a nonempty intersection with A. If the intersection with each neighborhood consists of points other than x, then x is a limit point of A by definition. Otherwise, the intersection with some neighborhood of x consists of x only and so $x \in A$.

    (b) Suppose A is closed. Let x be a limit point of A. If $x \notin A$, then $x \in X \setminus A$. Since $X \setminus A$ is open, there exists a neighborhood U of x such that $U \subset X \setminus A$. Hence $U \cap A = \varnothing$, which is contrary to the assumption that x is a limit point of A and so $x \in A$. Conversely, suppose that each limit point of A is in A. Then no point of $X \setminus A$ can be a limit point of A. That means each point of $X \setminus A$ has a neighborhood which is completely contained in $X \setminus A$. This shows that $X \setminus A$ is open or, equivalently, A is closed.

    (c) This is the same as (b).

    (d) Let A' denote the set of all limit points of A. (A' is very often called the *derived* set.) If $x \notin A \cup A'$, then there is an open neighborhood U of x such that $U \cap (A \cup A') = \varnothing$ because of part (a) and the fact that if U contains a limit point of A, then $U \cap A \neq \varnothing$. This shows that $X \setminus (A \cup A')$ is open because $U \subset X \setminus (A \cup A')$ and x is arbitrary. Hence $A \cup A'$ is closed.

    (e) Since $A \subset A \cup A'$ (A' being the derived set of A), $A \cup A'$ is closed and since $\bar{A}$ is the smallest closed set containing A, we have $\bar{A} \subset A \cup A'$. Since $A \subset \bar{A}$, to show that $\bar{A} = A \cup A'$ it is enough to show that $A' \subset \bar{A}$. Let $x \in A'$. For each neighborhood U of x, $U \cap A$ contains points of A other than x. Since $A \subset \bar{A}$, $U \cap \bar{A}$ contains points of $\bar{A}$ other than x. This shows that x is a limit point of $\bar{A}$. But since $\bar{A}$ is closed by definition, we have $x \in \bar{A}$ by (b). This completes the proof.

Definition. The *boundary* ∂A of a set A in a topological space $(X, \mathscr{E})$ is defined by $\partial A = \bar{A} \cap \overline{(X \setminus A)}$.

The following proposition contains the relationship between various notions defined above.

Proposition 7. Let A be a subset in a topological space $(X, \mathscr{T})$. Then:

(a) $A^\circ = \{y \in A : y \text{ is not a limit point of } X \backslash A\} = B$, say;

(b) $X \backslash A^\circ = \overline{(X \backslash A)}$ or $A^{\circ c} = \overline{A^c}$;

(c) ∂A is a closed set;

(d) $X \backslash \partial(A) = A^\circ \cup (X \backslash A)^\circ$;

(e) $\partial A = \bar{A} \backslash A^\circ \neq \overline{(X \backslash A)} \backslash (X \backslash A)^\circ$;

(f) ∂A is the set of all $x \in X$ such that $x \notin A^\circ$ and $x \notin (X \backslash A)^\circ$;

(g) $\bar{A} = A \cup \partial A$;

(h) $A^\circ = A \backslash \partial A$;

(i) A is closed iff $\partial A \subset A$;

(j) A is open iff $A \cap \partial A = \varnothing$;

(k) $A^{\circ c} = \overline{A^c}$;

(l) $A^\circ = \overline{(A^c)}{}^c$;

(m) $\bar{A} = A^{c \circ c}$.

Proof. (a) Let $x \in A^\circ$. If x is a limit point of $X \backslash A$, then since A° is an open neighborhood of x which is completely contained in A, it follows that $\varnothing \neq A^\circ \cap (X \backslash A) \subset A \cap (X \backslash A) = \varnothing$, an absurdity. Therefore, $x \in B$. Conversely, if $x \in B$, then there exists an open neighborhood U of x such that $U \cap (X \backslash A) = \varnothing$, or equivalently, $x \in U \subset A$. Since U is an open set contained in A and since A° is the largest open set contained in A [Proposition 3(a)], we have $U \subset A^\circ$. Hence $x \in A^\circ$ and (a) is proved.

(b) Let $x \in X$, $x \notin A^\circ$. Suppose $x \in A$ and $x \notin A^\circ$. Then for each open neighborhood U of x, $U \cap (X \backslash A) \neq \varnothing$. For, otherwise, for some open neighborhood U of x, $U \cap (X \backslash A) = \varnothing$, or $U \subset A$. Since A° is the largest open set contained in A, $x \in U \subset A^\circ$, a contradiction. Therefore, $x \in \overline{X \backslash A}$ by Proposition 6. Hence $X \backslash A^\circ \subset \overline{X \backslash A}$. Conversely, suppose $x \in \overline{X \backslash A}$. Then either $x \in X \backslash A$ or x is a limit point of $X \backslash A$. In the first case clearly $x \in X \backslash A^\circ$, because $A^\circ \subset A$. In the second case, also $x \notin A^\circ$. For, otherwise, A° being an open neighborhood of x has to have a nonempty intersection with $X \backslash A$. In other words, there has to be a common point in A° and $X \backslash A$. But this is impossible since A and $X \backslash A$ are disjoint and therefore A° and $X \backslash A$ are disjoint because $A^\circ \subset A$. This shows that $x \in X \backslash A^\circ$ or $X \backslash A \subset X \backslash A^\circ$. Combining the two inclusion relations, we obtain (b).

(c) ∂A is trivially closed by definition.

(d) It is clear that the complementation is an idempotent operator, viz., $X\setminus(X\setminus A) = A$. Or if $A^c = X\setminus A$ then $A^{cc} = A$. Using this, we have

$$A^\circ \cup (X\setminus A)^\circ = (A^\circ)^{cc} \cup ((X\setminus A)^\circ)^{cc}$$
$$= [A^{cc} \cap (X\setminus A)^{cc}]^c \qquad \text{(by De Morgan's law)}$$
$$= [\overline{(X\setminus A)} \cap \bar{A}]^c \qquad \text{[by (b)]}$$
$$= (\partial A)^c = X\setminus \partial A.$$

(e) Clearly $\partial A \subset \bar{A}$ by definition. Moreover, every point in ∂A is a limit point of A as well as that of $X\setminus A$. In view of (a), no point of A° can be a limit point of $X\setminus A$. Hence $\partial A \subset \bar{A}\setminus A^\circ$. Also $\bar{A}\setminus A^\circ \subset X\setminus A^\circ = \overline{(X\setminus A)}$ by (b). Hence $\bar{A}\setminus A^\circ \subset \overline{(X\setminus A)} \cap \bar{A} = \partial A$. Combining the two inclusions, we have (e). The fact that ∂A also equals $\overline{(X\setminus A)}\setminus(X\setminus A)^\circ$ follows by symmetry if we replace A by $X\setminus A$ in the above argument.

(f) This is another way of stating (d).

(g) Since $\partial A \subset \bar{A}$ and $A \subset \bar{A}$, $A \cup \partial A \subset \bar{A}$. Let $x \in \bar{A}$. Then either $x \in A$ or x is a limit point of A. If $x \notin A$ and x is a limit point of A, then every neighborhood U of x will contain points of A and $X\setminus A$. For, if some U contains only points of A then $x \in U \subset A^\circ \subset A$, but this leads to a contradiction. Hence if $x \notin A$ and $x \in \bar{A}$, then x is a limit point of A as well as of $X\setminus A$. In otherwords,

$$x \in \bar{A} \cap \overline{(X\setminus A)} = \partial A.$$

This shows that $\bar{A} \subset A \cup \partial A$ and (g) is proved.

(h) By (e), $\partial A = \bar{A}\setminus A^\circ = \bar{A} \cap A^{\circ c}$, where $A^{\circ c} = X\setminus A^\circ$. Hence

$$A\setminus \partial A = A\setminus(\bar{A}\setminus A^{\circ c}) = A \cap (\bar{A} \cap A^{\circ c})^c = A \cap (\bar{A}^c \cup A^\circ)$$
$$= (A \cap \bar{A}^c) \cup (A \cap A^\circ) = \varnothing \cup A^\circ = A^\circ,$$

because $A = \bar{A} \cap A$ implies

$$A \cap \bar{A}^c = (\bar{A} \cap A) \cap \bar{A}^c = A \cap \varnothing$$

and $A^\circ \subset A$ implies $A^\circ \cap A = A^\circ$.

(i) A is closed iff $A = \bar{A}$ (Proposition 4). Hence by (g), $\partial A \subset \bar{A} = A$ iff A is closed.

(j) A is open iff $A = A^\circ$ (Proposition 4). Hence by (h), $\partial A \cap A = \varnothing$ iff A is open.

(k) This is another statement of (b).

(l) By taking complements in (b), (l) follows.

(m) By replacing A by A^c in (l) we have $A^{c\circ c} = \overline{A^{cc}} = \bar{A}$.

§10. Bases and Subbases

Let X be a set and $\mathscr{F}$ a subfamily of the power set $\mathscr{P}(X)$ of X. A subfamily $\mathscr{B}$ of $\mathscr{F}$ is said to be a *pseudobase* of $\mathscr{F}$ if each member of $\mathscr{F}$ contains a member of $\mathscr{B}$. If for each $x \in X$ and each F in $\mathscr{F}$ such that $x \in F$ there exists a B in $\mathscr{B}$ such that $x \in B \subset F$, then $\mathscr{B}$ is said to be a *base* of $\mathscr{F}$. It is quite clear that every base is a pseudobase. The converse is not true, as will be seen later on. A subfamily $\mathscr{S}$ of $\mathscr{F}$ is said to be a *subbase* of $\mathscr{F}$ if the family consisting of finite intersections of sets in $\mathscr{S}$ is a base for $\mathscr{F}$.

A subfamily $\mathscr{C}$ of $\mathscr{P}(X)$ is said to be a *covering* of X if each $x \in X$ is contained in some member of $\mathscr{C}$. Clearly $\mathscr{P}(X)$ is a covering of X. Moreover, the family $\{\{x\}: x \in X\}$ of singletons $\{x\}$ is also a covering of X.

A covering $\mathscr{C}$ of a set X is said to be *finite* (or *countable*) if it consists of only a finite (or *countable*) family of subsets.

A covering $\mathscr{D}$ is said to be a *refinement* of a covering $\mathscr{C}$ if each member of $\mathscr{D}$ is contained in some member of $\mathscr{C}$. A subcollection of a covering which is itself a covering is called a *subcovering*. Clearly, if a covering $\mathscr{C}$ is a subcovering of a covering $\mathscr{D}$ then $\mathscr{D}$ is a refinement of $\mathscr{C}$. $\mathscr{D}$ is said to be *reducible* to $\mathscr{C}$.

Let $\mathscr{U} = \{\mathscr{U}_x\}_{x \in X}$ denote the totality of all neighborhood systems of a topological space $(X, \mathscr{T})$. Clearly $\mathscr{U}$ is a covering of X and the family $\mathscr{C}$ of all open sets in a topological space is a refinement of $\mathscr{U}$. A covering consisting of open sets is called an *open* covering.

A topological space $(X, \mathscr{T})$, is said to be a *Lindelöf* (respectively *compact*) space if every open covering is reducible to a countable (respectively finite) open subcovering.

In regard to bases, there are two questions which naturally arise in this connection: Can one characterize the bases of the family of open sets in a topological space? Does every family of subsets of a set form a base of a topology? The first question is answered in the following:

Proposition 8. A subfamily $\mathscr{B}$ of open sets of a topological space $(X, \mathscr{T})$ is a base if and only if each open set is the union of elements of $\mathscr{B}$.

Proof. Suppose $\mathscr{B}$ is a base. Then for an arbitrary open set A and for each $x \in A$ there exists a B_x in $\mathscr{B}$ such that $x \in B_x \subset U$. Clearly $\bigcup_{x \in U} B_x = U$. Conversely, suppose each open set is a union of members of $\mathscr{B}$. Let U be an open set in A and let $x \in U$. Then since $U = \bigcup_\alpha B_\alpha$, where $B_\alpha \in \mathscr{B}$, for some α_0, $x \in B_{\alpha_0} \subset U$, and so $\mathscr{B}$ is a base.

In the following proposition, we describe the situation where a given family of subsets on a set X *defines a topology* or forms a base for the family of open sets for a topology. In general, an arbitrary family need not define a topology, e.g., the collection of all intervals $(-\infty, a)$, (a, ∞), $a \in R$, does not define a topology.

Proposition 9. A covering $\mathscr{B}$ of subsets on a set X forms a base for a topology $\mathscr{T}$ on X if and only if for each pair U, V in $\mathscr{B}$ and for each $x \in U \cap V$ there exists a W in $\mathscr{B}$ such that $x \in W \subset U \cap V$.

Proof. Suppose $\mathscr{B}$ is a base for a topology $\mathscr{T}$. Let U and V be in $\mathscr{B}$ and $x \in U \cap V$. Since U and V are open sets, there exist open sets U_1 and V_1 in $\mathscr{B}$ such that $x \in U_1 \subset U$ and $x \in V_1 \subset V$. Since $U_1 \cap V_1$ is open, there exists a W in $\mathscr{B}$ such that $x \in W \subset U_1 \cap V_1 \subset U \cap V$. Conversely, let $\mathscr{B}$ be a covering of X with the given condition. Let $\mathscr{T}$ denote the totality of all unions of members of $\mathscr{B}$ plus the empty set. Since $\mathscr{B}$ is a covering of X, X is in $\mathscr{T}$. Clearly, arbitrary unions of members from $\mathscr{T}$ are in $\mathscr{T}$. Now to show that the intersection of any finite family of sets in $\mathscr{T}$ is in $\mathscr{T}$, it is sufficient to show this for two members. Let $U = \bigcap_\alpha B_\alpha'$, $V = \bigcap_\beta B_\beta'$, where B_α', $B_\beta' \in \mathscr{B}$. By assumption for each $x \in U \cap V$ there exists B_x' such that $x \in B_x' \subset B_\alpha' \cap B_\beta'$. As x runs over $U \cap V$, we see that $\cup B_x'$ equals $U \cap V$ and hence $U \cap V \in \mathscr{T}$. This shows that $\mathscr{T}$ is a topology.

A family $\mathscr{F}$ of subsets of a set X is said to be *locally finite (or locally σ-finite) with respect to a given covering $\mathscr{C}$ of X* if for each $x \in X$ there exists C in $\mathscr{C}$ such that $x \in C$ and C has nonempty intersections with at most a finite (or countable) number of members of $\mathscr{F}$.

If $(X, \mathscr{T})$ is a topological space and $\mathscr{C}$ is the covering of all open sets (i.e., $\mathscr{C} = \mathscr{T}$), then a family $\mathscr{F}$ which is locally finite (or locally σ-finite) with respect to $\mathscr{C}$ is called simply a *locally finite* (or *locally σ-finite*) family.

Proposition 10. Every finite open covering of a topological space is locally finite.

Proof. Let $(X, \mathscr{T})$ be a topological space and $\mathscr{F}$ a finite open covering. Since X is always an open neighborhood of x and since only a finite number of members in $\mathscr{F}$ cover X, we see that $\mathscr{F}$ is locally finite.

Proposition 11. A compact topological space $(X, \mathscr{T})$ is a Lindelöf space.

Proof.   Since a finite open covering is clearly a countable open covering, the proposition is obvious from the definitions.

Proposition 12.   Every nonempty covering $\mathscr{F}$ by subsets of a set X is a subbase for a unique topology $\mathscr{C}$.

Proof.   Let $\mathscr{B}$ denote the family of all finite intersections of members of $\mathscr{F}$. To show that $\mathscr{B}$ is a base for a topology $\mathscr{C}$, we use Proposition 9. Since the intersection $U \cap V$ of any two members U, V in $\mathscr{B}$ is the intersection of a finite number of members from $\mathscr{F}$, it follows that $U \cap V$ is in $\mathscr{B}$. This shows that $\mathscr{B}$ is a base for a topology $\mathscr{C}$ on X. It is quite clear that the toplogy $\mathscr{C}$ is unique.

§11. First and Second Countable Spaces

A topological space $(X, \mathscr{C})$ is said to be *first countable* if for each $x \in X$, the neighborhood system of x has a countable base. (A base of the neighborhood system $\mathscr{U}_x$ is also called a *fundamental system of neighborhoods of x.*)

A topological space $(X, \mathscr{C})$ is said to be *second countable* if the family $\mathscr{C}$ of open sets has a *countable base*.

Proposition 13.   Every topological space $(X, \mathscr{C})$ which is second countable is also first countable. But the converse does not hold.

Proof.   Since the subfamily $\mathscr{C}_x$ of all open sets in $\mathscr{C}$ containing x is the system of all open neighborhoods of x, the first statement follows immediately.

For the converse, consider the discrete uncountable space $(X, \mathscr{C})$. Then for each $x \in X$, $\{x\}$ is a neighborhood of X. Clearly each neighborhood of x contains $\{x\}$. Hence X is first countable. However, X, being uncountable, cannot be second countable.

Proposition 14.   Every topological space $(X, \mathscr{C})$ which is second countable is a Lindelöf space.

Proof.   Let $\mathscr{F}$ be an open covering of X and let $\mathscr{B} = \{U_n\}$ be a countable base of $\mathscr{C}$. For each $x \in X$ and each P in $\mathscr{F}$, there exists some $U_{n(x)}$ in $\mathscr{B}$ [where $n(x)$ depends on x] such that $x \in U_{n(x)} \subset P$. Since $\{U_{n(x)}\}$ is a subfamily of $\mathscr{B}$, $\{U_{n(x)}\}$ is a countable covering of X. Hence $(X, \mathscr{C})$ is a Lindelöf space.

Proposition 15. In a second countable topological space $(X, \mathscr{C})$, every uncountable subset A has a limit point in A.

Proof. Suppose $\mathscr{B} = \{U_n\}$ is a countable base of $\mathscr{C}$. Suppose A does not have a limit point in A. Then for each $a \in A$ there exists $U_{n(a)}$ in $\{U_n\}$ such that $a \in U_{n(a)}$ and $U_{n(a)}$ does not contain any other element of A. Now since $\{U_{n(a)}\}$ is a countable covering of A and since each $U_{n(a)}$ contains only one element (viz., $\{a\}$) of A, it follows that A is countable. But this is contrary to the assumption that A is uncountable. Hence A must have a limit point in A.

A subset A of a topological space $(X, \mathscr{C})$ is said to be *dense* in X if $\bar{A} = X$. A topological space $(X, \mathscr{C})$ is said to be *separable* if X contains a countable dense subset.

Proposition 16. A second countable topological space $(X, \mathscr{C})$ is separable. But the converse does not hold.

Proof. Let $\{U_n\}$ be a countable base of $\mathscr{C}$. Let $A = \{x_n\}$ denote the set formed by choosing x_n in U_n for each $n \geq 1$ if U_n is nonempty. Then clearly $\{x_n\}$ is countable. To show that A is dense in X, we observe that $X \setminus \bar{A}$ is open and contains no member of $\{U_n\}$ (because A and $X \setminus \bar{A}$ are disjoint). Since $\{U_n\}$ is a covering of X, it follows that $X \setminus \bar{A} = \emptyset$ or $X = \bar{A}$. For the converse let $X = \{(x, y) \in R^2: y \geq 0\}$, the upper half-plane, with open sets defined by open disks in $\{(x, y) \in R^2: y \geq 0\}$ and, for $x \in R$, the closed disks in X having R as their tangent line. Then R^2 is separable because the subset $\{(x, y): x, y > 0, \text{rational}\}$ is clearly dense and countable, but it is not second countable, because otherwise R would be a countable set in the induced topology.

§12. Metric Spaces

We saw that each topological space which is second countable is also first countable but the converse does not hold. (See Proposition 13.) However, in the following subclass of first countable spaces, the converse is true.

Let X be a set. Let d be a real-valued function defined on the product $X \times X$ such that

   (i) $d(x, y) \geq 0$, and $= 0$ if $x = y$;

   (ii) $d(x, y) = d(y, x)$;

   (iii) $d(x, y) \leq d(x, z) + d(z, y)$, $x, y, z \in X$ (triangular inequality).

Such a function d is said to be *semimetric* or *pseudometric* and X together with d written as (X, d) is called a *semimetric* or *pseudometric space*. If d satisfies only (i) and (iii) then d is called a *quasimetric* and (X, d) a *quasimetric* space.

If, in addition, $d(x, y) = 0$ iff $x = y$, then d is said to be a *metric*, and X together with d is called a *metric space*.

Let (X, d) be a semimetric space. Let $x_0 \in X$ and $r > 0$. The set

$$B_r(x_0) = \{x \in X : d(x, x_0) < r\}$$

is called an *open ball* centered at x_0 with radius r and the set

$$\bar{B}_r(x_0) = \{x \in X : d(x, x_0) \leq r\}$$

is called a *closed ball* centered at x_0 with radius r.

Proposition 17.   Let (X, d) be a semimetric space. Then the family of $\{B_r(x)\}$, where r runs over positive real numbers and x over X, is a base for a topology on X.

Proof.   Let $B_{r_1}(x_1)$ and $B_{r_2}(x_2)$ be two members of the family and let

$$x_3 \in B_{r_1}(x_1) \cap B_{r_2}(x_2).$$

Put

$$r_3 = \min(r_1 - d(x_1, x_3), r_2 - d(x_2, x_3)).$$

Then for any $x \in B_{r_3}(x_3)$, by the triangular inequality we have:

$$d(x, x_i) \leq d(x, x_3) + d(x_3, x_1) < r_3 + d(x_3, x_i) < r_i$$

for $i = 1,2$. Hence

$$B_{r_3}(x_3) \subset B_{r_1}(x_1) \cap B_{r_2}(x_2)$$

and by Proposition 9, the family $\{B_r(x)\}$ is a base for a topology.

The topology for which the family of all open balls in a semimetric (or metric) space forms a base is called the *semimetric (or metric) topology* and a topological space $(X, \mathscr{F})$ whose topology can be given by a metric (or pseudometric or quasimetric) is called a *metrizable (or pseudometrizable or quasimetrizable)* space.

Proposition 18.   A topological space (X, d) in which the topology is defined by a semimetric (or metric) d is a first countable space.

Proof. For each positive integer n and for each $x \in X$ the family $\{B_{1/n}(x)\}$, $n \geq 1$, is a countable family of open neighborhoods of x. To show that it forms a fundamental system of neighborhoods, let $x \in B_r(x)$ for some $r > 0$. Choose a positive integer n such that $n^{-1} < r$. Then we see that $B_{n^{-1}}(x) \subset B_r(x)$. Hence, (X, d) is first countable.

Proposition 19. A semimetric space (X, d) is second countable iff it is separable.

Proof. The fact that every second countable space is separable has already been proved (Proposition 16). For the "if" part, let $\{x_n\}$ be a countable dense subset of (X, d). Let Q denote the set of all positive rational numbers. Then the family $\{B_r(x_n): n \geq 1, r \in Q\}$ is a countable family of open balls in X. We show that this family is a base for the open sets of the semimetric topology. Let $x \in X$. For any open neighborhood P of x there exists a real number $\varepsilon > 0$ such that $B_\varepsilon(x) \subset P$. Since $\{x_n\}$ is dense in X, there exists x_n such that $x_n \in B_{\varepsilon/3}(x)$. Let r be an element in Q such that $\varepsilon/3 < r < 2\varepsilon/3$. Then $B_r(x_n) \subset B_\varepsilon(x)$. For if $y \in B_r(x_n)$, then

$$d(y, x) \leq d(y, x_n) + d(x_n, x) < r + \varepsilon/3 < 2\varepsilon/3 + \varepsilon/3 = \varepsilon.$$

This completes the proof.

Proposition 20. Every discrete space is metrizable. But the converse is not true.

Proof. Let $(X, \mathscr{T})$ be a discrete space. Define the following metric: For $x, y \in X$, let

$$d(x, y) = \begin{cases} 0 & \text{if } x = y; \\ 1 & \text{if } x \neq y. \end{cases}$$

It is easy to check that d is a metric. Since the discrete topology is the finest, it is sufficient to show that the metric topology defined by d is finer than the discrete topology. Now if P is any subset of X and if $x \in P$, then $B_\varepsilon(x) \subset P$, $0 < \varepsilon < 1$. Thus P is open in the metric topology and so the metric topology is finer than the discrete topology. Thus the two topologies coincide.

For the converse, the set of real numbers with the usual metric: $d(x, y) = |x-y|$ is a nondiscrete metric space.

We saw in the previous section that second countable spaces are separable as well as first countable. Also, it is clear that spaces having countable bases have countable pseudobases.

Proposition 21.   A topological space having a countable pseudobase is separable.

Proof.   Let $\{B_n\}$ be a countable pseudobase in a topological space $(X, \mathscr{T})$. For each n, choose $x_n \in B_n$. Then $A = \{x_n\}$ is a countable set. If $\bar{A} \neq X$, then $X \setminus \bar{A}$, being an open set, contains a $B_{n_0} \in \{B_n\}$. Hence $x_{n_0} \in X \setminus \bar{A}$, which is impossible.

Thus we see that spaces having a countable pseudobase lie in between those which have countable bases and those which are separable.

Proposition 22.   A separable topological space $(X, \mathscr{T})$ which is first countable has a countable pseudobase.

Proof.   Let $\{x_n\}$ be a countable dense subset of X and let $\{B_m(x)\}$ denote a countable base of the neighborhood system at x. We show that the countable family $\{B_m(x_n)\}$, when m and n are positive integers, is a pseudo-base. Let U be any open set; then there exists an $x_n \in \{x_n\}$ in U, because $\{x_n\}$ is dense in X. Since U is an open neighborhood of x_n, there exists a B_m in $\{B_m(x_n)\}$, $m \geq 1$, containing x_n and contained in U.

We have seen that a separable space which is first countable need not have a countable base (Proposition 16).

Theorem 4.   If (X, d) is a metric space, then the following statements are equivalent:

  (a) (X, d) is separable;
  (b) (X, d) has a countable pseudobase;
  (c) (X, d) has a countable base.

Proof.   By Propositions 21 and 22, (a) $\Leftrightarrow$ (c). Since (c) $\Rightarrow$ (b) $\Rightarrow$ (a) in general, it follows that (a), (b), and (c) are equivalent.

§13. Nets

The concept of nets is a generalization of real sequences, as the concept of a topological space is a generalization of the real line from the point of view of limit points.

A partially ordered nonempty set $(\Gamma, \geq)$ or $(\Gamma, \leq)$ is called a *directed set* if for $\alpha, \beta \in \Gamma$ there exists $\gamma \in \Gamma$ such that $\gamma \geq \alpha$ and $\gamma \geq \beta$ (or $\alpha \leq \gamma$ and $\beta \leq \gamma$).

A directed set is called a *net*. Also, one can define a net as a mapping with its domain as a directed set: Let X be a set and $(\Gamma, \leq)$ a directed set. Then a mapping $\alpha \to x_\alpha \in X$ of $(\Gamma, \leq)$ into X is said to be a net and lies in X. If the net $(\Gamma, \leq)$ lies in the set X, we shall denote it by $(x_\alpha, \alpha \in \Gamma)$, where Γ is a directed set. A subset B of a directed set $(\Gamma, \leq)$ is said to be *cofinal* if for each $\beta \in \Gamma$ there exists $\gamma \in B$ such that $\beta \leq \gamma$. Let $(x_\alpha, \alpha \in \Gamma)$ be a net in X and A any subset of X. $(x_\alpha, \alpha \in \Gamma)$ is said to be *eventually* in A if there exists $\alpha_0 \in \Gamma$ such that if $\alpha \in \Gamma$ and $\alpha \geq \alpha_0$, then $x_\alpha \in A$. A net $(x_\alpha, \alpha \in \Gamma)$ in X is said to be *frequently* in a subset A of X if for each $\alpha \in \Gamma$ there exists a $\beta \in \Gamma$ such that $\beta \geq \alpha$ and $x_\beta \in A$.

The set of real numbers is a directed set. So is the set of all integers. Both sets are directed by "$\geq$," meaning that a is greater than or equal to b whenever $a \geq b$.

Proposition 23. Let $(X, \mathcal{T})$ be a topological space. The neighborhood system $\mathcal{U}_x$ of each point $x \in X$ is a directed set.

Proof. We define ordering $V_x \leq U_x$ by inclusion, i.e., $U_x \subset V_x$. Then $\mathcal{U}_x$ is a partially ordered set. Furthermore, if $V_x \leq U_x$ then $W_x = V_x \cap U_x \in \mathcal{U}_x$ and $U_x \leq W_x$, $V_x \leq W_x$.

As is commonly known, a mapping from the set of all positive integers into any set X is said to be a *sequence*. Thus a net whose domain is the set of all positive integers is nothing but a sequence.

If B, a subset of a directed set $(\Gamma, \geq)$, with the induced ordering is directed, then $(B, \geq)$ is called a *directed subset* of $(\Gamma, \geq)$. A *subnet* $\{x_\beta, \beta \in B\}$ in A is a directed subset of a net $\{x_\alpha, \alpha \in \Gamma\}$ in A.

A subnet $\{x_\beta, \beta \in B\}$ of a net $\{x_\alpha, \alpha \in \Gamma\}$ is said to be *cofinal* if for each $\alpha \in \Gamma$ there exists $\beta \in B$ such that $\beta \geq \alpha$.

If a net $\{x_\alpha, \alpha \in \Gamma\}$ is frequently in A, then the subnet $\{x_\beta, \beta \in B\}$ where $B = \{\beta \in \Gamma, x_\beta \in A\}$ is cofinal.

Let $(X, \mathcal{T})$ be a topological space and let $\{x_\alpha, \alpha \in \Gamma\}$ be a net in X. $\{x_\alpha, \alpha \in \Gamma\}$ is said to *converge* to a point $x_0 \in X$ if it is eventually in each neighborhood of x_0.

Observe that the convergence of a net in a topological space depends on its topology. Thus a net converging to a point in one topology may fail to converge to the same point in another topology. Clearly if the topology $\mathcal{T}_1$ is coarser than the topology $\mathcal{T}_2$, then the convergence of a net to a point with respect to the topology $\mathcal{T}_2$ implies its convergence to the same point with respect to $\mathcal{T}_1$. If $\mathcal{T}_1$ and $\mathcal{T}_2$ are equivalent or equal, the convergence of a net to a point in $\mathcal{T}_1$ implies and is implied by its convergence to the same point in $\mathcal{T}_2$.

A net $\{x_\alpha, \alpha \in \Gamma\}$ in X is said to be *eventually constant* if there exists $\alpha_0 \in \Gamma$ such that for $\alpha \geq \alpha_0$, $x_\alpha = x_0$ for some $x_0 \in X$.

Proposition 24. Let $(X, \mathscr{E})$ be a discrete space. Then every convergent net (in particular, every convergent sequence) is eventually constant.

Proof. Let $\{x_\alpha, \alpha \in \Gamma\}$ be a convergent net in X converging to x_0. Since $(X, \mathscr{E})$ is a discrete space, $\{x_0\}$ being a subset of X is an open neighborhood of x_0. Therefore there exists $a_0 \in \Gamma$ such that $x_\alpha \in \{x_0\}$ whenever $\alpha \geq \alpha_0$, or $x_\alpha = x_0$ whenever $\alpha \geq \alpha_0$. Therefore $\{x_\alpha, \alpha \in \Gamma\}$ is eventually constant.

Corollary 1. Let $(X, \mathscr{E})$ be a discrete space. Then a sequence is convergent iff it is eventually constant.

Proof. By definition each eventually constant sequence is convergent. The converse follows from the above proposition.

Proposition 25. Let $(X, \mathscr{E})$ be an indiscrete space. Then every net and hence every sequence in X converges to each point of X.

· **Proof.** Let $\{x_\alpha, \alpha \in \Gamma\}$ be a net in X and let x_0 be any element in X. Then the only neighborhood of x_0 is X and hence each neighborhood of x_0 contains $\{x_\alpha, \alpha \in \Gamma\}$, i.e., the latter is eventually in each neighborhood of X. Hence it converges to x_0.

Proposition 26. If a net in a topological space converges to a point then each of its subnets also converges to the same point.

Proof. Obvious.

We observe that if a net $\{x_\alpha, \alpha \in \Gamma\}$ which is not eventually constant in a topological space $(X, \mathscr{E})$ converges to a point x_0 in X, then it is clear that x_0 is a limit point of the net $\{x_\alpha, \alpha \in \Gamma\}$ which is a subset of X. One is immediately motivated to find a connection between the limit points of a set and the limits of nets:

Theorem 5. Let $(X, \mathscr{E})$ be a topological space and A a subset of X. Then:

(a) A point $x_0 \in X$ is a limit point of A if and only if there is a net $\{x_\alpha, \alpha \in \Gamma\}$ in $A \setminus \{x_0\}$ converging to x_0;

(b) $x_0 \in \bar{A}$ if and only if there is a net in A converging to x_0;

(c) A is closed iff each limit point of each net in A is in A, i.e., iff no net in A converges to a point in $X \setminus A$.

Proof. (a) Suppose x_0 is a limit point of A. Then for each neighborhood $U \in \mathcal{U}_{x_0}$, there exists $x_u \in (A \setminus \{x_0\}) \cap U$. Since $\mathcal{U}_x$ is a directed set, $\{x_u; U \in \mathcal{U}_x\}$ is a net in $A \setminus \{x_0\}$ which clearly converges to x_0. Conversely, if there is a net $\{x_\alpha, \alpha \in \Gamma\}$ in $A \setminus \{x_0\}$ converging to x_0, then every neighborhood U of x_0 contains points of A other than x_0. Hence x_0 is a limit point of A.

(b) $x_0 \in \bar{A}$ iff $x_0 \in A$ or x_0 is a limit point of A. If $x_0 \in A$ then the constant net $\{x_\alpha: x_\alpha = x_0\}$ converges to x_0. If x_0 is a limit point, then by (a) there exists a net in A converging to x_0. Hence $x_0 \in \bar{A}$ iff there exists a net in A converging to x_0.

(c) A is closed iff $A = \bar{A}$ iff $A \supset \bar{A}$ by definition. Therefore, (c) follows from (b).

Corollary 2. Let $(X, \mathcal{T})$ be a first countable topological space (in particular, a metric space) and A a subset of X. Then:

(a) A point $x_0 \in X$ is a limit point of A iff there is a sequence $\{x_n, n \geq 1\}$ in $A \setminus \{x_0\}$ converging to x_0;

(b) $x_0 \in \bar{A}$ iff there is a sequence in A converging to x_0;

(c) A is closed iff each limit point of each sequence in A is in A or iff no sequence in A converges to a point in $X \setminus A$.

Proof. Let $\{U_n, n \geq 1\}$ be a countable base of $\mathcal{U}_x$. If we put $V_n = \bigcap_{k=1}^n V_k$, then it is easy to see that $\{V_n\}$ is also a base of $\mathcal{U}_x$ and $V_{n+1} \subset V_n$ for each n. Now the corollary follows from the above theorem.

Proposition 25 implies that the limits of convergent nets in an arbitrary topological space need not be unique. In the next chapter we see that limits of convergent nets are unique iff the topological space satisfies a particular property (viz., T_2-axiom).

Let $\{x_\alpha; \alpha \in \Gamma\}$ be a net in a topological space $(X, \mathcal{T})$. A point $x_0 \in X$ is said to be a *limit point* of $\{x_\alpha, \alpha \in \Gamma\}$ if the net is frequently in each neighborhood of x_0.

An important connection between the limit points of a net and the convergence of its subnets is given by the next proposition. But first we need the following:

Let $(\Gamma, \geqq)$ and $(B, \geqq)$ be two directed sets. We can define ordering on the Cartesian product $\Gamma \times B$ as follows: Let $(\alpha, \beta), (\alpha', \beta') \in \Gamma \times B$. Let $\geq$ be defined as:

$$(\alpha, \beta) \geq (\alpha', \beta') \text{ iff } \alpha \geq \alpha' \text{ and } \beta \geq \beta'.$$

It is easy to verify that $\Gamma \times B$ is a directed set under $\geq$. Similarly, if for $\alpha \in A$ (index set), Γ_α is a directed set with partial ordering $\geq_\alpha$, we define $\prod_{\alpha \in A} \Gamma_\alpha$ as the set of all mappings $f: A \to \bigcup_\alpha \Gamma_\alpha$ such that $f(\alpha) \in \Gamma_\alpha$ with the ordering: for $f, g \in \prod \Gamma_\alpha, f \geq g$ iff $f(\alpha) \geq g(\alpha)$ for each $\alpha \in A$. $\prod_\alpha \Gamma_\alpha$ is a directed set. In particular, Γ^A, if we put $\Gamma_\alpha = \Gamma$ for all $\alpha \in A$, is a directed set.

Proposition 27. A point x in a topological space $(X, \mathscr{E})$ is a limit point of a net $\{x_\alpha, \alpha \in \Gamma\}$ iff a subnet of $\{x_\alpha, \alpha \in \Gamma\}$ converges to x.

Proof. Suppose x is a limit point of a net $\{x_\alpha, \alpha \in \Gamma\}$. Let $\mathscr{U}_x$ be the neighborhood system of x. Then $\mathscr{U}_x$ is a directed set and the net $\{x_\alpha: x_\alpha \in U_\alpha \in \mathscr{U}_x\}$ is frequently in U for each U in $\mathscr{U}_x$. Let $\Gamma \times \mathscr{U}_x$ be the product set. For each $(\alpha, U) \in \Gamma \times \mathscr{U}_x$, define $p(\alpha, U) = \alpha$, the projection of $\Gamma \times \mathscr{U}_x$ on Γ. Then for (α_1, U_1) and (α_2, U_2) in $\Gamma \times \mathscr{U}_x$, where $(\alpha_1, U_1) \geq (\alpha_2, U_2)$ (i.e., $\alpha_1 \geq \alpha_2$ and $U_1 \subset U_2$), we have $p(\alpha_1, U_1) \geq p(\alpha_2, U_2)$. The range of p is cofinal in Γ because $\{x_\alpha, \alpha \in \Gamma\}$ is frequently in each U. The mapping $(\alpha, U) \to x_\alpha$ defines a subnet $\{x_\beta, \beta \in B\}$ of $\{x_\alpha, \alpha \in \Gamma\}$ which eventually is in each $U \in \mathscr{U}_x$. For, if $U \in \mathscr{U}_x$ and if $\alpha \in \Gamma$ is arbitrary in Γ such that $x_\alpha \in U$ and if (β, V) is in $\Gamma \times \mathscr{U}_x$ such that $(\beta, V) \geq (\alpha, U)$, then $x_\beta \in V \subset U$, i.e., $\{x_\beta, \beta \in B\}$ is eventually in U. Therefore $\{x_\beta, \beta \in B\}$ converges to x.

Conversely, suppose a subnet of $\{x_\alpha, \alpha \in \Gamma\}$ converges to x. If x is not a limit point of $\{x_\alpha, \alpha \in \Gamma\}$ then there is a neighborhood U of x such that $\{x_\alpha, \alpha \in \Gamma\}$ is not frequently in U. That means it is eventually in $X \backslash U$. Hence a subnet of $\{x_\alpha, \alpha \in \Gamma\}$ is in $X \backslash U$ and therefore it cannot converge to x. This completes the proof.

The following proposition is a generalization of Cantor's celebrated diagonalization process for real double sequences.

Proposition 28. Let Γ be a directed set and, for each $\alpha \in \Gamma$, B_α a directed set. For each $(\alpha, f) \in \Gamma \times \prod_{\alpha \in \Gamma} B_\alpha$, where $f \in \prod_{\alpha \in \Gamma} B_\alpha$, let us define

$$d_\alpha: (\alpha, f) \to (\alpha, f(\alpha)),$$

a mapping of $\Gamma \times \prod_{\alpha \in \Gamma} B_\alpha$ into $\Gamma \times B_\alpha$. Let $x_{\alpha, \gamma}$ be a member of a topological space $(X, \mathcal{E})$ for each $\alpha \in \Gamma$ and $\gamma \in B_\alpha$ such that $\lim_\alpha \lim_\gamma x_{\alpha,\gamma} = x$ exists in X. Then the net $\{x_{\alpha,f(\alpha)}\}$ converges to x.

Proof. Let U be an open neighborhood of x. Choose $\alpha_0 \in \Gamma$ such that $\lim_\gamma (x_{\alpha,\gamma}) \in U$ for each $\alpha \geq \alpha_0$. Also, for $\alpha \geq \alpha_0$ choose $f(\alpha)$ in B_α such that $x_{\alpha,\gamma} \in U$ for all $\gamma \geq f(\alpha) \in B_\alpha$. But if α is a member of Γ such that $\alpha \geq \alpha_0$, then let $f(\alpha)$ be an arbitrary element in B_α. Clearly if $(\alpha, g) \geq (\alpha_0, f)$, then $\alpha \geqq \alpha_0$. Hence $\lim_\gamma x_{\alpha,\gamma} \in U$ and therefore $x_{\alpha,g(\alpha)} \in U$ because $g(\alpha) \geqq f(\alpha)$. This proves that for any $(\alpha, g) \in \Gamma \times \prod_{\alpha \in \Gamma} B_\alpha$ such that $(\alpha, g) \geq (\alpha_0, f)$, $d_\alpha(\alpha, g) = x_{\alpha,g(\alpha)} \in U$. In other words, $\{x_{\alpha,f(\alpha)}, \alpha \in \Gamma\}, f \in \prod_\alpha B_\alpha$ converges to x.

From Theorem 5, we observe that it follows approximately that the closure of a set can be described by means of the convergence of a net. Hence a topology can be defined by assuming the convergence of nets. We investigate this phenomenon below and obtain necessary and sufficient conditions for the existence of a topology in terms of nets and their limits.

Let X be a set. Let $\mathcal{N}(X)$ denote a collection of pairs $(\{x_\alpha\}, x)$, where $\{x_\alpha\}$ is a net in X and $x \in X$.

We ask, when does there exist a topology $\mathcal{E}$ on X such that for any net $\{x_\alpha\}$, $(\{x_\alpha\}, x) \in \mathcal{N}(X)$ iff $\{x_\alpha\}$ converges to x? We prove the following:

Theorem 6. Let $(X, \mathcal{E})$ be a topological space and let $\mathcal{N}(X)$ denote the totality of all pairs $(\{x_\alpha\}, x)$ such that $\lim x_\alpha = x$. Then the following conditions are satisfied:

(a) If $\{x_\alpha\}$ is a constant net, i.e., $x_\alpha = x$ for all $\alpha \in \Gamma$, $\alpha \geq \alpha_0$, then $(\{x_\alpha\}, x) \in \mathcal{N}(X)$;

(b) if for any net $\{x_\alpha\}$, $\lim_\alpha x_\alpha = x$ then for any subnet $\{x_\beta\}$ of $\{x_\alpha\}$, $(\{x_\beta\}, x) \in \mathcal{N}(X)$.

(c) Let Γ be a directed set and assume that for each $\alpha \in \Gamma$, B_α is a directed set. Let (α, f) be an element in the product $\Gamma \times \prod_\alpha B_\alpha$. Define the mapping $(\alpha, f) \to (\alpha, f(\alpha))$. If $\lim_\alpha \lim_f (\alpha, f) = x$, then $(\{x_{\alpha,f(\alpha)}\}, x) \in \mathcal{N}(X)$.

(d) If for any net $\{x_\alpha\}$, $\lim_\alpha x_\alpha \neq x$ then there is a subnet $\{x_\beta\}$ of $\{x_\alpha\}$ such that for no subnet $\{x_\gamma\}$ of which, we have $\lim_\gamma x_\gamma = x$.

Conversely, given a family of pairs $(\{x_\alpha\}, x)$ where $\{x_\alpha\}$ is a net satisfying (a)–(d), there exists a unique topology $\mathcal{E}$ on X such that $(\{x_\alpha\}, x) \in \mathcal{N}(X)$ iff $\lim_\alpha x_\alpha = x$.

Proof. To prove the first part, we see that (a) is obvious by the definition of limits of constant nets. (b) This follows from Proposition 27. (c) This follows from Proposition 28 on limits. To show (d), let $\{x_\alpha\}$ be a net which does not converge to x; then $\{x_\alpha\}$ is frequently in $X \setminus U$ for some neighborhood U of X. Hence for a cofinal subnet B of Γ, $\{x_\beta, \beta \in B\}$ is in $X \setminus U$. But then it is quite clear that no subnet of $\{x_\beta, \beta \in B\}$ can converge to x.

To prove the converse, it is sufficient to assign an operator φ satisfying conditions (a)–(d) of Theorem 1, §7.

For each subset A of X, let $\varphi(A)$ denote the subset of X consisting of all x such that for some net $\{x_\alpha\}$ in A, $(\{x_\alpha\}, x) \in \mathcal{N}(X)$.

For the empty set $\varnothing$, $\varphi(\varnothing)$ is clearly empty. Since for each $x \in A$, $\{x_\alpha\}$, where $x_\alpha = x$, is a constant net, $(\{x_\alpha\}, x) \in \mathcal{N}(X)$ by (a). This shows that $A \subset \varphi(A)$. To show that $\varphi(\varphi(A)) = \varphi(A)$, we use condition (c). It is clear that $\varphi(A) \subset \varphi(\varphi(A))$ because $A \subset \varphi(A)$ as shown above. For the reverse inclusion relation, let $(\{x_\alpha\}, x) \in \mathcal{N}(X)$, $\alpha \in \Gamma$ be a net in $\varphi(A)$. Then for each $\alpha \in \Gamma$ there is a directed set B_α and a net $\{x_{\alpha,\gamma}, \gamma \in B_\alpha\}$, $x_{\alpha,\gamma} \in A$, such that $(\{x_{\alpha,\gamma}, \gamma \in B_\alpha\}, x_\alpha) \in \mathcal{N}(X)$. But by (c), there exists a net $\{(\alpha, f(\alpha))\}$ such that $(\{x_{\alpha, f(\alpha)}\}, x) \in \mathcal{N}(X)$. Hence $x \in \varphi(A)$. In other words, $\varphi(\varphi(A)) \subset \varphi(A)$. Therefore $\varphi(\varphi(A)) = \varphi(A)$. Finally, let A, B be any two subsets of X. If $x \in \varphi(A)$, then by the definition of φ, $x \in \varphi(A \cup B)$. Hence $\varphi(A) \subset \varphi(A \cup B)$. Similarly $\varphi(B) \subset \varphi(A \cup B)$. Therefore $\varphi(A) \cup \varphi(B) \subset \varphi(A \cup B)$. To show the reverse inclusion, let $\{x_\alpha, \alpha \in \Gamma\}$ be a net in $A \cup B$ and suppose $(\{x_\alpha\}, x) \in \mathcal{N}(X)$. Put $\Gamma_A = \{\alpha \in \Gamma, x_\alpha \in A\}$, $\Gamma_B = \{\alpha \in \Gamma, x_\alpha \in B\}$. Then clearly $\Gamma_A \cup \Gamma_B = \Gamma$. Therefore at least one of Γ_A, Γ_B is cofinal with Γ. Hence at least one of the subsets $\{x_\alpha, \alpha \in \Gamma_A\}$ and $\{x_\alpha, \alpha \in \Gamma_B\}$ of $\{x_\alpha, \alpha \in \Gamma\}$ has the property that $\{(x_\alpha, \alpha \in \Gamma_A), x\} \in \mathcal{N}(X)$ or $((\{x_\alpha, \alpha \in \Gamma_B\}), x) \in \mathcal{N}(X)$ by condition (b). Hence $x \in \varphi(A) \cup \varphi(B)$ and we have shown that $\varphi(A) \cup \varphi(B) = \varphi(A \cup B)$. Thus all the conditions of Theorem 1 are satisfied. Therefore there exists a unique topology $\mathcal{C}$ on X such that $\varphi(A) = \mathrm{Cl}_{\mathcal{C}} A$ for each $A \in \mathcal{P}(X)$. Finally, to show that $\lim_\alpha x_\alpha = x$ iff $(\{x_\alpha\}, x) \in \mathcal{N}(X)$, let $(\{x_\alpha\}, x) \in \mathcal{N}(X)$, and assume $\lim_\alpha x_\alpha \neq x$. Then there exists an open neighborhood U of x such that $\{x_\alpha, \alpha \in \Gamma\}$ is not eventually in U. Therefore, there is a cofinal subset B of Γ such that $x_\alpha \in X \setminus U$ for $\alpha \in B$. But the subnet $\{x_\alpha, \alpha \in B\}$ in $X \setminus U$ is such that $(\{x_\alpha, \alpha \in B\}, x) \in \mathcal{N}(X)$ by (b). This shows that $(X \setminus U) \neq \varphi(X \setminus U) = \overline{X \setminus U}$, proving that $X \setminus U$ is not closed or U is not open. This is a contradiction. Hence $\lim_\alpha x_\alpha = x$. To show the "only if" part, let $\{x_\alpha\}$ be a net such that $\lim_\alpha x_\alpha = x$ and assume $(\{x_\alpha\}, x) \notin \mathcal{N}(X)$. For each $\alpha \in \Gamma$, let $\Gamma_\alpha = \{\beta \in \Gamma: \beta \geq \alpha\}$. Since $\{x_\alpha, \alpha \in \Gamma\}$ converges to x, $x \in \bigcap_{\alpha \in \Gamma} \bar{F}_\alpha$, where $F_\alpha = \{x_\beta; \beta \in \Gamma_\alpha\}$ by Theorem 5. Hence for each $\alpha \in \Gamma$ there is a directed set Γ_α and a net $\{f_{z,\beta}; \beta \in \Gamma_\alpha\}$ such that $(\{x_{\alpha, f(\alpha)}\}, x) \in \mathcal{N}(X)$. Hence by (c), there is a subnet $\{x_{\beta, f(\beta)}\}$ converging to x. This contradicts (d) and the proof is complete.

§14. Filters

Let X be a nonempty set and let $\mathscr{F}$ be a nonempty subfamily of $\mathscr{P}(X)$. $\mathscr{F}$ is said to be a *filter* if the following conditions hold:

(a) $\varnothing$ does not belong to $\mathscr{F}$;

(b) for F and G in $\mathscr{F}$, $F \cap G \in \mathscr{F}$;

(c) if for any H in $\mathscr{P}(X)$, $F \subset H$, where $F \in \mathscr{F}$ then H is also in $\mathscr{F}$.

A nonempty subfamily $\mathscr{B}$ of $\mathscr{P}(X)$ is said to be a *filter base* if

(a') $\varnothing$ does not belong to $\mathscr{B}$;

(b') for any B_1 and B_2 in $\mathscr{B}$ there exists $B_3 \in \mathscr{B}$ such that $B_3 \subset B_1 \cap B_2$.

A filter $\mathscr{F}$ is said to be *finer* than a filter $\mathscr{G}$ or $\mathscr{G} \subset \mathscr{F}$ if each G in $\mathscr{G}$ is a member of $\mathscr{F}$. The family of all filters on a set X can be partially ordered by inclusion: $\mathscr{G} \subset \mathscr{F}$.

Proposition 29. Let $\mathscr{F}$ and $\mathscr{G}$ be two filters. Then the family $\mathscr{F} \cup \mathscr{G} = \{F \cup G : F \in \mathscr{F} \text{ and } G \in \mathscr{G}\}$ forms a filter.

Proof. Since neither $\mathscr{F}$ nor $\mathscr{G}$ contains the empty set, $\varnothing \notin \mathscr{F} \cup \mathscr{G}$. If $F_1 \cup G_1$ and $F_2 \cup G_2$ are two members in $\mathscr{F} \cup \mathscr{G}$, then

$$(F_1 \cup G_1) \cap (F_2 \cup G_2) \supset (F_1 \cap F_2) \cup (G_1 \cap G_2) \in \mathscr{F} \cup \mathscr{G},$$

because $F_1 \cap F_2 \in \mathscr{F}$ and $G_1 \cap G_2 \in \mathscr{G}$ owing to the fact that $\mathscr{F}$ and $\mathscr{G}$ are filters. Furthermore, let H be any set such that $F \cup G \subset H$, where $F \in \mathscr{F}$ and $G \in \mathscr{G}$. Since $F \cup G \subset H$ implies $F \subset H$ and $G \subset H$, it follows that $H \in \mathscr{F} \cup \mathscr{G}$.

Proposition 30. Let $\{\mathscr{F}_\alpha : \alpha \in \Gamma\}$ be a family of filters such that Γ is a linearly ordered set and $\mathscr{F}_\alpha \subset \mathscr{F}_\beta$ for $\alpha \leq \beta$. Then

$$\bigcup_{\alpha \in \Gamma} \mathscr{F}_\alpha = \{F : F \in \mathscr{F}_\alpha \text{ for some } \alpha \in \Gamma\}$$

is also a filter.

Proof. It is clear that $\varnothing \notin \bigcup_{\alpha \in \Gamma} \mathscr{F}_\alpha$. Moreover, if $A, B \in \bigcup_{\alpha \in \Gamma} \mathscr{F}_\alpha$, then $A \in \mathscr{F}_\alpha$ for some $\alpha \in \Gamma$ and $B \in \mathscr{F}_\beta$ for some $\beta \in \Gamma$. Since Γ is linearly ordered, either $\mathscr{F}_\alpha \subset \mathscr{F}_\beta$ or $\mathscr{F}_\beta \subset \mathscr{F}_\alpha$. Hence $A, B \in \mathscr{F}_\gamma$, where $\gamma = \max(\alpha, \beta)$. Hence $A \cap B \in \mathscr{F}_\gamma \subset \bigcup_{\alpha \in \Gamma} \mathscr{F}_\alpha$. Furthermore, if H is any set containing an $F \in \bigcup_{\alpha \in \Gamma} \mathscr{F}_\alpha$ then $H \in \mathscr{F}_\alpha$, where $F \in \mathscr{F}_\alpha$ for some α because $\mathscr{F}_\alpha$ is a filter. Therefore $\bigcup_{\alpha \in \Gamma} \mathscr{F}_\alpha$ is a filter.

In view of the above propositions, the partially ordered family of all filters on a set is inductive, i.e., every linearly ordered subfamily has an upper bound. Hence by Zorn's lemma there is a maximal element. This maximal element is called an *ultrafilter*.

Proposition 31.   Let $\mathscr{U}_x$ be the system of all neighborhoods of a point x in a topological space $(X, \mathscr{C})$. Then $\mathscr{U}_x$ is a filter. A base $\mathscr{B}_x$ of $\mathscr{U}_x$ is a filter base.

Proof.   Since $x \in U$ for each $U \in \mathscr{U}_x$, $\varnothing \notin \mathscr{U}_x$. Since the intersection of any two neighborhoods of x is a neighborhood of x, the intersection of any two members of $\mathscr{U}_x$ is in $\mathscr{U}_x$. By definition, any set containing a neighborhood of x is a neighborhood of x. Hence $\mathscr{U}_x$ is a filter. The other part is quite simple.

A filter $\mathscr{F}$ in a topological space $(X, \mathscr{C})$ is said to *converge* to $x \in X$ if each neighborhood U of x contains a member of $\mathscr{F}$. A point $x \in X$ is a *limit* or *accumulation* point of a filter $\mathscr{F}$ on X if $x \in \cap \{\bar{F} : F \in \mathscr{F}\}$.

Proposition 32.   If a filter $\mathscr{F}$ on a topological space $(X, \mathscr{C})$ converges to $x \in X$ then x is a *limit* or *accumulation* point of $\mathscr{F}$.

Proof.   Let U be a neighborhood of x. Then there exists an F in $\mathscr{F}$ such that $F \subset U$. Since $\mathscr{U}_x$ is a filter, it follows that $\mathscr{F}$ is finer than $\mathscr{U}_x$. Hence for each F and U in $\mathscr{U}_x$, $F \cap U \neq \varnothing$ and so $x \in \bar{F}$ for all $F \in \mathscr{F}$.

Proposition 33.   Let $\mathscr{F}$ be a finer filter than a filter $\mathscr{G}$ on a topological space $(X, \mathscr{C})$. If $\mathscr{G}$ converges to $x \in X$, then $\mathscr{F}$ also converges to x.

Proof.   For each neighborhood U of x there exists a G in $\mathscr{G}$ such that $G \subset U$. Since $\mathscr{F}$ is finer than $\mathscr{G}$, $G \in \mathscr{F}$ and hence $\mathscr{F}$ converges to x.

Proposition 34.   Let P be a subset of a topological space $(X, \mathscr{C})$. Then:

(a) P is open iff P is a member of each filter converging to a point of P;

(b)   a point $x \in X$ is a limit point of P iff $P \backslash \{x\}$ is a member of some filter converging to x.

Proof.   (a) Suppose P is open. Let $\mathscr{F}$ be an arbitrary filter converging to $x \in P$. Then, since P is an open neighborhood of x, there exists $F \in \mathscr{F}$ such that $F \subset P$. Hence $P \in \mathscr{F}$ because $\mathscr{F}$ is a filter. For the "only if" part, suppose P is in each filter converging to a point of P. If P is not open, there exists $x \in P$ such that x is a limit point of $X \backslash P$ or $x \in \partial P$ (Proposition 7, §8). For each neighborhood U of x, $U \cap (X \backslash) P \neq \varnothing$ and $U \cap P \neq \varnothing$. Clearly the filter $\mathscr{U}_x = \{U\}$ of neighborhoods of x converges to x. Hence $P \in \mathscr{U}_x$ by assumption and so $P \cap (X \backslash P) \neq \varnothing$, a contradiction. Hence P must be open.

(b) Suppose x is a limit point of P. Then for each neighborhood U of x, $U \cap (P \backslash \{x\}) \neq 0$. If $P \backslash \{x\}$ is not a member of any filter converging to x, then for some neighborhood U of x, $U \cap (P \backslash \{x\}) = \varnothing$, a contradiction. On the other hand, if $P \backslash \{x\}$ is a member of some filter converging to x, then for any neighborhood U of x, there exists $F \in \mathscr{F}$ such that $F \subset U$ and $(P \backslash \{x\}) \in \mathscr{F}$. Hence $(P \backslash \{x\}) \cap U \neq \varnothing$ and so x is a limit point of P.

We have seen the existence of ultrafilters before. They play an important role.

Proposition 35. An ultrafilter $\mathscr{F}$ in a topological space $(X, \mathscr{C})$ either converges to a point or has no limit point.

Proof. Suppose $\mathscr{F}$ has a limit point $x \in X$. Then $x \in \bar{F}$ for each F in $\mathscr{F}$. For each neighborhood U of x, $U \cap F \neq \varnothing$. Let $\mathscr{G} = \{U \cap F: U \in \mathscr{U}_x, F \in \mathscr{F}\}$. Then $\mathscr{G}$ is a filter-base finer than $\mathscr{F}$ and $\mathscr{U}_x$. Since $\mathscr{F}$ is an ultrafilter, $\mathscr{F} = \mathscr{G}$. Further, since $\mathscr{U}_x$ converges to x and since $\mathscr{F}$ is finer than $\mathscr{U}_x$, $\mathscr{F}$ converges to x.

Theorem 7. Let $\mathscr{F}$ be an ultrafilter on a topological space $(X, \mathscr{C})$. Let F and G be any two disjoint sets in X such that $F \cup G$ is in $\mathscr{F}$. Then either F or G is in $\mathscr{F}$.

Proof. We know that the power set $\mathscr{P}(X)$ is a distributive lattice, where $A \wedge B = A \cap B$ and $A \vee B = A \cup B$, and $A \leq B$ is defined by inclusion. Clearly, each filter in X is an ideal in $\mathscr{P}(X)$ and contains X. Now if $F, G \in \mathscr{F}$ then $F \wedge G = F \cap G = \varnothing \in \mathscr{F}$, which is a contradiction in view of the definition of a filter. Let $\mathscr{G}_1 = \{H: H \supset F \text{ and disjoint from } G, H \in \mathscr{F}\}$ and $\mathscr{G}_2 = \{H: H \supset G \text{ and disjoint from } F, H \in \mathscr{F}\}$. Then $\mathscr{G}_1$ and $\mathscr{G}_2$ are ideals and $\mathscr{G}_1 \cap \mathscr{G}_2 = \varnothing$. Hence by Theorem 1, §4, Chapter 1, there are disjoint maximal ideal and dual ideal $\mathscr{G}_1'$ and $\mathscr{G}_2'$ such that $\mathscr{G}_1' \cup \mathscr{G}_2' = \mathscr{P}(X)$. Clearly either F or G is in $\mathscr{G}_1'$. Suppose F is in $\mathscr{G}_1'$, which is a maximal ideal containing F and disjoint from G; it follows that $F \in \mathscr{F}$ and $G \notin \mathscr{F}$.

As one knows from the earlier results that nets can be taken to define a topology, so can a filter, since a filter is closed under finite intersections, arbitrary unions, and includes the whole set X provided one assumes the presence of the empty set.

On the other hand, it seems likely that filters can be taken to define nets and vice versa, as the following proposition demonstrates:

Proposition 36. Let $\{x_\alpha, \alpha \in \Gamma\}$ be a net in a nonempty set X. Then the family $\mathscr{F}$ of all A such that $\{x_\alpha, \alpha \in \Gamma\}$ is eventually in A forms a filter. Conversely, let $\mathscr{F}$ be a filter in X, and let Γ be the set of all pairs (x, F) such that $x \in F$ and F is in $\mathscr{F}$. Define the ordering in Γ by $(y, G) \geq (x, F)$ iff $G \subset F$. Let f be the mapping of Γ with $f(x, F) = x$. Then $\mathscr{F}$ is precisely the family of all sets A such that $\{f(x, F): (x, F) \in \Gamma\}$ is eventually in A.

Proof. Straightforward.

Proposition 37. Let $\{x_\alpha, \alpha \in \Gamma\}$ be a net in a topological space $(X, \mathscr{E})$ and let $F_\alpha = \{x_\beta, \beta > \alpha, \alpha, \beta \in \Gamma\}$. Then a point x_0 is a limit point of $\{x_\alpha, \alpha \in \Gamma\}$ iff $x_0 \in \cap \{\bar{F}_\alpha, \alpha \in \Gamma\}$.

Proof. Suppose x_0 is a limit point of $\{x_\alpha, \alpha \in \Gamma\}$. Since $\{x_\alpha, \alpha \in \Gamma\}$ is frequently in each neighborhood of x_0, each F_α has a nonempty intersection with each neighborhood of x_0. Hence $x_0 \in \bar{F}_\alpha$ for each $\alpha \in \Gamma$. On the other hand, suppose $x_0 \in \bar{F}_\alpha$ for each $\alpha \in \Gamma$. If x_0 is not a limit point of $\{x_\alpha, \alpha \in \Gamma\}$, there exists a neighborhood U of x_0 such that $\{x_\alpha, \alpha \in \Gamma\}$ is not frequently in U. Hence for some $\beta \in \Gamma$, if $\alpha \geq \beta$ then $x_\alpha \notin U$. That means $F_\beta \cap U = \varnothing$ and so $x_0 \notin \bar{F}_\beta$, a contradiction.

§15. Topologies Defined by Other Topologies

In the previous sections, we have seen that a topology on a set can be defined by several methods, e.g., via closures, interiors, nets and filters, etc. Now we want to show that sometimes a topology constructed out of given topologies may also be interesting and useful.

Proposition 38. Let $\mathscr{E}_1$ and $\mathscr{E}_2$ be two topologies on a set X. Let $\mathscr{E}_1 \wedge \mathscr{E}_2$ (respectively $\mathscr{E}_1 \vee \mathscr{E}_2$) denote the set of all subsets W such that $W \in \mathscr{E}_1$ and $W \in \mathscr{E}_2$ (respectively $W = U \cup V$) where U runs over $\mathscr{E}_1$ and V over $\mathscr{E}_2$ respectively. Then:

(a) $\mathscr{T}_1 \wedge \mathscr{T}_2$ is a topology which is the finest topology coarser than both $\mathscr{T}_1$ and $\mathscr{T}_2$;

(b) $\mathscr{T}_1 \vee \mathscr{T}_2$ is a topology which is the coarsest topology finer than $\mathscr{T}_1$ and $\mathscr{T}_2$.

Proof. (a) It is easy to verify that $\mathscr{T}_1 \wedge \mathscr{T}_2$ is a topology. Now let $\mathscr{T}$ be any topology which is coarser than both $\mathscr{T}_1$ and $\mathscr{T}_2$. Then each $\mathscr{T}$-open set W is $\mathscr{T}_1$-open and $\mathscr{T}_2$-open. That means that W is in $\mathscr{T}_1$ and also in $\mathscr{T}_2$ and hence in $\mathscr{T}_1 \wedge \mathscr{T}_2$ and so $\mathscr{T}$ is coarser than $\mathscr{T}_1 \wedge \mathscr{T}_2$.

(b) Similarly one proves that $\mathscr{T}_1 \vee \mathscr{T}_2$ is a topology and that it is the coarsest topology finer than $\mathscr{T}_1$ and $\mathscr{T}_2$.

Remark. One can extend the operations $\vee$ and $\wedge$ to any number of families $\{\mathscr{T}_\alpha\}$. Then $\vee_\alpha \mathscr{T}_\alpha$ and $\wedge_\alpha \mathscr{T}_\alpha$ are topologies.

Definition 1. Let $\{\mathscr{T}_\alpha\}$ be a family of topologies on a set X. We define:

$$\bigcap_\alpha \mathscr{T}_\alpha = \{W: W \in \mathscr{T}_\alpha \text{ for each } \alpha\};$$

$$\bigcup_\alpha \mathscr{T}_\alpha = \{W: W \in \mathscr{T}_\alpha \text{ for some } \alpha\}.$$

Proposition 39. (a) $\bigcap_\alpha \mathscr{T}_\alpha$ is a topology equal to $\wedge \mathscr{T}_\alpha$. (b) $\bigcup_\alpha \mathscr{T}_\alpha$ is not a topology in general, unless X is a set consisting of at most two points. However, it forms a subbase of the topology $\vee_\alpha \mathscr{T}_\alpha$.

Proof. (a) Since X and $\varnothing$ are in $\mathscr{T}_\alpha$ for each α, it follows that X and $\varnothing$ are in $\bigcap_\alpha \mathscr{T}_\alpha$. Similarly, the finite intersection and union of sets in $\bigcap_\alpha \mathscr{T}_\alpha$ are in $\mathscr{T}_\alpha$ for each α, it follows that $\bigcap_\alpha \mathscr{T}_\alpha$ is a topology.

(b) $\bigcup_\alpha \mathscr{T}_\alpha$ fails to satisfy the axiom of finite intersections. For instance, if U_α and V_α are open sets in each $\mathscr{T}_\alpha$, then $U_\alpha \cap V_\alpha$ nead not be in $\bigcup_\alpha \mathscr{T}_\alpha$, unless E is at most a two-point set. However, it forms a subbase of a topology, as is easy to see.

Theorem 8. The set of all topologies on a set is a complete lattice, where $\vee$ and $\wedge$ are lattice operations and the inclusion $\subset$ is the partial ordering.

Proof. It is clear that the relation $\subset$ defines a partial ordering on the set of all topologies. The rest follows from Proposition 39.

Definition 2. Let $\mathscr{C}_1$ and $\mathscr{C}_2$ be two topologies on a set X. Let $\mathscr{U}_1(x)$ and $\mathscr{U}_2(x)$ denote the neighborhood systems at each x with respect to topologies $\mathscr{C}_1$ and $\mathscr{C}_2$, respectively Let $\mathscr{C}_2(\mathscr{C}_1)$ denote the family of subsets $\mathrm{Cl}_{\mathscr{C}_2} U(x)$ for each $U \in \mathscr{U}_1(x)$.

Proposition 40. $\mathscr{C}_2(\mathscr{C}_1)$ defines a topology on X.

Proof. Let U_1 and U_2 be two members in $\mathscr{U}_1(x)$. Then there exists $U_3 \in \mathscr{U}_1(x)$ such that $x \in U_3 \subset U_1 \cap U_2$. But then

$$x \in \mathrm{Cl}_{\mathscr{C}_2} U_3 \subset \mathrm{Cl}_{\mathscr{C}_2}(U_1 \cap U_2) \subset \mathrm{Cl}_{\mathscr{C}_2} U_1 \cap \mathrm{Cl}_{\mathscr{C}_2} U_2.$$

This means that $\{\mathrm{Cl}_{\mathscr{C}_2} U\}$ forms a base of neighborhood system at x under the topology $\mathscr{C}_2(\mathscr{C}_1)$.

Remark. In general, the topology $\mathscr{C}_2(\mathscr{C}_1)$ need not satisfy a separation axiom even though $\mathscr{C}_1$ and $\mathscr{C}_2$ do. Clearly, $\mathscr{C}_2(\mathscr{C}_1)$ is coarser than $\mathscr{C}_1$.

Another method of defining a new topology is given by the following:

Definition 3. Let $\mathscr{C}$ be a topology on a set X. Let $\mathscr{C}$ be the family of all $\mathscr{C}$-compact subsets of X. Let $k(\mathscr{C}, \mathscr{C})$ denote the topology on X defined as follows: A subset V is $k(\mathscr{C}, \mathscr{C})$-open if and only if $V \cap C$ is relatively $\mathscr{C}$-open on each C in $\mathscr{C}$.

It is easily seen that $k(\mathscr{C}, \mathscr{C})$ is a topology and called the *k-extension* of $\mathscr{C}$ (see also §40).

Proposition 41. Each $\mathscr{C}$-open set is $k(\mathscr{C}, \mathscr{C})$-open, i.e., $k(\mathscr{C}, \mathscr{C})$ is finer than $\mathscr{C}$. $k(\mathscr{C}, \mathscr{C})$ and $\mathscr{C}$ have the same compact subsets.

Proof. Let U be $\mathscr{C}$-open and let C be $\mathscr{C}$-compact. Then clearly $U \cap C$ is relatively $\mathscr{C}$-open on C and hence U is $k(\mathscr{C}, \mathscr{C})$-open. In other words, $k(\mathscr{C}, \mathscr{C})$ in finer than $\mathscr{C}$. Thus every $k(\mathscr{C}, \mathscr{C})$-compact subset of X is $\mathscr{C}$-compact. Now let A be a $\mathscr{C}$-compact subset and let $\{P_\alpha\}$ be a $k(\mathscr{C}, \mathscr{C})$-open covering of A. Clearly for each α, $P_\alpha \cap A = P_\alpha$ is $\mathscr{C}$-open. Hence $\{P_\alpha \cap A\}$ is a $\mathscr{C}$-open covering of A and by $\mathscr{C}$-compactness of A only a finite number of sets of the covering $\{P_\alpha\}$ cover A and hence A is $k(\mathscr{C}, \mathscr{C})$-compact.

Examples and Exercises

1. (An infratopology which is not a topology.) Let R be the set of real numbers. The family of all closed sets under the metric topology (Exercise 5) is not closed under arbitrary unions, hence it does not define a topology, but this family contains R and $\emptyset$ and is closed under finite intersections and so it is an infratopology.

2. (Supratopology not a topology.) Again take the reals R with the usual topology. Now consider $R \cup \{\emptyset\} \cup \{R \backslash \{x\} : x \in R\}$. Then the family does not define a topology but it is a supratopology.

3. There is a countable-ultratopological space which is not an ultratopological space. In other words, there is a set with a family closed under countable intersections but not arbitrary intersections. Take the reals R together with $\emptyset$ and $\{A\}$, $A \subset R$ such that A^c is countable. Then $(\bigcap_{i=1}^{\infty} A_i)^c = \bigcup_{i=1}^{\infty} A_i^c$, which is countable. However, it is well known that an arbitrary union of countable sets is not necessarily countable; hence R, with this family, is a countable-ultratopological space but not an ultratopological space.

4. (A pseudobase which is not a base.) Consider the reals R with the metric topology and consider $\{A\}$, the family of compact sets A. This is a pseudobase because singletons are compact and any open set which is nonempty contains a singleton as a subset. This family is not a base.

5. Let R denote the set of all real numbers. Then $d(x, y) = |x - y|$, x, $y \in R$, defines a metric topology, sometimes called the *usual* or *Euclidean* topology.

6. Let C denote the set of all complex numbers. As in (1), $d(x, y) = |x - y|$, $x, y \in C$ defines a metric topology.

7. Let $n > 0$ be an integer. Let R denote the set of all n triples: i.e., $x \in R^n$ iff $x = (x_1, x_2, \ldots, x_n)$, $x_i \in R$, $1 \leq i \leq n$.

$$d_2(x, y) = \left[\sum_{i=1}^{n} (x_i - y_i)^2 \right]^{1/2}$$

defines a metric topology called the n-dimensional Euclidean topology; C^n is defined similarly.

8. Let X be any set. Let $\mathcal{C}$ consist of X and $\emptyset$. Then $(X, \mathcal{C})$ is a topological space. (This is called the indiscrete topology.)

9. Let X be any set. Let T consist of all subsets of X. Then (X, T) is a topological space. (This is called the discrete topology.)

10. Let X be a linearly ordered set. Let T denote the family of all subsets U such that $U_a = \{x : x \geq a \text{ for some } a\}$. Let $\mathcal{C} = \{U_a, a \in X\}$ denote

the totality of such U_a's. $(X, \mathcal{C})$ is a supratopological space but not a topological space.

11. (Example of Lindelöf space which is not second countable.) Let R be endowed with the topology which has the family $\mathcal{B}$ of all half-open and half-closed intervals $[a, b]$ as base. $(R, \mathcal{C})$ is not second countable. {For, let x be an arbitrary real number and let $[x, \infty)$ be a neighborhood of x. There exists a set B_x from $\mathcal{B}$ such that $x \in B_x \subset [x, \infty)$. Since R is uncountable, $\mathcal{B}$ contains at least as many sets as the power of the continuum. Hence $\mathcal{B}$ must be uncountable.}

But $(R, \mathcal{C})$ is a Lindelöf space. {Let $\mathcal{P}$ be an open covering of R. For each $x \in R$ there exists an open set B_x in $\mathcal{P}$ and rational numbers a_x, b_x such that $a_x < x < b_x$ and $[a_x, b_x] \subset B_x$. Clearly $\{(a_x, b_x): a_x, b_x$ rational$\}$ is a countable covering of R and so $\{B_x\}$, corresponding to $\{(a_x, b_x)\}$, is a countable open subcovering.}

12. (A separable space which is not second countable.) The space of Exercise 11 is not second countable, but it is separable. {Let Q be the set of all rational numbers. Then Q is a countable set. For each $x \in R$ and P a neighborhood of x there exists a real number $b \neq x$ such that $[x, b) \subset P$. Choose $r \in Q$ such that $r \in [x, b) \subset P$. This shows that Q is dense in R. Hence R is separable.}

13. (A separable space which contains a nonseparable subspace.)

   (a) Let $E = R^2$ and let $\mathcal{C}$ be the topology which has the following base: For each $(x, y) \in E$, $y \neq 0$, let $\mathcal{B}$ contain all open disks containing (x, y). For (x, y), $y = 0$, let $\mathcal{B}'$ contain the family of all closed disks which have a tangent at $(x, 0)$. Then R^2 is separable [the set of all points (x, y), x, y rational numbers, is dense in R^2]. But the set $\{(x, y), y = 0\}$ of the real line with the induced topology is not separable because the real line R with the induced topology is a discrete space and R is uncountable. Hence R cannot be separable.

   (b) Let $E = R^2$. Let $\mathcal{C}$ be the box topology, consisting of rectangles $[a, b) \times [c, d)$ as base. Then $(E, \mathcal{C})$ again is separable. But the subset $\{(x, y): x + y = 1\}$, endowed with the induced topology, is not separable. {The proof follows similarly as above.} $(E, \mathcal{C})$ is not a Lindelöf space.

14. Let $[0, \Omega)$ denote the set of all ordinals less than the first uncountable ordinal. Let ω_1, $\omega_2 \in [0, \Omega)$ and we order $\omega_1 \leq \omega_2$ if the ordinal ω_1 is less than or equal to the ordinal ω_2. Let $(\omega_1, \omega_2) = \{\omega \in [0, \Omega): \omega_1 < \omega < \omega_2\}$. Then the collection of all intervals (ω_1, ω_2) forms a base for the topology on $[0, \Omega)$, called the *order* topology.

   (a) Show that $[0, \Omega)$ is first countable but not second countable.

   (b) If $[0, \Omega]$ is the set of all ordinals less than or equal to the first
        uncountable ordinal Ω, then in the order topology on $[0, \Omega]$
        show that Ω is a limit point of $[0, \Omega)$ but is not a sequential
        limit of $[0, \Omega)$. Show that $[0, \Omega]$ is neither first nor second
        countable.

15. Let $\mathscr{F}$ be a filter on a set X. Then $\mathscr{F}$ and $\varnothing$ define a topology on X.
     Let $\mathscr{F}_x$ denote the set of all filters converging to $x \in X$. Show that
     $\cap \{\mathscr{G}: \mathscr{G} \in \mathscr{F}_x\}$ is a neighborhood system at x.

16. (a) Show that the topology $\mathscr{C}_2(\mathscr{C}_1)$ defined in Definition 2, §15, is
          not necessarily Hausdorff even if $\mathscr{C}_1$, $\mathscr{C}_2$ are.

     (b) If $\mathscr{C}_1 \supset \mathscr{C}_2$, show that $\mathscr{C}_1 \supset \mathscr{C}_2(\mathscr{C}_1) \supset \mathscr{C}_2$. If $\mathscr{C}_2$ is Hausdorff
          so is $\mathscr{C}_2(\mathscr{C}_1)$.

     (c) Is $\mathscr{C}_2(\mathscr{C}_1)$ Hausdorff if the diagonal set $\varDelta$ in the product space
          $(X, \mathscr{C}_1) \times (X, \mathscr{C}_2)$ is closed?

17. (On connected spaces.) A topological space E is said to be *connected*
     if E cannot be written as the union of two nonempty disjoint open (or
     closed) subsets of E. In other words, $E \neq P \cup Q$, where P, Q are
     nonempty disjoint and open. A subset A of a topological space E is
     said to be connected if it is connected in the induced topology. Show the
     following:

     (a) If A is connected, so is $\bar{A}$.

     (b) if $\{A_\alpha\}_{\alpha \in \Gamma}$ is a family of connected sets such that no two mem-
          bers have empty intersection then $\cup A_\alpha$ is connected.

     (c) Show that the set of all real numbers is a connected space.

     (d) A maximal connected subset of a topological space containing
          a subset A is called the *component* of A. Show that each com-
          ponent containing a point is a closed subset [use (a)].

18. (On locally connected spaces.) A topological space E is said to be
     locally connected if each neighborhood of a point of E contains a
     connected neighborhood.

     (a) E is locally connected iff for each $x \in E$ and each neighborhood
          U of x the component of U to which x belongs is a neighborhood
          of x iff the topology has a base of open connected subsets.

     (b) Each component of an open set in E is open iff E is locally
          connected.

     (c) Show that the subset

     $$\{(x, y): y = \sin(1/x), x \neq 0\} \cup \{(0, 0)\}$$

     of R^2 is connected but not locally connected.

III

Continuity and Separation Axioms

In this chapter, we first define continuous mappings and their various equivalent formulations. We then consider various separation axioms for topological spaces. We prove the existence (Urysohn's lemma) of continuous functions separating disjoint closed subsets of a normal space. Here and later we sometimes will denote a topological space E with a topology $\mathscr{E} = \{U\}$ by E_u.

§16. Continuous and Open Mappings

Definition 1. Let E_u and F_v be two topological spaces. A mapping f of E into F is said to be *continuous* if for each v-open set V in F, $f^{-1}(V)$ is an u-open set of E_u.

Remark. Observe that the continuity of a mapping is defined only if the spaces are *a priori* endowed with topologies.

Given a triple $f: E \to F$, i.e., two sets and a mapping f of E into F, there are three kinds of problems associated with this triple. First, given f and a topological space F_v, determine a topology u on E such that $f: E_u \to F_v$ is continuous. (This is a simple problem and will be easily solved.) Secondly, given f and a topological space E_u, determine a topology v on F such that $f: F_u \to F_v$ is continuous. Thirdly, given a mapping f from a topological space E_u into a topological space F_v, under what conditions is f continuous?

53

We shall discuss these problems in the sequel. It must be remarked that even though the solution of the first problem is simple, the solutions of the other two are not.

Before we discuss separation axioms, we collect a few propositions concerning continuous maps.

Proposition 1. Let $f: E_u \to F_v$ and $g: F_v \to G_w$ be two continuous mappings. Then the composition $g \circ f: E_u \to G_w$ is also continuous.

Proof. For each open set W in G_w,

$$(g \circ f)^{-1}(W) = f^{-1}(g^{-1}(W)).$$

Since f and g are continuous, $f^{-1}(g^{-1}(W))$ is an open set in E_u. Hence $g \circ f$ is continuous.

Corollary 1. The class of all topological spaces and continuous mappings is a category.

Proof. Immediate.

Some useful subcategories of the category of topological spaces will be considered later.

Proposition 2. Let f be a continuous mapping of a topological space E_u into a topological space F_v. Let H be a subset of E, endowed with the relative topology. Then the restriction $f \,|\, H: H \to F_v$ is continuous just as $f: E \to f(E)$ is continuous.

Proof. Let V be an open set in F_v. Then because f is continuous, $f^{-1}(V)$ is open in E_u. Hence by the definition of the relative topology on H, $f^{-1}(V) \cap H$ is open in H and so $f \,|\, H$ is continuous. Similarly one proves the other part.

Remark. It is not true that a mapping whose restriction is continuous on a subset as its domain is continuous. For example, $f(x) = 1$ or -1 according as $x \geq 0$ or $x < 0$ is continuous for $x \geq 0$ but not for all real x.

From this remark, it also follows that if the composition mapping $g \circ f$ of f and g is continuous then f and g both are not necessarily continuous. Take $f \,|\, H$ for g in the above proposition and consider the example in the preceding remark.

It is possible for a noncontinuous mapping to be continuous on a subset or only on certain points. This motivates the introduction of local continuity.

Definition 2. A mapping f of a topological space E_u into a topological space F_v is said to be *continuous at a point* $x_0 \in E$ if for each neighborhood V of $f(x_0)$, $f^{-1}(V)$ is a neighborhood of x_0.

Proposition 3. A mapping $f: E_u \to F_v$ is continuous if and only if f is continuous at each point of E.

Proof. It is simple to verify.

A most useful characterization of continuity is given by:

Theorem 1. Let $f: E_u \to F_v$ be a mapping. The following statements are equivalent:

(a) f is continuous;

(b) for each closed set C in F, $f^{-1}(C)$ is closed;

(c) for each member B of the subbase of the topology v in F_v, $f^{-1}(B)$ is open in E;

(d) f is continuous at each $x \in E$;

(e) for each $x \in E_u$ and for each neighborhood V of $f(x)$, there is a neighborhood U of x such that $f(U) \subset V$;

(f) for each $x \in E$ and for each net $\{x_\alpha, \alpha \in \Gamma\}$ converging to $x \in E$, $\{f(x_\alpha), \alpha \in \Gamma\}$ converges to $f(x)$;

(g) for each $x \in E$ and for each filter $\mathscr{F}$ converging to $x \in E$, $\{f(U): U \in \mathscr{F}\}$ converges to $f(x)$;

(h) for each subset A of E, $f(\bar{A}) \subset \overline{f(A)}$;

(i) for each subset B of F_v, $\overline{f^{-1}(B)} \subset f^{-1}(\bar{B})$.

Proof. (a) $\Leftrightarrow$ (b) For each set C of F_v, C is closed iff $F \backslash C$ is open. Therefore, by definition, (a) is true iff

$$f^{-1}(F \backslash C) = f^{-1}(F) \backslash f^{-1}(C) = E \backslash f^{-1}(C)$$

is open iff $f^{-1}(C)$ is closed.

(b) $\Rightarrow$ (c) For each member B of any subbase of the topology v in F, $f^{-1}(B)$ is open because B, being a member of the subbase, is open and because f is continuous.

(c) $\Rightarrow$ (d)   Let x be an arbitrary point of E. Let V be a neighborhood of $f(x)$. Then there exists a member B of the subbase of v such that $f(x) \in B \subset V$. By (c), $f^{-1}(B)$ is an open set containing x. Since $f(x) \in B \subset V$ implies $x \in f^{-1}(B) \subset f^{-1}(V)$, it follows that $f^{-1}(V)$ is a neighborhood of x.

(d) $\Rightarrow$ (e)   As in (c) $\Rightarrow$ (d), $U = f^{-1}(B)$, being open, is an open neighborhood of x. Hence $f(U) = f(f^{-1}(B)) = B \subset V$ proves (e).

(e) $\Rightarrow$ (f)   Let $x \in E$ and let V be a neighborhood of $f(x)$. By (d) there exists a neighborhood U of x such that $f(U) \subset V$. Since $\{x_\alpha, \alpha \in \Gamma\}$ converges to x, there exists α_0 such that for $\alpha \geq \alpha_0$, $x_\alpha \in U$, whence $f(x_\alpha) \in f(U) \subset V$ for all $\alpha \geq \alpha_0$. Since V is an arbitrary neighborhood of $f(x)$, it follows that $\{f(x_\alpha), \alpha \in \Gamma\}$ converges to $f(x)$.

(f) $\Rightarrow$ (g)   Let $\mathscr{F}$ be a filter converging to $x \in E$. Then the neighborhood filter $\mathscr{U}_x$ of x is finer than $\mathscr{F}$. Clearly $\{x_u$, an element chosen from each U, $U \in \mathscr{U}_x\}$ is a net converging to x and hence $\{f(x_u), U \in \mathscr{U}_x\}$ converges to $f(x)$ by (f). But since $\{f(U), U \in \mathscr{U}_x\}$ is a filter base containing $f(x)$ and since $f(x_u) \in f(U)$, it follows that $\{f(U), U \in \mathscr{U}_x\}$ converges to $f(x)$ and so does $\{f(F), F \in \mathscr{F}\}$ because it is a coarser filter base than $\{f(U), U \in \mathscr{U}_x\}$.

(g) $\Rightarrow$ (h)   Let $y \in f(\bar{A})$. Then $y = f(x)$, where $x \in \bar{A}$. There exists a filter $\mathscr{F}$ in A converging to x. By (g), $\{f(G), G \in \mathscr{F}\}$ converges to $f(x)$. Hence $f(x)$ is a limit point of $\{f(G), G \in \mathscr{F}\}$. Since $f(G) \cap f(A) \neq \varnothing$ for each $G \in \mathscr{F}$, $f(x) \in \overline{f(A)}$. Therefore $f(\bar{A}) \subset \overline{f(A)}$.

(h) $\Rightarrow$ (i)   Let $A = f^{-1}(B)$ for an arbitrary set B in F. Then by (h), $f(\bar{A}) \subset \overline{f(A)}$, or $f(\overline{f^{-1}(B)}) \subset \overline{f(f^{-1}(B))} = \bar{B}$. Hence $\overline{f^{-1}(B)} \subset f^{-1}(\bar{B})$.

(i) $\Rightarrow$ (b)   Let C be a closed subset of F. Then $\bar{C} = C$ and by (i), $f^{-1}(C) = f^{-1}(\bar{C}) \supset \overline{f^{-1}(C)}$. Hence $f^{-1}(C) = \overline{f^{-1}(C)}$ because $f^{-1}(C) \subset \overline{f^{-1}(C)}$, in general. Therefore $f^{-1}(C)$ is closed.

Exercise.   Show that $f: E \rightarrow F$ is continuous iff for each subset $B \subset F$, $f^{-1}(B^\circ) \subset [f^{-1}(B)]^\circ$.

Proposition 4.   Let (E, d) and (F, d') be two metric spaces. Let $f: E \rightarrow F$ be a mapping. The following statements are equivalent:

(a) f is continuous;

(b)   for each $x_0 \in E$ and each $\varepsilon > 0$ there exists $\delta = \delta(\varepsilon, x_0)$ depending upon ε and x_0 such that for $x \in E$, $d(x, x_0) < \delta$ implies $d'(f(x), f(x_0)) < \varepsilon$.

(c)   For each $x \in E$ and for each sequence $\{x_n\}$ converging to x, $\{f(x_n)\}$ converges to $f(x)$.

Proof. (a) $\Rightarrow$ (b)   Since open balls form a base for the metric topology, by part (c) of Theorem 1, $f^{-1}\{y \in F: d'(y, f(x_0)) < \varepsilon\}$ is open and hence there exists a $\delta > 0$ such that

$$\{x \in E: d(x, x_0) < \delta\} \subset f^{-1}\{y \in F: d'(y, f(x_0)) < \varepsilon\}.$$

In other words, for $x \in E$, $d(x, x_0) < \delta$ implies $d'(f(x), f(x_0)) < \varepsilon$.

(b) $\Rightarrow$ (c)   If $\{x_n\}$ is a sequence converging to $x \in E$, then there exists a positive interger n_0 such that $d(x_n, x) < \delta$ for $n \geq n_0$. Hence $d'(f(x_n), f(x)) < \varepsilon$ for all $n \geq n_0$ by (b). Since $\varepsilon > 0$ is arbitrary, it shows that $\{f(x_n)\}$ converges to $f(x)$.

(c) $\Rightarrow$ (a)   Suppose f is not continuous. Then for some $\varepsilon > 0$ and for all integers $n \geq 1$, there exists x_n such that $d(x_n, x_0) < 1/n$, but $d'(f(x_n), f(x_0)) > \varepsilon$. Thus we obtain a sequence $\{x_n\}$ converging to x_0 but $\{f(x_n)\}$ does not converge to $f(x_0)$. This contradicts (c).

Remark.   The statement (c) given in the above proposition is very often called the *sequential continuity*. For metric spaces it is equivalent to ordinary continuity. In general, however, this need not be the case. For example let $[0; \Omega)$ be the set of all ordinals less then the first uncountable ordinal Ω. $[0, \Omega)$ is an ordered space with the order topology defined by the order "$\leq$". It is clear that Ω is a limit point of $[0, \Omega)$ but it is not the sequential (cf. Exercise 15, Chapter II) limit. Thus the identity map: $[0, \Omega] \to [0, \Omega]$ is continuous but not sequentially continuous.

A notion somewhat dual to continuity is that of openness.

Definition 3.   A mapping $f: E_u \to F_v$ is said to be *open* if for each open set U in E, $f(U)$ is open in F. If for each closed set C in E, $f(C)$ is closed in F, then f is called a *closed* mapping.

Proposition 5.   Let $f: E_u \to F_v$ and $g: F_v \to G_w$. If f, g are open (or closed) then so is $g \circ f$.

Proof.   For each open (or closed) set U in E, $f(U)$ is open (or closed) in F and hence $g(f(U))$ is open (or closed) in G since g is open (or closed). Hence $g \circ f$ is open (or closed).

Proposition 6.   Let $f: E_u \to F_v$ be an injective mapping. Then the inverse mapping $f^{-1}: f(E_u) \to E_u$ exists. f is continuous iff f^{-1} is open.

Proof. For each $x \in E$, $f(x) \in F$. Since f is injective, for each $y \in f(E)$ there is only one $x \in E$ such that $f(x) = y$ and so $f^{-1}(y) = x$ is defined. If f is continuous, then clearly for each open set V in F, $f^{-1}(V)$ is open. But a set P in $f(E)$ is open iff for some open set V in F, $P = V \cap f(E)$. Hence

$$f^{-1}(P) = f^{-1}(V) \cap f^{-1}(f(E)) = f^{-1}(V) \cap E = f^{-1}(V)$$

is open. In other words, f^{-1} is open iff f is continuous.

In general, there is no connection between continuous, open, and closed mappings. For example, let R_d be the real line R with the discrete topology and R_u be R with the usual metric topology; then the identity mapping $i: R_d \to R_u$ is continuous but not open and not closed and so $i^{-1}: R_u \to R_d$ is open and closed but not continuous. Also the projection mapping $p: R^2 \to R$ defined by $p(x, y) = x$, $(x, y) \in R^2$ is open but not closed because the set $\{(x, 1/x), x \neq 0\}$ is closed in R^2 but its projection $R \setminus \{0\}$ is not closed in R.

Definition 4. A mapping $f: E_u \to f_v$ which is bijective, continuous, and open is called a *homeomorphism*.

Proposition 7. Let $f: E_u \to F_v$ be a mapping. Then f is open iff for any subset A of E, $f(A^\circ) \subset [f(A)]^\circ$.

Proof. Suppose f is open. Let A be any subset of E. Since A° is open, so is $f(A^\circ)$. But then from the inclusion $f(A^\circ) \subset f(A)$ and the fact that $[f(A)]^\circ$ is the largest open set contained in $f(A)$, we have $f(A^\circ) \subset [f(A)]^\circ$. Conversely, suppose for any subset A of E, $f(A^\circ) \subset [f(A)]^\circ$. Let P be an open subset in E. Then $P = P^\circ$ and so $f(P) = f(P^\circ) \subset [f(P)]^\circ$ by assumption. Since in general $[f(P)]^\circ \subset f(P)$, it follows that $f(P) = [f(P)]^\circ$, which is open and hence f is open.

Proposition 8. Let $f: E_u \to F_v$ be an injective mapping. If f is continuous then $[f(A)]^\circ \subset f(A^\circ)$ for any subset A of E.

Proof. Suppose f is continuous and let A be a subset of E. Since $[f(A)]^\circ$ is open and f continuous, $f^{-1}([f(A)]^\circ)$ is open in E. Clearly

$$f^{-1}([f(A)]^\circ) \subset f^{-1}(f(A)) = A$$

because f is injective. Since A° is the largest open set contained in A, we have $f^{-1}([f(A)]^\circ) \subset A^\circ$ or $[f(A)]^\circ \subset f(A^\circ)$.

Corollary 2. Let $f: E_u \to F_v$ be a continuous, open, and injective mapping. Then for any subset $A \subset E$, $f(A^\circ) = [f(A)]^\circ$.

Proof. Combine Propositions 7 and 8.

Proposition 9. Let $f: E_u \to F_v$ be a mapping. Then f is closed iff for each subset A of E, $\overline{f(A)} \subset f(\bar{A})$.

Proof. Suppose f is closed. If A is any subset of E, then $A \subset \bar{A}$ implies $f(A) \subset f(\bar{A})$. Since f is closed and $\bar{A}$ is a closed subset, $f(\bar{A})$ is closed and so $\overline{f(A)} \subset f(\bar{A})$. Conversely, if the condition holds for each subset A, then for any closed subset C of E, $\overline{f(C)} \subset f(\bar{C}) = f(C) \subset \overline{f(C)}$, because $C = \bar{C}$. This proves that $f(C)$ is closed and hence f is closed.

Proposition 10. Let $f: E_u \to F_v$ be a mapping. Then f is continuous and closed iff for each subset A of E, $f(\bar{A}) = \overline{f(A)}$.

Proof. f is continuous iff $f(\bar{A}) \subset \overline{f(A)}$ (Theorem 1) and f is closed iff $\overline{f(A)} \subset f(\bar{A})$ (Proposition 9). Combining the two inequalities we have $f(\bar{A}) = \overline{f(A)}$ iff f is continuous and closed.

A few useful facts about continuity are collected in the following:

Proposition 11. Let $f: E_u \to F_v$ be a mapping. Then:

(a) f is continuous if E_u is a discrete space;

(b) f is continuous if F_v is an indiscrete space;

(c) if f maps E onto a single point $y_0 \in F_v$, then f is continuous (in other words, each constant map is continuous);

(d) f is continuous and open if both E and F are discrete spaces;

(e) for discrete spaces E and F, f is a homeomorphism iff f is bijective.

Proof. (a) If E_u is a discrete space, then for each open set V in F, $f^{-1}(V)$, being a subset of E, is open. Hence f is continuous.

(b) This follows because the only nonempty open set in F is F and $f^{-1}(F) = E$, which is open.

(c) For each open set V in F, $f^{-1}(V) = E$ or $\emptyset$, according as $y_0 \in V$ or $y_0 \notin V$. Since E and $\emptyset$ are open, f is continuous.

(d) By (a), f is continuous when E is discrete. Similar arguments show that f is open when F is discrete.

(e) For discrete spaces E and F, f is always continuous and open by (d). Hence the map f is a homeomorphism iff f is bijective.

§17. Topologies Defined by Mappings

As mentioned in the first paragraph of §13, very often one wants to know if there exists a topology on a set E such that a mapping or a family of mappings of E into a topological space F_v is continuous. The following propositions answer this question.

Proposition 12. Let E be a set and F_v a topological space. Let f be a mapping of E into F_v. Then there exists a coarsest topology u on E such that $f: E_u \to F_v$ is continuous.

Proof. Let $\{V\}$ be the family of open sets in F_v. Then the family $\{f^{-1}(V)\}$ has the following property: For each finite family

$$f^{-1}(V_i), \quad 1 \le i \le n, \qquad \bigcap_{i=1}^{n} f^{-1}(V_i) = f^{-1}\left(\bigcap_{i=1}^{n} V_i\right)$$

is in $\{f^{-1}(V)\}$ because $\bigcap_{i=1}^{n} V_i$ is in $\{V\}$. Also, for any subfamily $\{f^{-1}(V_\alpha)\}$ in $\{f^{-1}(V)\}$, $\bigcup_\alpha f^{-1}(V_\alpha) = f^{-1}(\bigcup_\alpha V_\alpha)$ is in $\{f^{-1}(V)\}$ because $\bigcup_\alpha V_\alpha$ is in $\{V\}$. Hence $\{f^{-1}(V)\}$ defines a topology u on E. Since $f^{-1}(V)$ is u-open for each V in $\{V\}$, f is continuous. It is easy to check that u is the coarsest topology making $f: E_u \to F_v$ continuous.

Proposition 13. Let E be a set and F_v a topological space. Let $G = \{f_\alpha\}$ be a family of mappings of E into F_v. Then there exists a coarsest topology u on E such that each f_α in G is continuous.

Proof. As in the above proposition, if $\{V\}$ is the family of all open sets in F_v, then for each $f_\alpha \in G$, $\{f_\alpha^{-1}(V)\}$ defines a topology on E. Let u be the topology which has the family $\{f_\alpha^{-1}(V)\}$, where f_α runs over G and V over $\{V\}$, as its subbase. Then u is the coarsest topology such that E endowed with this topology has the property that each $f_\alpha: E_u \to F_v$ is continuous.

§18. Separation Axioms

In this section we describe the "Alexandroff–Hoff Trennungsaxioms" or the so-called separation axioms. There are six basic separation axioms that are frequently used in topology.

Axiom A_0.   For any two distinct points of a topological space, at least one of them has an open neighborhood which does not contain the other point.

Axiom A_1.   For any two distinct points of a topological space, each has an open neighborhood not containing the other point.

Axiom A_2.   For any two distinct points of a topological space, each has an open neighborhood which does not intersect the other.

Axiom A_3.   For any point x and any closed set C such that $x \notin C$ in a topological space there exist disjoint open sets U_i ($i = 1, 2$) such that $x \in U_1$ and $C \subset U_2$.

Axiom A_4.   Given below on page 64.

Axiom A_5.   Let C_i ($i = 1, 2$) be any two disjoint closed subsets of a topological space E_u. Then there exist two disjoint open subsets such that $C_i \subset U_i$, $i = 1, 2$.

Axiom A_6.   Let M_i ($i = 1, 2$) be any two subsets of a topological space E_u such that $(M_1 \cap \bar{M}_2) \cup (\bar{M}_1 \cap M_2) = \varnothing$. Then there exist disjoint open sets U_i such that $M_i \subset U_i$, $i = 1, 2$.

Proposition 14.   Axiom A_i implies Axiom A_{i-1} for $i = 1, 2$, and A_6 implies A_5. But the converses are not true.

Proof.   It is quite clear that A_i implies A_{i-1} for $i = 1, 2$. To show that A_6 implies A_5, let C_i be any two disjoint closed subsets of a topological space. Then $C_1 = \bar{C}_1$, $\bar{C}_2 = C_2$, and $C_1 \cap C_2 = \varnothing$ implies $(C_1 \cap \bar{C}_2) \cup (\bar{C}_1 \cap C_2) = \varnothing$. Hence A_5 follows from A_6. For the converses, see the examples at the end of this chapter.

Definition 5.   A topological space E_u satisfying the axiom A_i is called a T_i-space for $i = 0$–$3, 5, 6$. A T_2-space is also called a *Hausdorff space* (or a *separated* space elsewhere).

A T_1-space satisfying axiom A_i ($i = 3, 5, 6$) is respectively called *regular*, *normal*, and *completely normal*.

T_0-spaces do not play much role in general topological spaces for reasons pointed out in the sequel. It is, however, well known that in topological spaces which have additional algebraic structures, e.g., topological groups, topological linear spaces (cf., Husain [45], [46]), Axiom A_0 always implies A_1 and A_2.

Before we proceed further, we will characterize T_1-spaces.

Proposition 15. A topological space E_u is a T_1-space iff each singleton $\{x\}$ is a closed subset.

Proof. Suppose E_u is a T_1-space. Let $x, y \in E$, $x \neq y$. Since E_u is a T_1-space, there exists an open neighborhood U_y of y such that $x \notin U_y$. Let y run over $E \backslash \{x\}$, then $E \backslash \{x\} = \bigcup_y U_y$, which is an open set, and hence $\{x\}$ is a closed set. To prove the converse, let each singleton in E be a closed set. Let $x \neq y$, then $E \backslash \{x\}$ and $E \backslash \{y\}$ are open and $x \notin E \backslash \{x\}$, $y \notin E \backslash \{y\}$. Hence E_u is a T_1-space.

Proposition 16. Let E_u be a T_1-space and let A be a subset of E_u. Then the set A' of all limit points of A is closed.

Proof. Let $x \in \bar{A'}$. Let V be an open neighborhood of x. Then there exists $y \in A' \cap (V \backslash \{x\})$. Since E is a T_1-space, y has an open neighborhood U such that $x \notin U$. But then $U \cap V$ is a neighborhood of y which does not contain x. Since $U \cap V \subset V$ and $(U \cap V) \cap A \neq \varnothing$, it shows that $(V \backslash \{x\}) \cap A \neq \varnothing$. Thus $x \in A'$ and so A' is closed.

Proposition 17. Let E be any set. Then there exists a unique coarsest topology u on E such that E_u is a T_1-space.

Proof. Let $\mathscr{U}$ be the family of all subsets U of E such that $E \backslash U$ is finite. Clearly E is in $\mathscr{U}$. For each finite subfamily $\{U_i\}$, $i = 1, \ldots, n$, $\bigcup_{i=1}^{n}(E \backslash U_i)$ $= E \backslash \bigcap_{i=1}^{n} U_i$ being finite implies $\bigcap_{i=1}^{n} U_i$ is in $\mathscr{U}$. Also, for any family $\{U_\alpha\}$ in $\mathscr{U}$, $\bigcap_\alpha (E \backslash U_\alpha) = E \backslash \bigcup_\alpha U_\alpha$ is finite and so $\bigcup_\alpha U_\alpha$ is also in $\mathscr{U}$. Now by including $\varnothing$ in $\mathscr{U}$, we have a topology on E. Since for each $x \in E$, $\{x\}$ is finite, $E \backslash \{x\}$ is open and hence $\{x\}$ is closed. Therefore by Proposition 15, E is a T_1-space.

To show that u is the coarsest topology for which E_u is a T_1-space, let v be another topology such that E_v is a T_1-space. Let U be any u-open set, then $E \backslash U = \{x_i, i = 1, \ldots n\}$. Since E_v is a T_1-space, each $\{x_i\}$ is v-closed and hence $E \backslash U$ is v-closed. This shows that U is v-open and so u is the coarsest topology for which E_u is a T_1-space. The uniqueness is obvious.

Recall from calculus that the limit of a convergent sequence in $\mathbb{R}$ or $\mathbb{C}$ is unique. It is conceivable that the limits of nets or filters in a general topological space need not be unique. For example, this happens in indiscrete spaces. Such a situation is neither feasible nor desirable for applications in analysis. Hausdorff spaces avoid this anomaly.

Theorem 2. Let E_u be a topological space. The following statements are equivalent:

(a) E_u is a Hausdorff space;

(b) each convergent net of filter has a unique limit;

(c) the diagonal $\varDelta = \{(x, y): x = y, x, y \in E\}$ is a closed subset of the product space $E_u \times E_u$.

Proof. (a) $\Rightarrow$ (b) Suppose E_u is a Hausdorff space. Let $\{x_\alpha, \alpha \in \varGamma\}$ be a convergent net in E_u. Suppose $\{x_\alpha\}$ converges to x_1 and x_2. Suppose $x_1 \neq x_2$. Since E_u is a Hausdorff space, there exist two disjoint open sets U and V such that $x_1 \in U$ and $x_2 \in V$. That means that $\{x_\alpha, \alpha \in \varGamma\}$ is eventually in both U and V. But this is impossible. Hence (b) follows.

(b) $\Rightarrow$ (c) Let $(x, y) \in \bar{\varDelta}$. Then there exists a net $\{(x_\alpha, y_\alpha), \alpha \in \varGamma\}$ in $\varDelta$ such that (x_α, y_α) converges to (x, y). Since $(x_\alpha, y_\alpha) \in \varDelta$, $x_\alpha = y_\alpha$ for all $\alpha \in \varGamma$ and since $\{x_\alpha\}$ converges to x, x must be equal to y because otherwise (b) would be contradicted.

(c) $\Rightarrow$ (a) Let $x \neq y$, then $(x, y) \notin \varDelta$. Since $\varDelta$ is closed, there exists an open neighborhood $U \times V$ of (x, y) such that $(U \times V) \cap \varDelta = \varnothing$, where U is a neighborhood of x and V that of y. Whence $U \cap V = \varnothing$ and E_u is a Hausdorff space.

Corollary 3. Let E_u be a topological space and v any other coarser Hausdorff topology on E. Then E_u is also a Hausdorff space.

Proof. Let $x, y \in E$, $x \neq y$. Since E_v is Hausdorff, there exist disjoint v-open neighborhoods U, V of x and y, respectively. Since $u \supset v$, U and V are u-open and hence E_u is Hausdorff.

Theorem 3. (Characterization of regular spaces). Let E_u be a topological space. The following statements are equivalent:

(a) E_u is a T_3-space (or regular space);

(b) for each $x \in E$ and each open neighborhood U of x, there exists an open neighborhood V of x such that $\bar{V} \subset U$; in other words, each point in E has a fundamental system of closed neighborhoods (if in addition E is a T_1-space).

Proof. (a) $\Rightarrow$ (b) Let U be an open neighborhood of x. Then $E \backslash U$ is closed and $x \notin E \backslash U$. By (a), there exist disjoint open sets V and W such that $x \in V$ and $E \backslash U \subset W$. Since W is open, $E \backslash W$ is closed and

$V \subset E \setminus W$ (because $V \cap W = \varnothing$) and so $\bar{V} \subset E \setminus W$. Thus the relations: $x \in V \subset \bar{V} \subset E \setminus W \subset U$ establish (b).

(b) $\Rightarrow$ (a)   Let C be a closed set and $x \notin C$. Then $E \setminus C$ is an open neighborhood of x. By (b), there exists an open neighborhood V of x such that $x \in V \subset \bar{V} \subset E \setminus C$. Clearly $E \setminus \bar{V} \supset C$, $x \in V$ and $(E \setminus \bar{V}) \cap V = \varnothing$. Since V and $E \setminus \bar{V}$ are open sets, (a) is proved.

Proposition 18.   Let E_u be a regular space. Then for any closed subset C and any element $x \in E \setminus C$, there exist open sets U and V containing x and C respectively such that $\bar{U} \cap \bar{V} = \varnothing$.

Proof.   By Theorem 3 above, there exists an open neighborhood W of x such that $x \in W \subset \bar{W} \subset E \setminus C$. Repeating the same argument, there exists an open neighborhood U of x such that $x \in U \subset \bar{U} \subset W$. Putting $V = E \setminus \bar{W}$, we see that $C \subset V \subset \bar{V} \subset E \setminus W$, because $E \setminus W$ is closed. Clearly $\bar{U} \cap \bar{V} = \varnothing$.

Proposition 19.   Every regular space E_u is a Hausdorff space.

Proof.   Let $x, y \in E$, $x \neq y$. By hypothesis x has an open neighborhood U which does not contain y. Then by the regularity (Theorem 3) there exists an open neighborhood V of x such that $x \in V \subset \bar{V} \subset U$. Since $y \notin U$, it follows that $y \notin \bar{V}$. This shows that $y \in E \setminus \bar{V}$. Since $V \cap (E \setminus \bar{V}) = \varnothing$ and V, $E \setminus V$ are open neighborhoods of x and y respectively, this proves that E is a Hausdorff space.

Definition 6.   A T_1 topological space E_u is said to be *completely regular* if the following holds:

Axiom A_4.   For any closed subset C of E and any element $x \in E \setminus C$, there exists a continuous function from E_u into the closed unit interval $[0, 1]$ such that $f(x) = 0$ and $f(y) = 1$ for all $y \in C$, or $f(x) = 0$ and $f(C) = 1$.

Proposition 20.   Every completely regular space E_u is regular.

Proof.   Let C be a closed subset of E_u and $x \in E \setminus C$. There exists a continuous function f from E into $I = [0, 1]$ such that $f(x) = 0, f(C) = 1$. The intervals $[0, \frac{1}{3})$ and $(\frac{2}{3}, 1]$ are open in I, hence by the continuity of f, $U = f^{-1}([0, \frac{1}{3}))$ and $V = f^{-1}((\frac{2}{3}, 1])$ are open in E_u. Clearly $U \cap V = \varnothing$, $x \in U$ and $C \subset V$. This proves that E_u is regular.

The converse of the above proposition is not true (see Exercise 16).

Remark. It is possible to define a topological space E_u to be completely regular without satisfying the T_1-axiom. But then such spaces need not be Hausdorff.

However, the following is true:

Proposition 21. Every completely regular space is a Hausdorff space.

Proof. This follows from Propositions 19 and 20.

Remark. A completely regular space is sometimes called a *Tychonoff space* and the space defined in Definition 6 without the T_1-axiom is called a T_4-space.

Proposition 22. On each completely regular space (or a Tychonoff space) E, the set $C(E)$ of all continuous real-valued functions separates points, i.e., for each pair $x, y, x \neq y$, there exists an $f \in C(E)$ with $f(x) \neq f(y)$.

Proof. Let $x, y \in E$, $x \neq y$. Since E is a T_1-space, $\{y\}$ is a closed subset of E. By the definition of complete regularity of E, there exists a continuous function $f: E \to [0, 1]$ such that $f(x) = 0$, $f(y) = 1$. Hence f separates x and y.

Theorem 4. Let E_u be a T_1-topological space. The following statements are equivalent:

(a) E is completely regular;

(b)   for each x and each open neighborhood P of x, there exists a family $\{U_\alpha, \alpha \in D\}$ of open sets such that $x \in U_\alpha \subset P$ for all α and for $\alpha < \beta$, $U_\alpha \subset \bar{U}_\alpha \subset U_\beta$, where D is the set of all dyadic rationals in $[0, 1]$.

Proof. (a) $\Rightarrow$ (b)   Suppose E is completely regular. Let $x \in P$, where P is an open neighborhood of x. Then by (a), there exists a continuous function f of E into $[0, 1]$ such that $f(x) = 0$ and $f(E \backslash P) = 1$. For each $\alpha \in D$, define

$$U_\alpha = \{y \in E: f(y) < \alpha\}.$$

Clearly,

$$\bar{U}_\alpha \subset \{y \in E: f(y) \leq \alpha\} \subset \{y \in E: f(y) < \beta\},$$

if $\alpha < \beta$ for $\alpha, \beta \in D$. Thus for $\alpha < \beta$ we have $U_\alpha \subset \bar{U}_\alpha \subset U_\beta$. Moreover, each U_α is open because f is continuous and $U_\alpha \subset P$ because $0 \leq \alpha < 1$ for $\alpha \in D$.

(b) $\Rightarrow$ (a)   Let C be a closed subset of E and let $x \in E \backslash C$. By (b) there exists a family $\{U_\alpha, \alpha \in D\}$ of open sets U_α such that $x \in U_\alpha \subset E \backslash C$, and for $\alpha < \beta, \alpha, \beta \in D, \bar{U}_\alpha \subset U_\beta$. We define the following function:

$$f(y) = \begin{cases} \inf\{\alpha: y \in U_\alpha, \alpha \in D\}; \\ 1, \ y \notin \bigcup_{\alpha \in D} U_\alpha. \end{cases}$$

It is clear that $0 \leq f(y) \leq 1$ for all $y \in E$. Since $x \in U_\alpha$ for all $\alpha \in D, f(x) = 0$. Furthermore, if $y \in C$ then $y \notin \bigcup_{\alpha \in D} U_\alpha$ because $U_\alpha \subset E \backslash C$ for each α and hence $f(y) = 1$ for all $y \in C$.

To show that f is continuous, first let $0 \leq \alpha \leq 1$; then $f(y) < \alpha$ if and only if $y \in U_\gamma$ for some $\gamma < \alpha$ and hence

$$\{y \in E: f(y) < \alpha\} = \bigcup_{\gamma < \alpha} U_\gamma,$$

which is open.

Secondly, let $0 \leq \alpha < 1$, then $f(y) > \alpha$ if and only if $y \notin \bar{U}_\gamma$ for some $\gamma > \alpha$ and hence

$$\{y \in E: f(y) > \alpha\} = \bigcup_{\gamma > \alpha} (E \backslash \bar{U}_\gamma)$$

is also an open set. Since the intervals $[0, \alpha)$ and $(\alpha, 1]$ form a subbase of the topology of $[0, 1]$, we have established that f is continuous by Theorem 1, §16.

Theorem 5.   Let E_u be a T_1-topological space. The following statements are equivalent:

(a) E is normal;

(b)   for each closed set C and any open set U containing C there exists an open set V such that $C \subset V \subset \bar{V} \subset U$;

(c)   for each closed set C and any open set U containing C there exists a subset A of E such that $C \subset A° \subset \bar{A} \subset U$.

Proof.   (a) $\Rightarrow$ (b)   Let C be a closed set and U an open set such that $C \subset U$. Since $E \backslash U$ is closed and $C \cap (E \backslash U) = \varnothing$, by (a) there exist disjoint open sets V and W such that $C \subset V$ and $E \backslash U \subset W$. Since $V \cap W = \varnothing$ implies $V \subset E \backslash W$, we have $\bar{V} \subset E \backslash W$ because W is open. But $(E \backslash U) \subset W$ implies $(E \backslash W) \subset U$ and so we have $C \subset V \subset \bar{V} \subset U$. This proves (b).

(b) $\Rightarrow$ (a)   Let C_i ($i = 1, 2$) be two disjoint closed subsets in E. Clearly $E \setminus C_i$ ($i = 1, 2$) are open and $C_1 \subset E \setminus C_2$, $C_2 \subset E \setminus C_1$. By (b), there exists an open set V such that $C_1 \subset V \subset \bar{V} \subset E \setminus C_2$. Putting $U = E \setminus \bar{V}$, which is open and disjoint with $V \subset \bar{V} \subset E \setminus C_2$, it follows that $U \supset C_2$. Thus U and V are the desired disjoint open sets separating C_1 and C_2. This proves (a).

(c) $\Leftrightarrow$ (a)   This is obtained by putting $V = A$ in (b) and by observing that $V = V^\circ = A^\circ = A$, because V is open.

Definition 7.   A topological space E is said to be *perfectly normal* if it is normal and each closed subset of E is the intersection of a countable family of open sets (i.e., each closed subset is a G_δ-set}.

Proposition 23.

(a)   Each completely normal space is normal;

(b)   each perfectly normal space is normal.

Proof.   Clearly (a) and (b) follow from the definitions.

Remark.   The class of normal spaces contains those of completely normal and perfectly normal spaces. We show that the class of all normal spaces also contains the class of all metrizable spaces and the latter lies in the intersection of the classes of completely normal and perfectly normal spaces.

Theorem 6.   Every metrizable space E_u is completely normal as well as perfectly normal.

Proof.   Let E_u be a topological space, the topology of which is defined by a metric d. First we show that E_u is completely normal. Let F_1 and F_2 be two separated subsets of E such that $(F_1 \cap \bar{F}_2) \cup (\bar{F}_1 \cap F_2) = \varnothing$. Let $U_1 = \{x \in E : d(x, F_1) < d(x, F_2)\}$ and $U_2 = \{x \in E : d(x, F_2) < d(x, F_1)\}$, where $d(x, A) = \inf_{y \in A} d(x, y)$, for any subset A of E and any element $x \in E$. Since for any subset $A \neq \varnothing$, the mapping $x \to d(x, A)$ is continuous (Exercise 19). Also, for any $x \in F_1$, $d(x, F_1) = 0$ and since $F_1 \cap \bar{F}_2 = \varnothing$, $d(x, F_2) > 0$. Therefore $F_1 \subset U_1$. Similarly $F_2 \subset U_2$. Also $U_1 \cap U_2 = \varnothing$, because for no x, can we have $d(x, F_1) < f(x, F_2)$ and $d(x, F_1) > d(x, F_2)$ at the same time. Thus E is completely normal and hence normal (Proposition 23).

To show that E is also perfectly normal, let C be a closed subset of E. Then for each positive integer n, $U_n = \{x: d(x, C) < 1/n\}$ is open. Clearly $C \subset \bigcap_{n=1}^{\infty} U_n$. For the reverse inclusion, let $x \in \bigcap_{n=1}^{\infty} U_n$. Then $d(x, C) < 1/n$ for all $n \geq 1$. Hence $d(x, C) = 0 \Leftrightarrow x \in \bar{C} = C$ (Corollary 2, §12). This proves that C is a G_δ-set.

Proposition 24. Let E be a normal space and F a T_1-topological space. If $f: E \to F$ is a continuous closed surjective map, then F is also normal.

Proof. Let C be a closed set and U open in F such that $C \subset U$. Then $f^{-1}(C) \subset f^{-1}(U)$. Since f is continuous, $f^{-1}(C)$ is closed and $f^{-1}(U)$ open in E. By normality of E, there exists a subset B of E such that $f^{-1}(C) \subset B^\circ \subset B \subset f^{-1}(U)$ (Theorem 5). Since f is continuous and closed, $\overline{f(B)} = f(\bar{B})$ (Proposition 10, §16). Thus we have $C \subset f(B^\circ) \subset \overline{f(B)} = f(\bar{B}) \subset U$. Now we show that $f(B^\circ) \subset (f(B))^\circ$ or $B^\circ \subset f^{-1}((f(B))^\circ)$. By Proposition 7, §9, since f is surjective, we have

$$f^{-1}((f(B))^\circ) = f^{-1}(\overline{[f(B)]^c}^c) = (f^{-1}\overline{(f(B))^c})^c \supset (\overline{f^{-1}(f(B))^c})^c$$
$$= ((\overline{f^{-1}f(B))^c})^c \supset (\overline{B^c})^c = B^\circ.$$

Thus we have shown that

$$C \subset f(B^\circ) \subset (f(B))^\circ \subset \overline{f(B)} \subset U.$$

By putting $f(B) = A$ and using Theorem 5, we see that F is a normal space.

§19. Continuous Functions on Normal Spaces

It is obvious that each constant mapping of a topological space E_u into another topological space F_v is continuous. Now the question is: On what topological spaces do there exist nonconstant continuous mappings? Since the continuity of a mapping depends upon the topologies of its domain and range spaces, it is relevant to fix the range space once and for all, e.g., the closed unit interval or the real line, and then ask for the domain spaces on which there exist nonconstant continuous functions. We have seen that on completely regular spaces there exist continuous functions into [0, 1] such that the values of f on a closed set and at any point outside the closed set are different. We shall see that not even on regular Hausdorff spaces do there exist nonconstant continuous functions. In this section, we prove the Urysohn lemma, which deals with the existence of nonconstant continuous functions with a stronger property, viz., separating disjoint closed sets. We also prove Tietze's extension theorem.

Theorem 7. (Urysohn's lemma). Let E_u be a normal space and C_i ($i = 0, 1$) two closed disjoint subsets of E_u. Then there exists a continuous function f on E_u into the finite closed interval $[a, b]$ such that $f(C_0) = a$ and $f(C_1) = b$, where a and b are real numbers and $a \neq b$.

Proof. Observe that $t \to a + (b - a)t$ is a continuous bijective mapping of the closed unit interval $[0, 1]$ onto $[a, b]$. Moreover $g^{-1}(t) = (t-a)/(b-a)$ is a continuous $1 : 1$ mapping of $[a, b]$ onto $[0, 1]$. Hence g is a homeomorphism of $[0, 1]$ onto $[a, b]$. Therefore, for the proof of the theorem it is sufficient to prove the existence of a continuous mapping from E_u into $[0, 1]$ such that $f(C_0) = 0$ and $f(C_1) = 1$.

Let D denote the set of all dyadic rational numbers in $[0, 1]$. Each $r \in D$ is for the form: $r = m/2^n$ where m and n are nonnegative integers such that $m \leq 2^n$. Clearly D is countable.

Put $U_1 = E \backslash C_1$; then $C_0 \subset U_1$ because $C_0 \cap C_1 = \emptyset$. Since E is normal, there exists (Theorem 5, §18) an open set $U_{1/2}$ such that $C_0 \subset U_{1/2} \subset \bar{U}_{1/2} \subset U_1$. Similarly, by normality of E there exist open sets $U_{1/4}$, $U_{3/4}$ such that $C_0 \subset U_{1/4} \subset \bar{U}_{1/4} \subset U_{1/2} \subset \bar{U}_{1/2} \subset U_{3/4} \subset \bar{U}_{3/4} \subset U_1$. Continuing this process, by induction we define U_r for each $r \in D$ such that $C_0 \subset U_r \subset \bar{U}_r \subset U_1$ and $\bar{U}_r \subset U_s$ if $r < s$ for $r, s \in D$.

Now define the function f as follows:

$$f(x) = \begin{cases} \inf\{r : x \in U_r, r \in D\}; \\ 1, \quad \text{if } x \notin U_r \text{ for } r \in D. \end{cases}$$

It is clear that $f(x) = 1$ if $x \in C_1$ and $f(x) = 0$ if $x \in C_0$. Also by definition $0 \leq f(x) \leq 1$ for all $x \in E$. We have to show only that f is continuous. For this we observe that the family of intervals $[0, a)$, $(b, 1]$, $a, b \in (0, 1)$ form a subbase of the topology on $[0, 1]$. Now for $0 \leq a \leq 1$, $f(x) < a$ if and only if $x \in U_r$ for some $r \in D$, $r < a$. Thus

$$\{x : 0 \leq f(x) < a\} = \bigcup_{r < a} \{U_r : r \in D\},$$

being the union of open sets, is open. Similarly, for $0 \leq b < 1$, $f(x) > b$ if and only if $x \notin \bar{U}_r$ for some $r > b$, $r \in D$. Thus

$$\{x : f(x) > b\} = \bigcup_{r > b} \{(E \backslash \bar{U}_r) : r \in D\}$$

is again an open set. Hence f is continuous.

Corollary 4.   Every normal space is completely regular.

Proof.   Since each singleton in a T_1-space is a closed set, the corollary follows from the above theorem combined with the definition of complete regularity.

The converse is not true (see Example 5).

Corollary 5.   On each normal space E and $A \subsetneq E$ closed, there exists a continuous function $f \colon E \to [0, 1]$ such that $f(A) = 0$ or $A \subset f^{-1}(0)$.

Proof.   Let E be a normal space and $x \in E \backslash A$. Since $\{x\}$, A are disjoint closed sets, by the above theorem there exists a continuous function f on E into $[0, 1]$ such that $0 = f(A) \neq f(x) = 1$.

Definition 8.   Let F be a subset of a set E and G any set. A mapping f of E into G is said to be an *extension* of a mapping g of F into G if $f(x) = g(x)$ for all $x \in F$. Alternatively, we say f *extends* g from F to E.

If f extends g uniquely, i.e., there is only one $f \colon E \to G$ which extends a given $g \colon F \to G$, then f is said to be the *unique extension*. If f and g are continuous, f is said to be a *continuous extension* of g.

Another useful characterization of normal spaces is given by the following:

Theorem 8.   Let E_u be a T_1-topological space. The following statements are equivalent:

(a) E is normal.

(b)   For each pair of disjoint closed subsets A and B of E, there exists a continuous function f on E into $[a, b]$ for any real number $a, b, a \neq b$, such that $f(A) = a$ and $f(B) = b$.

(c)   (Tietze's extension theorem)   For each closed subset F of E, each continuous real-valued function (bounded or not) on F has a continuous extension.

Proof.   We have already proved that (a) $\Rightarrow$ (b) (cf., Theorem 7, Urysohn's lemma).

(b) $\Rightarrow$ (a)   Let A and B be two disjoint closed subsets of F. Then by (b) there exists a continuous function f on E such that $f(A) = 0$ and $f(B) = 1$. Now the sets $U = \{x \colon f(x) < \tfrac{1}{2}\}$ and $V = \{x \colon f(x) > \tfrac{1}{2}\}$ are disjoint open sets in E and $A \subset U$, $B \subset V$. This proves (a).

(b) ⇒ (c)   Let F be a closed subset of a normal space E and let g be a continuous function on F. Since R and $R \cup \{\infty, -\infty\} = \bar{R}$ are homeomorphic with (a, b) and $[a, b]$ respectively for some a, b, we can regard g to be a continuous map from F into $[0, 1]$ in view of the fact that $[a, b]$ is homeomorphic with $[0, 1]$ and $(a, b) \subset [a, b]$.

If g is unbounded and $g(x) = \infty$ for some $x \in F$, we consider the extended real line $\bar{R} = R \cup \{\infty, -\infty\}$, which is homeomorphic with $[0, 1]$. In any case it is sufficient to prove the theorem under the additional condition that g is a continuous mapping of F into the closed interval $[0, 1]$.

The main idea is to approximate g uniformly by a sequence of continuous functions on E. The limit function of the sequence will work.

The subsets

$$C_1 = \{x \in F: g(x) \leq \tfrac{1}{3}\} \quad \text{and} \quad D_1 = \{x \in F: g(x) \geq \tfrac{2}{3}\}$$

are disjoint closed sets in F and hence closed in E because F is a closed subset of E. Therefore, by (b), there exists a continuous function f_1 from E into $[0, \tfrac{1}{3}]$ such that $f_1(C_1) = 0$ and $f_1(D_1) = \tfrac{1}{3}$. But then it is clear that $0 \leq g - f_1 \leq \tfrac{2}{3}$ on F. Suppose we have determined continuous functions f_k from E into $[0, \tfrac{1}{3}(\tfrac{2}{3})^{k-1}]$, $k = 1, \ldots, n$, such that

$$0 \leq g - \sum_{k=1}^{n} f_k \leq \left(\frac{2}{3}\right)^n$$

on F.

Putting

$$g_{n+1} = g - \sum_{k=1}^{n} f_k,$$

we note again that

$$C_{n+1} = \{x \in F: g_{n+1}(x) < \tfrac{1}{3}(\tfrac{2}{3})^{n-1}\}$$

and

$$D_{n+1} = \{x \in F: g_{n+1}(x) \geq (\tfrac{2}{3})^{n+1}\}$$

are closed disjoint subsets of E. Hence, by (b), there exists a continuous function f_{n+1} from E into $[0, \tfrac{1}{3}(\tfrac{2}{3})^n]$ such that

$$f_{n+1}(C_{n+1}) = 0 \quad \text{and} \quad f_{n+1}(D_{n+1}) = \tfrac{1}{3}(\tfrac{2}{3})^n.$$

Thus on F we have

$$0 \leq g - \sum_{k=1}^{n+1} f_k = g_{n+1} - f_{n+1} \leq \left(\frac{2}{3}\right)^{n+1}$$

Hence by induction for each positive integer n, we have a continuous function f_n from E into $[0, \frac{1}{3}(\frac{2}{3})^{n-1}]$ such that

$$0 \le g - \sum_{k=1}^{n} f_k \le \left(\frac{2}{3}\right)^n$$

on F. Since $0 \le g \le 1$ on F by supposition, it follows that for all $x \in E$,

$$\sum_{k=1}^{\infty} f_k(x) \le \sum_{k=1}^{\infty} \frac{1}{3} \left(\frac{2}{3}\right)^{k-1} = M < \infty$$

for some finite positive number M. Hence by the well-known theorem (§50, Corollary 12) on uniform convergence,

$$f \equiv \sum_{k=1}^{\infty} f_k$$

is a continuous function on E. Moreover, for each $x \in F$,

$$|g(x) - f(x)| = \lim_{n \to \infty} |g(x) - \sum_{k=1}^{n} f_k(x)| \le \lim_{n \to \infty} \left(\frac{2}{3}\right)^n = 0.$$

This shows that f is an extension of g and the proof is complete.

(c) $\Rightarrow$ (b)   Let A, B be two disjoint closed subsets of E. Then $A \cup B$ is also a closed subset of E. The function $g(x) = 0$ or 1 according as $x \in A$ or $x \in B$ is a continuous function of $A \cup B$ into $[0, 1]$. By (c), there exists a continuous function f from E into $[0, 1]$ which is an extension of g. This proves (b) in view of the early remark in the proof of (b) $\Rightarrow$ (c).

Theorem 9.   Every normal second countable space is homeomorphic to a subspace of the Hilbert cube l_2^1, where

$$l_2^1 = \left\{ \{x_i\}_{i=1}^{\infty} : \sum_{i=1}^{\infty} |x_i|^2 \le 1 \right\},$$

with the induced l_2-norm topology.

Proof.   Let $\{U_n\}$ be a countable base of open sets in a normal space E. By Theorem 5, §18, we may assume that for each U_n there exists $U_m \in \{U_n\}$ such that $\bar{U}_m \subset U_n$. Observe that the sequence $\{U_n\}$ covers E. Now for each pair U_m, $U_n \in \{U_n\}$ with $\bar{U}_m \subset U_n$, there exists a continuous function $f_{nm}: E \to [0, 1]$ such that $f_{nm}(\bar{U}_m) = 0$ and $f_{nm}(E \setminus U_n) = 1$ by Theorem 7. We rearrange $\{f_{nm}: n \ge 1, m \ge 1\}$ in a sequence and denote it by $\{f_n\}$.

If U_n consists of a single point, i.e., it is an isolated point, then we take $f_n = 0$ for all large n in the rearrangement. Now define the map φ of E into the Hilbert space l_2 by

$$\varphi(x) = \left\{\frac{f_n(x)}{n}\right\}, \qquad n \geq 1.$$

Since $0 \leq f_n(x) \leq 1$ for all $n \geq 1$, we have

$$\sum_{n=1}^{\infty} |f_n(x)|^2\, n^{-2} \leq \sum_{n=1}^{\infty} n^{-2} < \infty.$$

Hence $\varphi(x) \in l_2$. Now we prove that φ is a homeomorphism into.

To prove that φ is injective, let $x, y \in E$, $x \neq y$. Certainly there exist $U_m, U_n \in \{U_n\}$ such that $x \in U_m \subset \bar{U}_m \subset U_n$ and $y \in E \setminus \bar{U}_n$, because $\{U_n\}$ is the base and E is normal. Hence the function f_n corresponding to U_n has the property

$$f_n(x) = 0 \neq 1 = f_n(y).$$

Hence

$$\left\{\frac{f_n(x)}{n}\right\} \neq \left\{\frac{f_n(y)}{n}\right\}$$

and so φ is $1 : 1$.

To show that φ is continuous, let $\varepsilon > 0$ be given and let $\varphi(x_0) \in \varphi(E) \subset l_2$. For any $x \in E$, $0 \leq f_n(x) \leq 1$ for all $n \geq 1$ and so there exists a positive integer n_1 such that

$$\sum_{n=n_1}^{\infty} \left|\frac{f_n(x) - f_n(x_0)}{n}\right|^2 \leq \sum_{n=n_1}^{\infty} \frac{4}{n^2} < \frac{\varepsilon^2}{2}.$$

Since each f_n is continuous, there exists a neighborhood U_n of x_0, $1 \leq n < n_1$ such that for all $x \in U_n$ $(1 \leq n < n_1)$

$$\frac{|f_n(x) - f_n(x_0)|^2}{n^2} < \frac{\varepsilon^2}{2n_1}.$$

Hence for all $x \in \bigcap_{n=1}^{n_1-1} U_n = U$ (which is a neighborhood of x_0), we have

$$\sum_{n=1}^{\infty} n^{-2}|f_n(x) - f(x_0)|^2 = \left(\sum_{n=1}^{n_1} + \sum_{n_1+1}^{\infty}\right)\left|\frac{f_n(x) - f_n(x_0)}{n}\right|^2 \leq \frac{n_1}{2n_1}\varepsilon^2 + \frac{\varepsilon^2}{2} < \varepsilon^2.$$

In other words, for all $x \in U$,

$$d\big(\varphi(x), \varphi(x_0)\big) = \|\varphi(x) - \varphi(x_0)\|_2 = \left[\sum_{n=1}^{\infty} \frac{|f_n(x) - f_n(x_0)|^2}{n^2}\right]^{1/2} < \varepsilon,$$

and so φ is continuous at x_0. Since x_0 is arbitrary, φ is continuous for all $x \in E$.

To show that φ is open onto $\varphi(E)$, let $x_0 \in E$. We show that for any open neighborhood U of x_0, $\varphi(U)$ is an open subset of $\varphi(E) \subset l_2$. Since $x_0 \in U$, there exist open sets U_{m_0}, $U_{n_0} \in \{U_n\}$ such that $x_0 \in U_{m_0} \subset \bar{U}_{m_0} \subset U_{n_0} \subset U$ and hence there exists a continuous function $f_{n_0} \colon E \to [0, 1]$ such that $f_{n_0}(x_0) = 0$ and $f_{n_0}(E \setminus U) = 1$. Now for any $x \in E \setminus U$,

$$\frac{|f_{n_0}(x) - f_{n_0}(x_0)|^2}{n_0^2} = n_0^{-2}.$$

Hence for all $x \in E \setminus U$,

$$d\big(\varphi(x), \varphi(x_0)\big) = \left[\sum_{n=1}^{\infty} \frac{|f_n(x) - f_n(x_0)|^2}{n^2} \right]^{1/2} \geq \left[\frac{|f_{n_0}(x) - f_{n_0}(x_0)|^2}{n_0^2} \right]^{1/2} = \frac{1}{n_0}.$$

Thus $\varphi(U)$ contains the intersection of a ball $B_{1/n_0}\big(f(x_0)\big)$ of radius $1/n_0$ with $\varphi(E)$. In other words, $\varphi(U)$ is relatively open in $\varphi(E)$. This completes the proof.

Examples and Exercises

1. (A topological space which is not a T_0-space.) Let $X = \{a, b\}$ be a 2-point set and $a \neq b$. The collection $\mathscr{E} = (X, \varnothing, \{a\})$ defines a topology on X which is not a T_0-space. If $a = 0$, $b = 1$, then $X' = \{0, 1\}$ with the topology is not a T_1-space but a T_0-space. $(X', \mathscr{E})$ is called *Sierpiński space*.

2. (A topological space which is a T_1-space but not Hausdorff.) Consider X, an infinite set with the topology $\mathscr{E} = \{\varnothing$, and $U \subset X$ such that $X \setminus U$ is finite$\}$. Then $(X, \mathscr{E})$ is a T_1-space but not Hausdorff.

3. (A T_3-space need not be a T_1-space.) Consider $X = \{a, b, c\}$ and $\mathscr{E} = \{\{a\}, \{b, c\}, \{a, b, c\}, \varnothing\}$. Then $(X, \mathscr{E})$ is a T_3-space but not a T_1-space.

4. (A Hausdorff space which is not a T_3-space.) Consider $E = \mathbb{R}$, reals with the topology $\mathscr{E}$ defined by open intervals and the rationals as subbasis. Then $(\mathbb{R}, \mathscr{E})$ is a Hausdorff space because $\mathscr{E}$ is finer than the Euclidean metric topology but it is not a regular space.

5. Let $[0, \Omega]$ denote the set of all ordinals less than or equal to the first uncountable ordinal Ω and $[0, \omega]$ denote the set of all ordinals less than or equal to the first infinite ordinal, both with the order topology. Then $[0, \Omega] \times [0, \omega]$ is a normal space and is called the *Tychonoff plank*. Let $X = [0, \Omega] \times [0, \omega] \setminus (\Omega, \omega)$ with the corner point (Ω, ω) of the plank removed. Then X is a completely regular space but not normal.

6. Let A be an uncountable set and $I = [0, 1]$. Then I^A is a normal space but not perfectly normal.

7. Let $f: E \to F$ be continuous and $g: F \to G$. Show that if $g \circ f$ is continuous then g is not necessarily continuous.

8. Show that the only Hausdorff topology on a finite set is the discrete topology.

9. Show that the discrete space is normal (and thus a T_5-space).

10. An indiscrete space is a T_5-space but not a T_0-space.

11. Show that $[0, 1]$ is completely normal.

12. Show that an open mapping need not be continuous.

13. Show that a continuous mapping need not be open.

14. Show that the continuous image of a connected space is connected.

15. Let $(E, \mathscr{T})$ be connected and $\mathscr{T}'$ be a topology coarser than $\mathscr{T}$. Then $(E, \mathscr{T}')$ is connected.

16. A regular space need not be completely regular (see also Novak [69]). Let $I = (0, 1)$ and $Q_2 = \{(r, s) \in I^2: r, s \text{ rational in } (0, 1)\}$. Let

$$X = Q_2 \cup \{(0, 0)\} \cup \{(1, 0)\}.$$

    Put

$$C = \{(\tfrac{1}{4}, r\sqrt{2}); \ r \text{ rational and } 0 < r\sqrt{2} < 1\};$$
$$D = \{(\tfrac{1}{2}, r\sqrt{3}); \ r \text{ rational and } 0 < r\sqrt{3} < 1\};$$
$$E = \{(\tfrac{3}{4}, r\sqrt{5}); \ r \text{ rational and } 0 < r\sqrt{5} < 1\}.$$

    Let $X \backslash [D \cup \{(0, 0), (1, 0)\}]$ be endowed with the product topology induced from I^2 and the neighborhoods of other points be given by

$$U_n(0, 0) = [0, \tfrac{1}{4}) \times [0, 1/n),$$
$$U_n(1, 0) = (\tfrac{3}{4}, 1] \times [0, 1/n),$$
$$U_n(\tfrac{1}{2}, r\sqrt{3}) = (\tfrac{1}{4}, \tfrac{3}{4}) \times (r\sqrt{3} - 1/n, r\sqrt{3} + 1/n)$$

    for $n \geq 1$. Show that X is regular but not completely regular.

17. Show that each continuous real-valued function f on the space $[0, \Omega)$ of ordinals less than the first uncountable ordinal Ω endowed with the order topology is eventually constant, i.e., there exists $\beta < \Omega$ such that f is constant on $[\beta, \Omega)$. [*Hint:* For each positive integer n, there is $\alpha_n < \Omega$ such that for all $\beta > \alpha_n$, $|f(\beta) - f(\alpha)| < 1/n$. For, otherwise, there exists a positive integer n such that for all $\alpha < \Omega$ there exists $\beta > \alpha$ with $|f(\alpha) - f(\beta)| \geq 1/n$. By induction there exists a sequence $\{\beta_i\}$ such that $\beta_i < \beta_{i+1}$ for $i \geq 1$ and $|f(\beta_i) - f(\beta_{i+1})| \geq 1/n$. If $\beta = \sup \beta_i$, then f cannot be continuous at β. Put $\alpha = \sup \alpha_n < \Omega$ and f is constant on $[\alpha, \Omega)$.]

Show that $[0, \Omega]$ is Lindelöf but not second countable.

18. Show that on the set $\mathscr{C}$ of all topologies on a nonempty set X, the operations $\vee$ and $\wedge$ defined in Definition 2 make $\mathscr{C}$ a complete lattice.

19. If (X, d) is a metric space, show that for any nonempty subset $A \subset X$, the mapping $x \to d(x, A)$ is continuous.

20. If A is a subset of a topological space X and χ_A is the characteristic function of A (i.e. $\chi_A(x) = 1$ or 0 according to whether $x \in A$ or $x \notin A$). Then $\chi_A \colon x \to R$ is continuous iff A is both closed and open in X.

21. Show by an example that there exists a topological space X, A a subset of X and f a real-valued continuous function on A which does not extend to a continuous function on X.

22. Let f be a mapping of X into Y and $A \subsetneq X$. Show by an example that $f \mid A$ is continuous but f is not continuous at any point of A.

23. Let $E = R$, the real line with the usual metric topology $d(x, y) = |x{-}y|$ and $F = R_d$, with the discrete topology. Then the identity map: $F \to E$ is continuous but not open (cf. Ex. 13).

24. Let Q be the subset of all rational numbers of the set R of real numbers, endowed with the induced topology. Then the injection $Q \to R$ is continuous but not open.

25. Show by an example that if $f \colon X \to Y$ maps connected subsets into connected subsets, then f is not necessarily continuous.

26. Let E be a normal space and $A \subset E$ a closed subset. A necessary and sufficient condition for the existence of a continuous function $f \colon E \to [0, 1]$ with $A = f^{-1}(0)$ is that A be a G_δ-set.

27. Let E be a normal space and A, B two disjoint closed subsets of E. A necessary and sufficient condition for the existence of a continuous function $f \colon E \to [0, 1]$ with $f^{-1}(0) = A$ and $f^{-1}(1) = B$ is that both A and B are G_δ-sets.

28. Let E be a normal and F any topological space. F is called an *absolute retract* (normal) or briefly AR if for each closed subspace A of E and any continuous function $f \colon A \to F$, there exists a continuous extension $f^\sim \colon E \to F$ of f. Show that $I^n = \{(x, \ldots, x_n) \colon x_i \in [0, 1], 1 \leq i \leq n\}$ is an AR-space and so is $\mathbb{R}^n$.

29. (Characterization of normality via covering.) Show that a topological space E is normal iff for any point-finite open covering $\{P_\alpha\}_{\alpha \in \Gamma}$ of E there exists an open covering $\{Q_\alpha\}_{\alpha \in \Gamma}$ of E such that $\bar{Q}_\alpha \subset P_A$ for each $\alpha \in \Gamma$.

30. Show that a topological space E is completely regular iff its topology is the coarsest topology which makes each map of E into $[0, 1]$ continuous.

IV

Methods for Constructing New Topological Spaces from Old

It is extremely useful to know whether or not certain topological spaces constructed out of given topological spaces maintain the properties of the given ones. There are several methods for constructing new topological spaces from old ones, e.g., subspaces, products, sums, and inductive and projective limits. In this chapter we discuss all these methods and show whether or not the spaces discussed in the previous chapters remain invariant under these construction methods.

§20. Subspaces

The method of endowing a subset of a topological space with an induced topology has already been encountered. Recall that if $\{U\}$ is the family of all open sets of a topological space E and F a subset of E, then the family $\{U \cap F\}$, where U runs over $\{U\}$, satisfies all the axioms of a topological space. The topology on F defined by $\{U \cap F\}$ is called the *relative* or *induced* topology on F. A subset F of E endowed with the relative topology is called a *subspace* of E. The mapping $i: F \to E$ is called the *inclusion mapping*.

The following statements are immediate:

Proposition 1.   Let F_u be a subspace of E_u. Then

(a)   the inclusion mapping $i: F_u \to E_u$ is continuous, but not open in general;

(b) i is open if and only if F is an open subset of E;

(c) i is closed if and only if F is a closed subset of E.

Proof.   (a)   The first part of (a) is obvious in view of the definitions involved. For the second part of (a), see the counterexample (Example and Exercise 24, Chap. III).

(b) i is obviously $1 : 1$ and continuous. If F is open, then for each open set U in E, $U \cap F$ is an open set in E. Since all the open sets of F are of the type $U \cap F$, when U is open in E, (a) follows.

(c)   If i is closed, clearly F is a closed subset of E. If F is a closed subset of E, then for each closed subset C of F, $F \backslash C$ being open in F, we have $F \backslash C = U \cap F$ for some open set U in E. This implies $C = F \backslash (U \cap F) = (E \backslash U) \cap F$. Since $E \backslash U$ and F are closed in E, C is closed in E.

Theorem 1.   Let F be a subspace of a topological space E.

(a)   If E is a T_0-space, so is F;

(b)   if E is a T_1-space, so is F;

(c)   if E is a T_2-(or Hausdorff) space, so is F;

(d)   if E is a T_3-space (or regular), so is F;

(e)   if E is a T_4-space (or completely regular), so is F;

(f)   if E is a T_5-space (or normal), so is F, provided F is closed;

(g)   if E is second countable, so is F;

(h)   if E is first countable, so is F;

(i)   if E is metrizable, so is F;

(j)   if E is a complete metric space, so is F, provided F is closed;

(k)   if E is a Lindelöf space, so is F, provided F is closed;

(l)   if E is compact, so is F, provided F is closed;

(m)   if E is locally compact, so is F, provided F is closed (or open, see Proposition 22, §38).

Proof.   (a) and (b) are obvious. For (c), let $x \neq y$, $x, y \in F$. Then there exist disjoint open neighborhoods U and V of x and y respectively in E. But then $U \cap F$ and $V \cap F$ are respectively disjoint open neighborhoods of x and y in F.

(d) Let P be an open neighborhood of $x \in F$. Then there is an open set U in E such that $P = U \cap F$. There exists an open neighborhood V of x in E such that $x \in V \subset \bar{V} \subset U$ (Theorem 3, §18, applies here since it is true for T_3-spaces as well) because E is a T_3-space. Since $V \cap F$ is an open neighborhood of x in F and $\overline{V \cap F} = \bar{V} \cap F \subset U \cap F = P$, it proves that F is a T_3-space. This combined with (b) gives the statement for regular spaces.

(e) Let E be a T_4-space and F a subspace of E. Let B be a closed subset of F such that $x \in F \backslash B$. There exists a closed subset C of E such that $B = C \cap F$ and $x \notin C$. Since E is a T_4-space, there exists a continuous function f on E into $[0, 1]$ such that $f(C) = 1$ and $f(x) = 0$. But then the restriction $f \mid F$ is also continuous and $f(C \cap F) = f(B) = 1$, $f(x) = 0$. This proves that F is a T_4-space. The fact that F is completely regular when E is, follows by combining (b) and the case for T_4-spaces.

(f) Let F be a closed subspace of a T_5-space E. Let A_0 and B_0 be two disjoint closed subsets of F. Since F is a closed subspace of E, A_0 and B_0 are closed subsets of E and hence there exist disjoint open sets U and V in E such that $A_0 \subset U$, $B_0 \subset V$ because E is a T_5-space. But then $U \cap F$ and $V \cap F$ are clearly disjoint open subsets of F and $A_0 \subset U \cap F$, $B_0 \subset V \cap F$. Therefore, F is a T_5-space. Normality of F follows similarly, when E is normal.

(g) Let $\{B_n\}$ be a countable base of the topology and let F be any subspace of E. Then $\{B_n \cap F\}$ is a countable family of open sets in F. To show that $\{B_n \cap F\}$ is a base, let V be any open set in F. Then there exists an open set U in E such that $U \cap F = V$. Let x be any arbitrary member of V. Then $x \in U$ and there exists a B_n such that $x \in B_n \subset U$. Hence $x \in B_n \cap F \subset U \cap F = V$.

(h) The proof of (h) is similar to that of (g).

(i) Obvious.

(j) Let F be a closed subset of a complete metric space E. By (i), F is a metric space. Let $\{x_n\}$ be a Cauchy sequence in F, i.e., for a given arbitrary $\varepsilon > 0$ there exists $n_0 = n_0(\varepsilon)$ such that $d(x_n, x_m) < \varepsilon$ for all n, $m \geq n_0$, where d is the metric for F induced from E. Since E is complete, there exists $x_0 \in E$ such that $d(x_n, x_0) \to 0$ as $n \to \infty$. It is clear that $x_0 \in \overline{\{x_n\}} \subset \bar{F} = F$, because F is closed and $\{x_n\} \subset F$. This shows that F is complete.

(k) Let F be a closed subspace of a Lindelöf space E. Let $\{P_\alpha\}$ be an open covering of F. Then for each α there exists an open set U_α in E such that $P_\alpha = U_\alpha \cap F$. It is clear that $\{U_\alpha\} \cup (E \backslash F)$ is an open covering of E. Hence there exists a countable subcovering $\{U_{\alpha_n}\} \cup (E \backslash F)$ of E. But then it is clear that $\{P_{\alpha_n}\}$ is a countable open subcovering of F.

(l)   This follows by using similar arguments to those in (k). In this case an open covering $\{U_\alpha\} \cup (E\backslash F)$ of E contains a finite subcovering $\{U_{\alpha_i}\}_{i=1}^n \cup (E\backslash F)$ of E. Hence $\{P_{\alpha_i}\}_{i=1}^n$ is a finite subcovering of F.

(m)   Let $x \in F$. There exists a neighborhood U of x in E such that $\bar{U}$ is compact. But $U \cap F$ is a neighborhood of x in F and $\overline{U \cap F} \subset \bar{U} \cap F$. Since by (l), $\overline{U \cap F}$ is compact, we have established (m).

Remark.   Parts (f), (j), (k), and (l) are not true in general if F is not closed, as follows from the examples given at the end of this chapter.

§21. Topological Sums

Let E_α $(\alpha \in A)$ be a family of topological spaces with the topology u_α and let $\bigcup_\alpha E_\alpha = E$. We define a topology on E. Assume that the induced topology on $E_\alpha \cap E_\beta$ from E_α and E_β are equal. Also either $E_\alpha \cap E_\beta$ is open in both E_α and E_β or closed in both E_α and E_β. We say that a subset U in E is open if and only if for each α, $E_\alpha \cap U$ is u_α-open in E_α. Let $\{U\}$ denote the family of all such open sets.

It is easy to check that $\{U\}$ defines a topology on E, and it is called the *weak topology*. (There is another topology also called the weak topology in topological vector spaces, which will be treated later. The confusion between the names will be cleared up by the context.) The weak topology induces u_α on E_α.

If, in addition, $E_\alpha \cap E_\beta = \varnothing$ for $a \neq \beta$, and $E = \bigcup_\alpha E_\alpha$, then E with the weak topology is called the *topological sum* of the E_α and is sometimes denoted by $\sum E_\alpha = E$.

Proposition 2.   Let E be the space with weak topology or, in particular, the topological sum of topological spaces $(E_\alpha)_{\alpha \in A}$ with topologies $\{u_\alpha\}$ and let F be a topological space. A mapping $f: E \rightarrow F$ is continuous if and only if $f \mid E_\alpha$ is continuous for each α.

Proof.   Suppose f is continuous. Then for each open set V of F, $f^{-1}(V)$ is open in E and hence $f^{-1}(V) \cap E_\alpha$ is open for each α. Hence $f \mid E_\alpha$ is continuous. Conversely, if V is open in F, then $(f \mid E_\alpha)^{-1}(V)$ is open for each α by assumption. But then $\bigcup_\alpha (f \mid E_\alpha)^{-1}(V) = f^{-1}(V)$ is open in E. Hence f is continuous.

It is worthwhile to note that each u_α-open set of E_α, in particular E_α, is open in $\sum E_\alpha$. Moreover each subset U of $\sum E_\alpha$ is open if and only if $U = \bigcup_{\alpha \in A} U_\alpha$, where U_α is u_α-open in E_α.

Theorem 2. Let $\{E_\alpha\}_{\alpha \in A}$ be a family of topological spaces E_α with the topology u_α, $\alpha \in A$, and E its topological sum.

(a) If each E_α is a T_0-space, so is E;

(b) if each E_α is a T_1-space, so is E;

(c) if each E_α is a Hausdorff space, so is E;

(d) if each E_α is a T_3-space (or regular), so is E;

(e) if each E_α is a T_4-space (or completely regular), so is E;

(f) if each E_α is a T_5-space (or normal), so is E;

(g) if each E_α is second countable and if A is countable, then E is also second countable;

(h) if each E_α is first countable so is E;

(i) if each E_α is metrizable, so is E;

(j) if each E_α is a complete metrizable space, so is E;

(k) if each E_α is a Lindelöf space and if A is countable, then E is also a Lindelöf space;

(l) if each E_α is a compact space and if A is a finite set, then E is compact;

(m) if each E_α is locally compact, so is E.

Proof. (a)–(c) are obvious.

(d) Let C be a closed subset of E and let $x \in E \setminus C$. It is clear that $C_\alpha = C \cap E_\alpha$ is a closed subset of E_α for each α. Since $E = \bigcup_\alpha E_\alpha$, there exists α such that $x \in E_\alpha$ and $x \in E_\alpha \setminus C_\alpha$ because $x \notin C$. Since each E_α is a T_3-space, there exist disjoint open sets U_α and V_α in E_α such that $x \in U_\alpha$ and $C_\alpha \subset V_\alpha$. For $\beta \neq \alpha$, put $V_\beta = E_\beta$ and let $V = \bigcup_\beta V_\beta$. Then it is clear that V is open in E and $C \subset V$. Moreover, V is an open set in E and $U_\alpha \cap V = \varnothing$ because E_α's are mutually disjoint by definition. This proves that E is a T_3-space. The regularity of E when each E_α is regular follows similarly.

(e) Let each E_α be a T_4-space and let C be a closed subset of E and $x \in E \setminus C$. Suppose $x \in E_\alpha \setminus C_\alpha$, where $C_\alpha = C \cap E_\alpha$. Let f_α be a continuous function on E_α into $[0, 1]$ such that $f_\alpha(C_\alpha) = 1$ and $f_\alpha(x) = 0$. Define f as follows:

$$f(y) = \begin{cases} f_\alpha(y) & \text{if } y \in E_\alpha; \\ 1 & \text{if } y \in E_\beta, \; \beta \neq \alpha. \end{cases}$$

Then $f(C) = 1$, because $y \in C$ implies either $y \in C_\alpha$ or $y \in E_\beta$, $\beta \neq \alpha$. In each case $f(y) = 1$. To show that f is continuous it is enough to observe in view of Proposition 2 that the restriction of f on each E_α is continuous, which is the case because $f \mid E_\alpha$ is continuous and $f \mid E_\beta$, $\beta \neq \alpha$ being constant, is equal to 1, and each E_α is open, we see that $f \mid E_\beta$ for $\beta \neq \alpha$ is continuous. Hence f is continuous.

(f)   Let each E_α be a T_5-space and let A and B be two disjoint closed subsets of E. Then $A \cap E_\alpha$ and $B \cap E_\alpha$ are disjoint closed subsets of E_α for each α. Since each E_α is a T_5-space, there exist disjoint open sets U_α and V_α in E_α such that $A \cap E_\alpha \subset U_\alpha$, $B \cap E_\alpha \subset V_\alpha$ for each α. But then $U = \bigcup_\alpha U_\alpha$, $V = \bigcup_\alpha V_\alpha$ are also open in E. Since U_α and V_α are disjoint for each α and so are E_α and E_β $(\alpha \neq \beta)$, it follows that U and V are disjoint open subsets of E. It is clear that $A \subset U$ and $B \subset V$. The normality of E when each E_α is normal follows similarly.

(g)   Suppose A is a countable set. Let $\{U_n{}^{(\alpha)}\}$ denote a countable base of the topology of E_α. Then it is clear that the family $\{U_n{}^{(\alpha)}\}$, where α runs over A and n over positive integers, is a countable family. Let $x \in E$. Then there exists α such that $x \in E_\alpha$. If U is any open set in E containing x, then $U \cap E_\alpha$ is an open set in E_α containing x. Hence there exists $U_n{}^{(\alpha)}$ such that $x \in U_n{}^{(\alpha)} \subset U \cap E_\alpha \subset U$. This proves that $\{U_n{}^{(\alpha)}\}$ is a base of the topology of the topological sum E.

(h)   Since each element x of E is an element of E_α for some α and since each x in E_α has a countable base of neighborhoods (because each E_α is first countable by assumption), it follows that E is first countable.

(i)   Let d_α be the metric for the metric space E_α, for each α. We can assume that $d_\alpha \leq 1$ for all α. For, if d' is any metric on a set, then $d = d'/(1 + d')$ is also a metric giving an equivalent topology and $d \leq 1$. With this assumption, we define the following: For $x, y \in E = \bigcup_\alpha E_\alpha$,

$$d(x, y) = \begin{cases} d_\alpha(x, y) & \text{if } x, y \in E_\alpha, \\ 1 & \text{if } x \in E_\alpha, y \in E_\beta \ (\beta \neq \alpha). \end{cases}$$

Then it is easy to check that d is a metric on E, it coincides with d_α for each α, and defines the weak topology on $\bigcup_\alpha E_\alpha$.

(j)   By (i), E is a metric space. As in (i), if the metric on each E_α is not bounded we can replace it by an equivalent bounded metric, and it is clear that E_α remains complete under this new bounded metric. Now we show that E is complete. Let $\{x_n\}$ be a Cauchy sequence in E. Then for certain n_0, $x_n \in E_\alpha$ for some α and $n \geq n_0$. For, suppose $x_n \in E_{\alpha_n}$ and $x_m \in E_{\alpha_m}$, $\alpha_n \neq \alpha_m$ when $n \neq m$. Then $d(x_n, x_m) = 1$ for all n, $m \geq n_0$, contradicting the assumption that $\{x_n\}$ is a Cauchy sequence. Therefore, for $n \geq n_0$, $\{x_n\} \subset E_\alpha$ for some α. Since E_α is complete, there exists $x_0 \in E_\alpha$ such that $\{x_n\}$ converges to $x_0 \in E_\alpha \subset E$. Hence E is complete.

(k)   Let each E_α be a Lindelöf space and let A be a countable set. Let $\{P\}$ be an open covering of E. Then $\{P \cap E_\alpha\}$ is an open covering of E_α for each α. Hence there is a countable subcovering $\{P_n \cap E_\alpha\}$ of E_α. Since A is countable, $\bigcup_{\alpha \in A}\{P_n \cap E_\alpha : n \geq 1\}$ is a countable open covering of E. Hence E is a Lindelöf space.

(l)   Let A be a finite set and let each E_α be a compact space for $\alpha \in A$. Let $\{P\}$ be an open covering of E. Then $\{P \cap E_\alpha\}$ is an open covering of E_α. Hence there is a finite subcovering $\{P_i \cap E_\alpha\}$ $(1 \leq i \leq n)$ of E_α. But then $\{P_i \cap E_\alpha\}$, $1 \leq i \leq n$ and where α runs over A, is a finite open subcovering of E. Hence E is compact.

(m)   Let $x \in E$. Then $x \in E_\alpha$ for some α. There exists a compact neighborhood U of x in E_α. Since a neighborhood of a point in E_α is also a neighborhood of x in E, and a compact subset in E_α is compact in E, it follows that E is locally compact. The proof is completed.

Given any family $\{E_\alpha\}_{\alpha \in A}$ of topological spaces with the topology $\{u_\alpha\}_{\alpha \in A}$. For each α, put $\{\alpha\} \times E_\alpha = E_\alpha'$ so that E_α' is homeomorphic with E_α, where $\{\alpha\}$ has the discrete topology. $E_\alpha' \cap E_\beta' = \emptyset$ for $\alpha \neq \beta$. The *free* union of $\{E_\alpha\}_{\alpha \in A}$ is the set $\bigcup_{\alpha \in A} E_\alpha'$. The set $\bigcup_{\alpha \in A} E_\alpha'$ with the weak topology defined by E_α' is called the *coproduct* or sum of E_α's and is denoted by $\prod_{\alpha \in A} E_\alpha'$.

§22. Topological Products

Let $E_\alpha(\alpha \in A)$ be a family of topological spaces. The set E of all mappings x of A into $\bigcup_\alpha E_\alpha$ for each α is called the *Cartesian product* of E_α's and is denoted by $\prod_{\alpha \in A} E_\alpha$.

If $x \in E$, then $x(\alpha) = x_\alpha \in E_\alpha$ for each $\alpha \in A$. For a fixed α, the mapping $p_\alpha : x \to x_\alpha$ is called the *projection mapping* or more precisely, the αth *projection mapping* or the αth *projection*. We shall write $x = (x_\alpha)$.

Since each E_α is a topological space, in view of §17, Chap. III, we can define the following topology on the set E: Let $\{U_\alpha\}$ be a base of the given topology on E_α for each α. Let E be endowed with the topology which has the family $\{p_\alpha^{-1}(U_\alpha)\}$, where α runs over A, as its subbase. This topology is called the *direct product* or briefly, the *product topology* on E. E endowed with the product topology is called the *topological product* or simply the product of E_α's and written as $E = \prod_{\alpha \in A} A_\alpha$.

Proposition 3.   Let $E = \prod_{\alpha \in A} E_\alpha$ be the topological product of topological spaces $E_\alpha(\alpha \in A)$ and let u be a topology on E. The following statements are equivalent:

 (a) u is the product topology on E;

 (b) u is the coarsest topology on E for which each projection p_α is continuous;

 (c) the family of subsets $\{U\}$, $U = \prod_{\beta \in A} U_\beta$, where U_β is an open set in E_β for $\beta = \alpha$ and $U_\beta = E_\beta$ for $\beta \neq \alpha$ in E, forms a subbase.

 Proof. (a) $\Leftrightarrow$ (b) Let v be the topology on E such that $p_\alpha \colon E_v \to E_\alpha$ is continuous for each α. Let U_α be an open set in E_α, then $p_\alpha^{-1}(U_\alpha)$ is a v-open set in E. But by the definition, the family $\{p_\alpha^{-1}(U_\alpha)\}$, where α runs over A, forms a subbase of the product topology. Thus if B is u-open, then for each $x \in B$, there exists a finite family $\{\alpha_i\}_{i=1}^n$ such that $x \in \prod_{\alpha \in A} U_\alpha \subset B$, where $U_\alpha = U_{\alpha_i}$ for $\alpha = \alpha_i$ ($i = 1, \ldots, n$) and $U_\alpha = E_\alpha$ for $\alpha \neq \alpha_i$. It is clear that

$$\prod_{\alpha \in A} U_\alpha = \bigcap_{i=1}^n p_{\alpha_i}^{-1}(U_{\alpha_i})$$

and hence v-open. This shows that B is v-open, since x is an arbitrary element of B. This proves that u is coarser than v. Further, if U_α is open in E_α then $p_\alpha^{-1}(U_\alpha)$ is open in E for any topology implies $p_\alpha^{-1}(U_\alpha)$ is u-open. Since u is the coarsest topology which makes each p_α continuous, u is the product topology.

 (a) $\Leftrightarrow$ (c) Observe that for each $\alpha \in A$ and each open set U_α in E_α, $p_\alpha^{-1}(U_\alpha) = \prod_{\beta \in A} U_\beta$, where $U_\beta = U_\alpha$ for $\beta = \alpha$ and $U_\beta = E_\beta$ for all $\beta \neq \alpha$. The result is clear.

 Proposition 3′. Let $E = \prod_\alpha E_\alpha$ be the product space of E_α's and let $p_\alpha \colon E \to E_\alpha$ be the projection map. Then each p_α is continuous and open.

 Proof. By definition of the product topology, $p_\alpha \colon E \to E_\alpha$ is continuous for each $\alpha \in A$. To show that p_α is open, let B be an open set in E. Let $x \in B$ and let $p_\alpha(x) = x_\alpha \in E_\alpha$. There exists a finite set $\{\alpha_i\}_{i=1}^n$ of indices such that $x \in \prod_{\alpha \in A} U_\alpha \subset B$, where $U_\alpha = U_{\alpha_i}$ (open set in E_{α_i}) for $\alpha = \alpha_i$ ($1 \leq i \leq n$) and $U_\alpha = E_\alpha$ for $\alpha \neq \alpha_i$. But then $p_\alpha(B) = E_\alpha$ for $\alpha \neq \alpha_i$ and $p_\alpha(B) = U_{\alpha_i}$ for $\alpha = \alpha_i$, $i = 1, \ldots, n$. In each case $p_\alpha(B)$ is open.

 Remark. Although from the above proposition, it follows that each p_α is continuous and open, it is not true that p_α is closed. For example, let $E = \mathbb{R}^2$ (plane) and $A = \{(x, 1/x) \colon x \neq 0,\, x \in \mathbb{R}\}$. Then A is closed in $\mathbb{R}^2$ but $p_1(A) = \{x \in \mathbb{R} \colon x > 0\}$ is not closed in $\mathbb{R}$.

Proposition 4. Let $E = \prod_{\alpha \in A} E_\alpha$. Let A_α be an arbitrary subset of E_α for each α. Then:

(a) A filter $\mathscr{F}$ in E converges to $x \in E$ iff $p_\alpha(\mathscr{F})$ converges to $p_\alpha(x)$ for each α. A net $\{x(\gamma),\ \gamma \in D\}$ in $E = \prod_{\alpha \in A} E_\alpha$ converges to $x \in E$ iff $p_\alpha(x(\gamma))$ converges to $p_\alpha(x)$ for each α.

(b) $\overline{\prod_{\alpha \in A} A_\alpha} = \prod_{\alpha \in A} \bar{A}_\alpha$.

(c) A subset $\prod_{\alpha \in A} A_\alpha$ of E is closed iff each A_α is closed in $E_\alpha \neq \varnothing$.

Proof. (a) Since each p_α is continuous, the convergence of the net $\{x(\gamma): \gamma \in D\}$ to x implies that $\{p_\alpha(x(\gamma))\}$ converges to $p_\alpha(x)$. Conversely, let $\{x(\gamma),\ \gamma \in D\}$ be a net in E such that $\{p_\alpha(x(\gamma)),\ \gamma \in D\}$ converges to $x_\alpha \in E_\alpha$ for each α. Then for each neighborhood U_α of x_α, $\{p_\alpha(x(\gamma)),\ \gamma \in D\}$ is eventually in U_α. This implies that $\{x(\gamma),\ \gamma \in D\}$ is eventually in $P_\alpha^{-1}(U_\alpha)$ for each $\alpha \in A$ and hence in each neighborhood of $x = (x_\alpha)$ because the family $\{p_\alpha^{-1}(U_\alpha)\}$ forms a subbase of the product topology. Therefore $\{x(\gamma),\ \gamma \in D\}$ converges to x. The arguments for filters follow similarly.

(b) Let $x \in \overline{\prod_{\alpha \in A} A_\alpha}$. Since $p_\alpha \colon E \to E_\alpha$ is continuous for each α and since x is a limit point of $\prod_{\alpha \in A} A_\alpha$, it follows that $p_\alpha(x) = x_\alpha$ is a limit point of A_α for each α. That is, $x_\alpha \in \bar{A}_\alpha$ for each α. Hence $x \in \prod_{\alpha \in A} \bar{A}_\alpha$. On the other hand, if $y \in \prod_{\alpha \in A} \bar{A}_\alpha$, then for any basic neighborhood $U = \prod_{\alpha \in A} U_\alpha$ of y [where U is obtained by a finite set $\{\alpha_i\}$, i.e., $U_\alpha = U_{\alpha_i}$ (open in E_{α_i}), $U_\alpha = E_\alpha$ for $\alpha \neq \alpha_i$], we have $U_\alpha \cap A_\alpha \neq \varnothing$. Hence $U \cap \prod_{\alpha \in A} A_\alpha \neq \varnothing$. This shows that $y \in \overline{\prod_{\alpha \in A} A_\alpha}$, and (b) is established.

(c) This is a particular case of (b).

Remark. Part (b) in the above proposition is not true for open sets A_α. However, if A_α is open for a finite subset $\{\alpha_i\}_{i=1}^n$ of A and $A_\alpha = E_\alpha$ for $\alpha \in A \backslash \{\alpha_i\}$ then by the definition of the product topology $\prod_{\alpha \in A} A_\alpha$ is open. We shall discuss later the conditions on the topology under which $\prod_{\alpha \in A} A_\alpha$ is open, whenever each A_α is open.

Proposition 5. Let $E = \prod_{\alpha \in A} E_\alpha$ and let F be a topological space. Then a mapping f of F into E is continuous iff $p_\alpha \circ f$ is continuous for each α, where $p_\alpha \colon E \to E_\alpha$ is the projection mapping.

Proof. Let u denote the product topology on E. If $f \colon F \to E_u$ is continuous, then the composition $p_\alpha \circ f$ of two continuous mappings f and $p_\alpha \colon E_u \to E_\alpha$ is continuous. On the other hand, assume $p_\alpha \circ f$ is continuous for each $\alpha \in A$. Let U be a neighborhood of $f(x) \in E$, $x \in F$. Then there is a finite subset $\{\alpha_i\}_{i=1}^n$ of A such that $f(x) \in \prod_{\alpha \in A} U_\alpha \subset U$, where U_α is open in E_α for $\alpha \in \{\alpha_i\}$ and $U_\alpha = E_\alpha$ for all $\alpha \in A \setminus \{\alpha_i\}$. Clearly

$$f^{-1}(U) \supset \bigcap_{i=1}^n f^{-1}\left(p_{\alpha_i}^{-1}(U_{\alpha_i})\right).$$

Since each $f^{-1}\left(p_{\alpha_i}^{-1}(U_{\alpha_i})\right)$ is a neighborhood of $x \in F$, by the assumption on continuity of $p_\alpha \circ f$, $f^{-1}(U)$ is a neighborhood of x. Hence f is continuous.

Theorem 3. Let E_α $(\alpha \in A)$ be a family of topological spaces and $E = \prod_{\alpha \in A} E_\alpha$.

(a) E is a T_0-space iff each E_α is;

(b) E is a T_1-space iff each E_α is;

(c) E is a Hausdorff space iff each E_α is;

(d) E is a T_3-space (or regular) iff each E_α is;

(e)   if each E_α is a T_4-space (or completely regular), so is E;

(f)   if each E_α is second countable and if A is countable, then E is also second countable. Conversely, if E is second countable, then so is E_α;

(g)   if each E_α is first countable and if A is countable, then E is also first countable. Moreover, if E is first countable so is E_α;

(h)   if each E_α is metrizable and if A is countable, then E is also metrizable. Conversely, if E is metrizable, so is each E_α.

(i)   (Tychonoff theorem) E is a compact space iff each E_α is;

(j)   if each E_α is locally compact for all $\alpha \in B$, where B is a finite subset of A and for all $\alpha \in A \setminus B$, E_α is compact, then E is locally compact;

(k) E is connected iff each E_α is.

Proof.   (a)   Let $x \neq y$, x, $y \in E$. Then for at least one α, $x_\alpha \neq y_\alpha$, where $x = (x_\alpha)$, $y = (y_\alpha)$. Since each E_α is a T_0-space at least one of x_α, y_α, say x_α, has an open neighborhood U_α such that $y_\alpha \notin U_\alpha$. Let $U = \prod_{\beta \in A} U_\beta$ where $U_\beta = U_\alpha$ for $\beta = \alpha$ and $U_\beta = E_\beta$ for $\beta \neq \alpha$. Then U is a neighborhood of x and $y \notin U$. The other part follows from Theorem 1(a).

(b)   The proof of (b) is similar to (a).

(c)   As in (a), let $x \neq y$, $x = (x_\alpha)$, $y = (y_\alpha)$. For at least one α, $x_\alpha \neq y_\alpha$. Since E_α is a Hausdorff space, there exist disjoint open neighborhoods U_α and V_α of x_α and y_α, respectively. Now putting $U = p_\alpha^{-1}(U_\alpha) = \prod_{\beta \in A} U_\beta$ (where $U_\beta = U_\alpha$ for $\beta = \alpha$ and $U_\beta = E_\beta$ for $\beta \neq \alpha$) and $V = \prod_{\beta \in B} V_\beta$ (defined similarly to U), we see that U and V are disjoint open neighborhoods of

x and y, respectively. Hence E is a Hausdorff space. The "only if" part follows from Theorem 1(c).

(d) Let $x \in E$ and let U be a neighborhood of x in E. First we assume that U is a member of the subbase of the product topology. Then $p_\alpha(U) = U_\alpha$ is an open set in E_α and $p_\beta(U) = E_\beta$ for all $\beta \neq \alpha$. Clearly U_α is a neighborhood of $p_\alpha(x) = x_\alpha$. Since E_α is a T_3-space, there exists a neighborhood V_α of x_α such that $x_\alpha \in V_\alpha \subset \bar{V}_\alpha \subset U_\alpha$. Hence $x \in p_\alpha^{-1}(V_\alpha) \subset p_\alpha^{-1}(\bar{V}_\alpha) \subset p_\alpha^{-1}(U_\alpha) = U$. Since each p_α is continuous and since $\bar{V}_\alpha$ is a closed subset, $p_\alpha^{-1}(\bar{V}_\alpha)$ is a closed subset of E. Hence $\overline{p_\alpha^{-1}(V_\alpha)} \subset p_\alpha^{-1}(\bar{V}_\alpha)$. But since $p_\alpha^{-1}(V_\alpha)$ is a neighborhood of x, it follows from a characterization of T_3-spaces (Theorem 3, Chap. 3) that E is a T_3-space if one considers the members of the subbase of the product topology on E. Observing that each open set in E contains a member of the base of the product topology which is the intersection of a finite number of members of its subbase, and that the inverse images of intersections are the intersections of the inverse images under a mapping, (d) is proved. The other part follows from part (d), Theorem 1.

(e) Let each E_α be completely regular. Let C be a closed subset of E and let $x \in E \backslash C$. Since $E \backslash C$ is an open neighborhood of $x = (x_\alpha)$, there are a finite number of open sets U_{α_i} $(1 \leq i \leq n)$ such that $x \in \prod_\alpha U_\alpha \subset E \backslash C$, where $U_\alpha = U_{\alpha_i}$ for $\alpha = \alpha_i$ $(1 \leq i \leq n)$ and $U_\alpha = E_\alpha$ for $\alpha \neq \alpha_i$. Let f_i be a continuous map of E_{α_i} into $[0, 1]$ such that $f_i(x_{\alpha_i}) = 1$ and $f_i(U_{\alpha_i}^c) = 0$. Now define $f \colon \prod_\alpha E_\alpha \to [0, 1]$ by $f(x) = \min\{f_i \circ p_{\alpha_i}(x) \colon 1 \leq i \leq n\}$. Then f is continuous because each $f_i \circ p_{\alpha_i}$ is continuous and $f(x) = 1$, $f(C) = 0$. Hence E is a T_4-space. The other part follows from Theorem 1(c).

(f) Let $\{U_n^{(\alpha)}\}$ be a countable base of the topology of E_α, $\alpha \in A$, where A is a countable set of indices. For a fixed n, let

$$U_{\alpha_n} = \prod_{\beta \in A} U_n^{(\beta)},$$

where $U_n^{(\beta)} = E_\beta$ for all $\beta \neq \alpha$, and $U_n^{(\beta)} = U_n^{(\alpha)}$ for $\beta = \alpha$. Then clearly U_{α_n} is an open set in $E = \prod_\alpha E_\alpha$ and is actually a member of a subbase of the product topology on E. Since A is countable, as α and n run over their respective sets of indices, we see that $\{U_{\alpha_n}\}$ is countable. Hence it is clear that the base, consisting of the family of finite intersections of members from $\{U_{\alpha_n}\}$, is a countable family and so E is second countable. The converse is obvious.

(g) This follows similarly to (f).

(h) Let A be a countable set of indices, say the set of positive integers, and let E_n be a metric space with metric d_n for each $n \geq 1$.

For each pair $x, y \in E = \prod_{n \geq 1} E_n$, define

$$d(x, y) = \sum_{n \geq 1} \frac{1}{2^n} \frac{d_n(x_n, y_n)}{1 + d_n(x_n, y_n)}.$$

It is clear that $d(x, y) \geq 0$ and $d(x, y) = d(y, x)$. To establish the triangular inequality, we see that for $x, y, z \in E = \prod_{n \geq 1} E_n, x = (x_n), y = (y_n), z = (z_n)$,

$$d_n(x_n, y_n) + d_n(y_n, z_n) \geq d(x_n, z_n) \qquad \text{for } n \geq 1,$$

and so

$d(x, y) + d(y, z)$

$$= \sum_{n \in A} \frac{1}{2^n} \left[\frac{d_n(x_n, y_n)}{1 + d_n(x_n, y_n)} + \frac{d_n(y_n, z_n)}{1 + d_n(y_n, z_n)} \right]$$

$$= \sum_{n \in A} \frac{1}{2^n} \left\{ \left[1 - \frac{1}{1 + d_n(x_n, y_n)} \right] + \left[1 - \frac{1}{1 + d_n(y_n, z_n)} \right] \right\}$$

$$= \sum_{n \in A} \frac{1}{2^n} \left\{ 2 - \left[\frac{2 + d_n(x_n, y_n) + d_n(y_n, z_n)}{1 + d_n(x_n, y_n) + d_n(y_n, z_n) + d_n(x_n, y_n) d_n(y_n, z_n)} \right] \right\}$$

$$\geq \sum_{n \in A} \frac{1}{2^n} \left\{ 2 - \left[\frac{2 + d_n(x_n, y_n) + d_n(y_n, z_n)}{1 + d_n(x_n, y_n) + d_n(y_n, z_n)} \right] \right\}$$

$$= \sum_{n \in A} \frac{1}{2^n} \left[1 - \frac{1}{1 + d_n(x_n, y_n) + d_n(y_n, z_n)} \right]$$

$$\geq \sum_{n \in A} \frac{1}{2^n} \left[1 - \frac{1}{1 + d_n(x_n, z_n)} \right]$$

$$= \sum_{n \in A} \frac{1}{2^n} \left[\frac{d_n(x_n, z_n)}{1 + d_n(x_n, z_n)} \right] = d(x, z).$$

Now to complete the proof of (h), we must show that the metric d induces the product topology on E.

If a sequence $\{x^{(n)}\}$ in E converges to x in E, i.e., $d(x^{(n)}, x) \to 0$, then $d_m(x_m^{(n)}, x_m) \to 0$ as $n \to \infty$ for each $m \geq 1$ and vice versa. This proves each projection P_n is continuous. Hence the metric topology induced by d is finer than the product topology on E (Proposition 3).

To show that the product topology is finer than the metric topology, let m be a positive integer and let $B_m(x) = \{y \in E: d(y, x) < 1/2^m\}$ be an open ball with center x. Then for each $y \in B_\varepsilon(x)$,

$$d(y, x) = \sum_{n \in A} \frac{1}{2^n} \left[\frac{d_n(y_n, x_n)}{1 + d_n(y_n, x_n)} \right] < \frac{1}{2^m}.$$

Let

$$U = \{y \in E: d_n(y_n, x_n) < 1/2^{m+2}, \qquad n \leq m + 2\}.$$

Clearly U is a neighborhood of x in the product topology. We show that $U \subset B_m(x)$. For each $y \in U$,

$$d(y, x) = \sum_{n=1}^{\infty} \frac{1}{2^n} \left[\frac{d_n(y_n, x_n)}{1 + d_n(y_n, x_n)} \right]$$

$$\leq \sum_{n=1}^{m+2} \frac{1}{2^{n+m+2}} + \sum_{n=m+3}^{\infty} \frac{1}{2^n}$$

$$< \frac{1}{2^{m+1}} + \frac{1}{2^{m+1}} = \frac{1}{2^m}.$$

Hence $y \in B_m(x)$. This shows that the product topology is finer than the metric topology and hence the proof is complete. The other part is obvious.

(i) If E is compact, then E_α, being the continuous image of a compact space E under the continuous projection mapping $P_\alpha: E \to E_\alpha$, is also compact.

We defer the proof of the other part, viz., Tychonoff's theorem until later (cf. §35, Theorem 3).

(j) In order to prove (j), it is clearly sufficient to show that a finite product of locally compact spaces is locally compact.

Let E_i ($1 \leq i \leq n$) be locally compact and let $E = \prod_{i=1}^{n} E_i$. Let $x \in E$, $x = (x_i)$ and let U_i be a neighborhood of x_i such that $\bar{U}_i$ is compact. Such a neighborhood exists, because E_i is locally compact. Then clearly $\prod_{i=1}^{n} U_i$ is a neighborhood of x and $\overline{\prod_{i=1}^{n} U_i} = \prod_{i=1}^{n} \bar{U}_i$ is compact by Tychonoff's theorem. Hence E is locally compact.

(k) Let $E = \prod_{\alpha \in A} E_\alpha$, where each E_α is connected. Suppose $E = P \cup Q$ where P, Q are disjoint open nonempty sets. Let f be a real function from E to the disconnected discrete space $\{0, 1\}$ consisting of 0 and 1, defined as follows:

$$f(x) = \begin{cases} 1 & \text{if } x \in P; \\ 0 & \text{if } x \in Q. \end{cases}$$

Since P and Q are disjoint open sets and $E = P \cup Q$, f is a continuous mapping of E onto $\{0, 1\}$. Let $a = (a_\alpha)$ be a fixed point of E. Define a mapping g_α of E_α into E by: $g_\alpha(x_\alpha) = \{y_\beta\}$, where $y_\beta = a_\alpha$ for all $\beta \neq \alpha$ and $y_\alpha = x_\alpha$. Then g_α is clearly continuous and hence the composition $f \circ g_\alpha$ is a continuous mapping of E_α into $\{0, 1\}$. If $f \circ g_\alpha$ were onto, then since E is connected, it would follow that $\{0, 1\}$ is connected, which is clearly not true. Hence

$f \circ g_\alpha$ takes only one value, i.e., $f \circ g_\alpha$ is constant or $f(g_\alpha(x_\alpha)) = f(a)$ for each $x_\alpha \in E$. Hence $f(x) = f(a)$ for all x in E which are equal to a in all coordinates except possibly the αth coordinate. Repeating this argument a finite number of times, we see that $f(x) = f(a)$ for all x in E which are equal to a in all coordinates except for a finite number. Since the set of all x's in E which differ from a in only a finite number of coordinates is dense (this is left as an exercise for the reader), we see that $f(x) = f(a)$ on a dense subset of E. Since f is continuous and constant on a dense subset of E, $f(x) = f(a)$ for all $x \in E$. This contradicts that f is two-valued. Hence E is connected. For the other part, we see that $p_\alpha \colon E \to E_\alpha$ is continuous and surjective. Hence if E is connected, so is E_α.

§23. Quotient Topology and Quotient Spaces

Apart from subspaces, sums, and products, another important method of constructing new spaces from the old is that of forming quotients. Let E be a topological space, F a set, and f a mapping of E onto F. It is interesting to find the finest topology on F such that $f \colon E \to F$ is continuous.

Definition 1.   Let f be a mapping of a topological space E onto a set F. The topology on F for which a subset $A \subset F$ is open iff $f^{-1}(A)$ is open in E is called the *quotient topology*. (It is easy to check that the sets A define a topology.)

Remark.   It is clear that the quotient topology depends upon the topology of E and upon f. Moreover, a subset $B \subset F$ is closed in the quotient topology iff $f^{-1}(B)$ is closed in E.

The following proposition characterizes the quotient topology.

Proposition 6.   Let f be a mapping of a topological space E_u onto a topological space F_v, where v is the quotient topology on F. Then f is continuous and open. Moreover, v is the finest topology on F which makes $f \colon E_u \to F_v$ continuous.

Proof.   The continuity of f follows from the definition of v. Let w be a topology on F such that $f \colon E_u \to F_w$ is continuous. Let W be a w-open set in F. By continuity of f, $f^{-1}(W)$ is open in E. But then by the definition of the quotient topology on F, W is v-open and so is $v \supset w$. This completes the proof. For each open set $U \subset E$, $U \subset f^{-1}(f(U))$ shows that f is open by the definition of the quotient topology.

Suppose $f: E_u \to F_w$ is continuous for some topology w on F. Then we know that w is coarser than the quotient topology on F. With an additional condition of f, we can say that w is the quotient topology. We have:

Corollary 1. If $f: E_u \to F_w$ is a continuous open and onto mapping, then w is the quotient topology.

Proof. Let v denote the quotient topology on F. By the continuity of f, $v \supset w$ (Proposition 6). If U is v-open, then $f^{-1}(U)$ is u-open (by the definition of v). Since $f: E_u \to F_w$ is open, we have $U = f(f^{-1}(U))$ to be w-open and so $w \supset v$. Hence $v = w$.

Remark. The converse of Corollary 1 is not true without an extra condition. For example, consider the characteristic function $\chi_A: [0, 1] \to \{0, 1\}$, where $A = [\frac{1}{2}, 1]$. With the quotient topology, $\{0, 1\}$ is the Sierpinski space but χ_A is not open (Example 1, Chap. III).

Proposition 7. Let F_v be a topological space with the quotient topology defined by a topological space E_u and the mapping $f: E_u \to F_v$. Then a mapping $g: F_v \to G_w$, where G_w is any topological space, is continuous iff the composition $g \circ f: E_u \to G_w$ is continuous.

Proof. Assume g is continuous. Since $f: E_u \to F_v$ is always continuous, it follows that the composition $g \circ f$ is continuous. Conversely, if $g \circ f$ is continuous, then for any open set W in G_w, $f^{-1}(g^{-1}(W))$ is u-open in E_u and hence $g^{-1}(W)$ is v-open in F_v because of the definition of the quotient topology. Hence g is continuous.

A most significant property of the quotient topology on the range space F is that it determines the range space completely. That is, given f and E_u we can determine a homeomorphic copy of F by means of f and the elements of the domain space E_u.

Proposition 8. Let $f: E_u \to F_v$ be a continuous mapping of a topological space E_u onto F_v and let v be the quotient topology on F. Then there exists a topological space G_w and a mapping g of G_w onto F_v such that g is a homeomorphism, where G is a set of subsets of E.

Proof.   For each $y \in F$, let $f^{-1}(y)$ be the inverse image of y. Let G denote the collection of all $f^{-1}(y)$, where y runs over F. Since for $y_1, y_2 \in F$, $y_1 \neq y_2$ implies $f^{-1}(y_1) \cap f^{-1}(y_2) = \varnothing$, G is a collection of pairwise disjoint subsets of E. To each $y \in F$, put $g(y) = f^{-1}(y) \in G$. Then g is a well-defined mapping of F onto G. It is clearly one-to-one. Hence $g^{-1}: G \to F$ is also defined.

For $x \in E$, define the mapping $q(x) = f^{-1}(f(x)) = g(f(x))$ of E into G. If $f^{-1}(y) \in G$ with y in F then there exists $x \in E$ (because f is onto) such that $y = f(x)$. Hence $f^{-1}(y) = f^{-1}(f(x)) = g(f(x)) = q(x)$. Therefore q is onto. Now, we endow G with the quotient topology w, defined by q and E_u. Then $q: E_u \to G_w$ is continuous. From the definition of q, we see that $q = g \circ f$. Since q and f are continuous, by Proposition 7 above, it follows that g is continuous. Also $f = g^{-1} \circ q$. Again, since w is the quotient topology on G, the continuity of f implies the continuity of g^{-1} by Proposition 7. Hence g is a homeomorphism.

Corollary 2.   Let $f: E_u \to F_v$ be a continuous mapping of E_u onto F_v, where v is the quotient topology. Then there exists a topological space G_w (see Proposition 8) and a continuous mapping q (see Proposition 8 again) of E_u onto G_w such that $f = g^{-1} \circ q$ and $q = g \circ f$. In other words, each continuous mapping f of E_u onto F_v can be factored into a continuous mapping followed by a homeomorphism.

Proof.   It is contained in the proof of Proposition 8.

After having introduced the quotient topology, we now introduce the concept of quotient space, which is defined by an equivalence relation.

Let E be a set and let $\sim$ be an equivalence relation (cf., Chapter 1, §2) between the elements of E. Then for any element x of E, $x^{\sim}$ is called the equivalence class of an element $x \in E$, i.e., $y \in x^{\sim}$ iff $y \sim x$. Thus we can decompose E into equivalence classes $\{x^{\sim}\}$, $x \in E$, such that either $x^{\sim} \cap y^{\sim} = \varnothing$ (in that case $x \not\sim y$) or $x^{\sim} = y^{\sim}$ (in the case $x \sim y$). The family of all disjoint equivalence classes can thus be regarded as a set with distinct elements and is called a *quotient* set of E. Thus each $x \in E$ belongs to a unique equivalence class $x^{\sim}$. The mapping $x \to x^{\sim}$ is sometimes called the *projection* or *injection*. In order to distinguish it from the projection mapping of the product and injections of subspaces, we shall call it the *quotient* or *canonical mapping*.

Definition 2.   Let E be a topological space with $\sim$ as an equivalence relation. The quotient set $E^{\sim}$, endowed with the quotient topology, for which the canonical mapping: $q: x \to x^{\sim}$ is continuous is called the *quotient space*.

Remark. By definition of the quotient topology, the canonical mapping $x \to x^\sim$ is continuous and open.

There is a different but equivalent method of defining the equivalence relation, namely, by means of a decomposition. *A partition* or *decomposition $\mathscr{D}$* of a set E is a nonempty family of pairwise disjoint subsets of E whose union is E. Let p denote the mapping which to each $x \in E$ assigns a unique element of $\mathscr{D}$ to which x belongs. Then p is called a *projection map* or *quotient map* of E onto $\mathscr{D}$.

Define the relation $x \sim y$ in E if and only if x and y belong to the same member of $\mathscr{D}$, that is, $p(x) = p(y)$. It is not hard to show that $x \sim y$ is an equivalence relation associated with the decomposition $\mathscr{D}$. Thus a decomposition always defines an equivalence relation and, conversely, an equivalence relation gives rise to a decomposition of a set.

Sometimes a mapping is identified with its graph. More precisely, given a mapping $f: E \to F$, the subset $G_f = \{(x, y) \in E \times F: y = f(x)\}$ is called the graph of f and, conversely, given a graph we can define a function whose graph coincides with the given one. A similar description works for relations. Given an equivalence relation $\sim$ associated with a decomposition of a set, the graph of $\sim$ is $\cup \{A \times \{A\}: A \in \mathscr{D}\}$, i.e., the relation $\sim$ is identified with a subset of $E \times E$.

Proposition 9. Let E be a set with an equivalence relation $\sim$. Then for any subset $A \subset E$, $A^\sim = \cup\{x \in E: x^\sim \in E^\sim \text{ and } x^\sim \cap A \neq \varnothing\}$, where $A^\sim$ is the image of A in $E^\sim$ under $\sim$.

Proof. If $a^\sim \in A^\sim$ then $a^\sim$ is a subset of E and $a^\sim \in E^\sim$. Clearly $a \in a^\sim \cap A$. Therefore $a^\sim \subset \cup \{x \in E: x^\sim \in E^\sim \text{ and } x^\sim \cap A \neq \varnothing\} = B$, say. Suppose $y \in B$. Then $y^\sim \in E^\sim$ and $y^\sim \cap A \neq \varnothing$. Let $c \in y^\sim \cap A$ so that $c \sim y$ and $c \in A$. Clearly $y \in c^\sim$ and so $y \in A^\sim$. Thus $B \subset A^\sim$ and we have shown that $A^\sim = B$.

Proposition 10. Let E be a topological space with an equivalence relation $\sim$ and $E^\sim$ the quotient space. The following statements are equivalent:

(i) $q: x \to x^\sim$ is an open mapping of E onto $E^\sim$;

(ii) for each open subset A of E, $A^\sim$ is open in $E^\sim$;

Proof. (i) $\Leftrightarrow$ (ii) follow immediately in view of the remark following the definition of a quotient space.

Theorem 4. Let E be a topological space with an equivalence relation $\sim$ and $E^\sim$ its quotient space.

(i) If E is either compact, connected, locally connected, separable, or Lindelöf, so is $E^\sim$;

(ii) if E is locally compact, so is $E^\sim$;

(iii) if E is a T_1-space, then $E^\sim$ is a T_1-space if the canonical mapping $q: x \to x^\sim$ is closed, i.e., for each $x \in E$, $q(x)$ is closed in $E^\sim$;

(iv) if E is a Hausdorff space and if the graph of q is closed in $E \times E^\sim$, then $E^\sim$ is a Hausdorff space.

Proof. (i) Since the canonical mapping $q: x \to x^\sim$ is continuous and onto, such a mapping preserves all the spaces mentioned in (i) under the continuous and surjective maps.

(ii) Let $x^\sim \in E^\sim$. Then there exists $x \in E$ such that $q(x) = x^\sim$. Since E is locally compact, there exists an open neighborhood U of x such that $\bar{U}$ is compact. Hence $q(\bar{U})$ is a compact neighborhood of $q(x) = x^\sim$, because q is continuous and open. But then $\overline{q(U)} \subset q(\bar{U})$ shows that $E^\sim$ is locally compact.

(iii) Let $x^\sim \in E^\sim$. Then $x^\sim = q(x)$, $x \in E$. Since q is closed by hypothesis and since $\{x\}$ is a closed set in E, $q(x) = \{x^\sim\}$ is a closed subset of $E^\sim$. Thus it follows that $E^\sim$ is a T_1-space if $q(x)$ is closed for each $x \in E$.

(iv) Let $x^\sim \neq y^\sim$, $x^\sim$, $y^\sim \in E^\sim$. Then x is not equivalent to y and hence there exist disjoint open sets U and V such that $x \in U$, $y \in V$ and no point of U is related to any point of V because $E \times E$ is Hausdorff and the graph of the relation $\sim$ is closed. Hence $U^\sim \cap V^\sim = \varnothing$. Since $x \to x^\sim$ is open, $U^\sim$ and $V^\sim$ are open and $x^\sim \in U^\sim$, $y^\sim \in V^\sim$. This proves that $E^\sim$ is Hausdorff.

Remark. In general the quotient space of a Hausdorff space is not a Hausdorff space (cf. Exercises 21).

There is another method of defining quotient spaces of topological spaces, namely, by identifying those points whose neighborhoods are exactly the same. More precisely, we say that x, y in a topological space E are equivalent and write $x \sim y$ iff $\mathscr{U}_x = \mathscr{U}_y$, where $\mathscr{U}_x$ is the neighborhood system at x. Equivalently $\overline{\{x\}} = \overline{\{y\}}$. The relation $\sim$ is an equivalence relation. Moreover, for each $x \in E$, $\dot{x} = \{\bar{x}\}$ is the equivalence class containing x.

Proposition 11. The mapping $\dot{q}: x \to \dot{x}$ is open and the quotient space $\dot{E} = E/\!\sim$ is a T_0-space if E is. If E is regular, so is $\dot{E}$.

Proof. It is clear that if A is an open set in E, and if $A \cap \dot{x} \neq \varnothing$, then $\{y \in E: y \in \dot{x}\} \subset A$. Hence by Proposition 9, $\dot{A} = \cup\{y \in E: y \in \dot{x}, \dot{x} \cap A \neq \varnothing\}$ and A° is open. Hence $\dot{q}$ is open.

To show that $\dot{E}$ is a T_0-space, let $\dot{x} \neq \dot{y}$, where $x \in \dot{x}$, $y \in \dot{y}$. Then x is not related to y, i.e., $\overline{\{x\}} \neq \overline{\{y\}}$ and so $x \neq y$. There exists an open neighborhood P of x or y which does not contain the other, say $y \notin P$ because E is a T_0-space. Since $\dot{q}$ is open, $\dot{q}(P) = \dot{P}$ is an open neighborhood of $\dot{x}$ and $\dot{y} \notin \dot{P}$. Hence $\dot{E}$ is a T_0-space.

Similar arguments work for regularity of $\dot{E}$.

As every topological space can be made Hausdorff by identifying the closures of points, we can similarly make pseudo-metric spaces into metric spaces as follows:

Proposition 12. Let (E, d) be a pseudometric space. Define the relation: $x \sim y$ iff $d(x, y) = 0$. Then $\sim$ is an equivalence relation and $E^\sim$ is a metric space.

Proof. It is easy to see that $\sim$ is an equivalence relation. Let $x \to x^\sim$ be the canonical mapping of E onto $E^\sim$. Define the metric $d^\sim(x^\sim, y^\sim) = \{d(x, y): \text{ for some } x \in x^\sim, y \in y^\sim\}$. To show that $d^\sim$ is a metric, we show first that $d^\sim$ does not depend upon the particular choice of x, y in $x^\sim, y^\sim$ respectively. If x_1, y_1 are other members in $x^\sim, y^\sim$ respectively, then

$$d(x, y) \leq d(x, x_1) + d(x_1, y_1) + d(y_1, y) = d(x_1, y_1),$$

since $d(x_1, x) = d(y_1, y) = 0$ because $x_1 \in x^\sim$, $y_1 \in y^\sim$. Similarly $d(x_1, y_1) \leq d(x, y)$ and hence $d(x_1, y_1) = d(x, y)$. Now it is quite clear that $d^\sim$ is a metric which defines the quotient topology on $E^\sim$.

§24. Projective and Inductive Limits

Inverse Limits. Let $(\Gamma, \leq)$ be a preordered set. For each $\alpha \in \Gamma$, let E_α be a topological space. For $\alpha, \beta \in \Gamma$, $\alpha \leq \beta$, there is a continuous map $f_{\alpha\beta}$ from E_β into E_α such that:

(i) $f_{\alpha\alpha} = $ identity map on E_α;

(ii) for $\alpha \leq \beta \leq \gamma$, $f_{\alpha\beta} \circ f_{\beta\gamma} = f_{\alpha\gamma}$, i.e., for each $x \in E_\gamma$, $f_{\alpha\beta}(f_{\beta\gamma}(x)) = f_{\alpha\gamma}(x)$.

Then $(E_\alpha, \Gamma, f_{\alpha\beta})$ is called an *inverse system* of topological spaces E_α.

Definition 3. Let $(E_\alpha, \Gamma, f_{\alpha\beta})$ be an inverse system of topological spaces. Let $E = \prod_{\alpha \in \Gamma} E_\alpha$ be the product space. The subset $\lim_{\leftarrow} E_\alpha = \{x = (x_\alpha) \in \prod_{\alpha \in \Gamma} E : \text{for all } \alpha \leq \beta, f_{\alpha\beta}(x_\beta) = x_\alpha\}$ of E is called the *inverse limit* of E_α's or of the inverse system $(E_\alpha, \Gamma, f_{\alpha\beta})$. Sometimes $\lim_{\leftarrow} E_\alpha$ is called the *projective limit* of the inverse system $\{E_\alpha, \Gamma, f_{\alpha\beta}\}$ when endowed with the topology induced from $\prod_\alpha E_\alpha$.

There is no reason to believe that the projective limit of an arbitrary inverse system always exists as a nonempty set. Since $\lim_{\leftarrow} E_\alpha$ is a subset of $\prod_\alpha E_\alpha$, $\lim_{\leftarrow} E_\alpha$ is respectively Hausdorff, regular, and completely regular if each E_α is. Moreover, we have:

Proposition 13. Let $\lim_{\leftarrow} E_\alpha$ be the inverse limit of nonempty Hausdorff spaces E_α's of an inverse system $(E_\alpha, \Gamma, f_{\alpha\beta})$. Then $\lim_{\leftarrow} E_\alpha$ is a closed subset of $\prod_{\alpha \in \Gamma} E_\alpha$.

Proof. We know from set theory that if E_α's are nonempty then $\prod_\alpha E_\alpha$ is also nonempty. By definition, $\lim_{\leftarrow} E_\alpha$ is a subset of $\prod_\alpha E_\alpha$. To show that it is closed, let $x \in \prod_{\alpha \in \Gamma} E \setminus \lim_{\leftarrow} E_\alpha$. Then for some $\alpha, \beta \in \Gamma, \alpha \leq \beta$, $f_{\alpha\beta}(x_\beta) \neq x_\alpha$. Since E_α is Hausdorff, there are disjoint open neighborhoods U_α and V_α such that $x_\alpha \in U_\alpha$ and $f_{\alpha\beta}(x_\beta) \in V_\alpha$. Let $U_\beta = f_{\alpha\beta}^{-1}(V_\alpha)$. Then U_β is an open set because $f_{\alpha\beta}$ is continuous. Now put $U = \prod_{\gamma \in \Gamma} U_\gamma$, where $U_\gamma = U_\alpha$ or U_β if $\gamma = \alpha$ or β and $U_\gamma = E_\gamma$ if $\gamma \neq \alpha, \beta$. Then U is an open neighborhood of x and $U \cap \lim_{\leftarrow} E_\alpha = \varnothing$. This proves that $\lim_{\leftarrow} E_\alpha$ is closed.

Corollary 3. Let $\lim_{\leftarrow} E_\alpha$ be the inverse limit of the inverse system $(E_\alpha, \Gamma, f_{\alpha\beta})$ of Hausdorff compact spaces E_α's. Then $\lim_{\leftarrow} E_\alpha$ is a compact Hausdorff space.

Proof. Since by Tychonoff's theorem (Theorem 3, §35), $\prod_{\alpha \in \Gamma} E_\alpha$ is compact and $\lim_{\leftarrow} E_\alpha$ is a closed subset of $\prod_{\alpha \in \Gamma} E_\alpha$ (Proposition 13), the corollary follows.

Remark. Observe that by definition, $\lim_{\leftarrow} E_\alpha$ is a subset of $\prod_{\alpha \in \Gamma} E_\alpha$. The mapping $p_\alpha : \prod_{\alpha \in \Gamma} E_\alpha \to E_\alpha$ is called the projection mapping and its restriction on $\lim_{\leftarrow} E_\alpha$ to E_α is called the canonical mapping and is sometimes denoted by

$$f_\alpha : \lim_{\leftarrow} E_\alpha \to E_\alpha.$$

Clearly f_α is continuous because each p_α is. By the definition of $\lim_{\leftarrow} E_\alpha$, it follows that $f_\alpha = f_{\alpha\beta} \circ f_\beta$ whenever $\alpha \leq \beta$. If Γ is directed, the sets $\{f_\alpha^{-1}(U_\alpha) : U_\alpha \text{ open in } E_\alpha\}$ form a base of the topology of $\lim_{\leftarrow} E_\alpha$.

Proposition 14. Let $(E_\alpha, \Gamma, f_{\alpha\beta})$ be an inverse system and $E = \lim_{\leftarrow} E_\alpha$. For each α, f_α is the canonical mapping of E into E_α. Let F be any topological space and g_α a continuous mapping of F into E_α for each $\alpha \in \Gamma$ such that $f_{\alpha\beta} \circ g_\beta = g_\alpha$ whenever $\alpha \leq \beta$. Then there exists a unique continuous mapping g of F into E such that $f_\alpha \circ g = g_\alpha$ for all $\alpha \in \Gamma$.

Proof. For each $x \in F$, by hypothesis $g_\alpha(x) = f_{\alpha\beta}(g_\beta(x))$, wherever $\alpha \leq \beta$. In other words, we can express the last equation by $p_\alpha(g(x)) = f_{\alpha\beta}(p_\beta(g(x)))$, where g is the unique mapping of F into $\prod_{\alpha \in \Gamma} E$ given by $g(x) = (g_\alpha(x))_{\alpha \in \Gamma}$. But then it follows from $p_\alpha(g(x)) = f_{\alpha\beta}(p_\beta(g(x)))$ that $g(x) \in \lim_{\leftarrow} E_\alpha$. Obviously g is continuous and $f_\alpha \circ g = g_\alpha$ for all $\alpha \in \Gamma$.

Proposition 15. Let $(E_\alpha, \Gamma, f_{\alpha\beta})$ and $(F_\alpha, \Gamma, g_{\alpha\beta})$ be two inverse systems, and $E = \lim_{\leftarrow} E_\alpha$, $F = \lim_{\leftarrow} F_\alpha$. Let f_α, g_α be the canonical mappings of E and F into E_α and F_α, respectively. If for each $\alpha \in \Gamma$ there is a continuous mapping u_α of E_α into F_α such that the diagram

$$
\begin{array}{ccc}
E_\beta & \xrightarrow{\ u_\beta\ } & F_\beta \\
{\scriptstyle f_{\alpha\beta}}\big\downarrow & & \big\downarrow{\scriptstyle g_{\alpha\beta}} \\
E_\alpha & \xrightarrow[\ u_\alpha\ ]{} & F_\alpha
\end{array}
$$

is commutative whenever $\alpha \leq \beta$, then exists a unique continuous mapping $u \colon E \to F$ such that the diagram

$$
\begin{array}{ccc}
E & \xrightarrow{\ u\ } & F \\
{\scriptstyle f_\alpha}\big\downarrow & & \big\downarrow{\scriptstyle g_\alpha} \\
E_\alpha & \xrightarrow[\ u_\alpha\ ]{} & F_\alpha
\end{array}
$$

is commutative for each α.

Proof. Since for $\alpha \leq \beta$, we have $g_{\alpha\beta} \circ u_\beta \circ f_\beta = u_\alpha \circ f_{\alpha\beta} \circ f_\beta = u_\alpha \circ f_\alpha$, by Proposition 14 there exists a mapping $h_\alpha \colon E_\alpha \to F$ such that $f_\alpha \circ h_\alpha = u_\alpha$ for each α. This proves the proposition if we take $u = \{h_\alpha\} \mid E$ as the restriction of $\{h_\alpha\} \colon \prod_\alpha E_\alpha \to F$. The uniqueness of u is easy.

Direct or inductive limits. Let $\{E_\alpha\}_{\alpha \in \Gamma}$ be a family of topological spaces, with Γ a directed set. Let $\{f_{\alpha\beta}\}$ be a family of mappings such that $f_{\alpha\beta} \colon E_\alpha \to E_\beta$ is continuous whenever $\alpha \leq \beta$ and the following conditions hold:

(i)   For $\alpha \leq \beta \leq \gamma$, $f_{\alpha\gamma} = f_{\beta\gamma} \circ f_{\alpha\beta}$;

(ii)   for each $\alpha \in \Gamma$, $f_{\alpha\alpha}$ is the identity. Then $(E_\alpha, \Gamma, f_{\beta\alpha})$ is called a *direct system*.

Let G be the free union of E_α's (see §21), in which each E_α is identified with its canonical image in G. We may introduce an equivalence relation between the elements of G as follows: For x, $y \in G$, we write $x \sim y$ if for any connecting maps $f_{\alpha\gamma}$ and $f_{\beta\sigma}$ of E_α and E_β, where $x \in E_\alpha$ and $y \in E_\beta$, $f_{\alpha\gamma}(x) = f_{\beta\sigma}(y)$. The directedness of Γ is needed to check that $\sim$ is an equivalence relation. The quotient space $E = G/\sim$ is called the *direct or inductive limit* of $(E_\alpha, \Gamma, f_{\alpha\beta})$ and is written as: $E = \lim_{\rightarrow} E_\alpha$ or $\lim \mathrm{Dir}_\Gamma E_\alpha$.

It is clear that the direct limit is nonempty if at least one E_α is nonempty. Let f denote the canonical mapping of G onto $G/\sim = E$. The restriction of f to E_α is called the canonical mapping f_α of E_α into E, which is clearly continuous and

$$f_\beta = f_\alpha \circ f_{\alpha\beta},$$

whenever $\alpha \leq \beta$. Clearly $E = \bigcup_{\alpha \in \Gamma} f_\alpha(E_\alpha)$.

Proposition 16.   Let $(E_\alpha, \Gamma, f_{\alpha\beta})$ be a direct system and $E = \lim_{\rightarrow} E_\alpha$. For each $\alpha \in \Gamma$, let f_α denote the canonical mapping of E_α into E. If u_α is a continuous mapping of E_α into a topological space F for each α such that $u_\beta \circ f_{\alpha\beta} = u_\alpha$ whenever $\alpha \leq \beta$, then there exists a unique continuous mapping u of E into F such that $u_\alpha = u \circ f_\alpha$ for all $\alpha \in \Gamma$.

Proof.   Let v be the mapping of G into F which coincides with u_α for each $\alpha \in \Gamma$. By the definition of direct sum topology on G, v is continuous from G into F. By hypothesis v is compatible with the equivalence relation. Hence there exists a unique mapping u of $E = G/\sim$ into F. It is clear that u is continuous and $u_\alpha = u \circ f_\alpha$ for $\alpha \in \Gamma$.

Corollary 4.   Let $(E_\alpha, \Gamma, f_{\beta\alpha})$ and $(F_\alpha, \Gamma, g_{\beta\alpha})$ be two direct systems and $E = \lim_{\rightarrow} E_\alpha$, $F = \lim_{\rightarrow} F_\alpha$. Let f_α (or g_α) denote the canonical mappings of E_α (or F_α) into E (or F) respectively. If for each $\alpha \in \Gamma$ there is a continuous mapping u_α of E_α into F_α such that the following diagram:

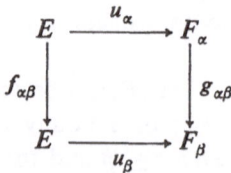

is commutative, whenever $\alpha \leq \beta$, then there exists a unique continuous mapping $u: E \to F$ such that for each $\alpha \in \Gamma$, the diagram

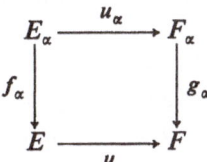

is commutative.

Proof. The proof is similar to that of Proposition 15.

The construction of inductive (or direct) limits may sometimes be described as follows:

Let $(E_\alpha)_{\alpha \in \Gamma}$ be a family of topological spaces and E a set. Let f_α be a mapping of E_α into E for each $\alpha \in \Gamma$ such that $E = \bigcup_{\alpha \in \Gamma} f_\alpha(E_\alpha)$. We may endow E with the topology $\mathcal{E}$, which is the *finest* topology such that for each $\alpha \in \Gamma$, each mapping $f_\alpha: E_\alpha \to E$ is continuous. The topological space $(E, \mathcal{E})$ is sometimes called the *generalized inductive limit* of E_α's. If each f_α is an injection and for every pair $\alpha, \beta \in \Gamma$, $\alpha \leq \beta$ implies $E_\alpha \subset E_\beta$ and the topology $\mathcal{E}_\beta$ on E_β induces the given topology on E_α and if Γ is countable, then $(E, \mathcal{E})$ is called the *strict inductive limit* of E_α's.

Proposition 17. Let $(E, \mathcal{E})$ be the generalized inductive limit of the system $(E_\alpha, \Gamma, f_\alpha)$ and F a topological space. A mapping $f: E \to F$ is continuous iff $f \circ f_\alpha$ is continuous for each $\alpha \in \Gamma$.

Proof. We know that $f_\alpha: E_\alpha \to E$ is continuous for each α. If $f: E \to F$ is continuous, then the composition $f \circ f_\alpha$ is certainly continuous. On the other hand, if V is an open neighborhood of $f_\alpha(x) \in F$ for some $x \in E$, then $f_\alpha^{-1}(V)$ is an open neighborhood of x, since f_α is continuous for each α. But then $f^{-1}(V)$ is a set in E such that $f^{-1}(V) \cap E_\alpha = f_\alpha^{-1}(V)$ is open in E_α for each $\alpha \in \Gamma$ which shows that $f^{-1}(V)$ is open in E by the definition of the generalized inductive limit topology. This proves that f is continuous.

Examples and Exercises

1. If F is a subspace of E, then a subset A of F is closed (or open) in E if F is closed (or open).
2. Show that a subspace of a separable space need not be separable (cf. Exercise 13, Chapter II).

3. Show that a subspace of a Lindelöf space need not be Lindelöf (cf. Exercise 13, Chapter II).

4. Show that the open interval $(0, 1)$ of the reals is not compact while the closed interval $[0, 1]$ is.

5. Show that a subspace of a normal space need not be normal.

7. Let $E = E_1 \cup E_2$ such that E_1, E_2 are closed subspaces of E. Let $A \subset E_1$ be such that $A \cap E_2$ is contained in an open subset B of E_2, then $A \subset (E_1 \cap B)^\circ$.

8. Let $E_u = R$, real line with the topology u defined by the family of half-closed and half-open intervals $[a, b)$. Show that this topology is finer than the usual Euclidean topology defined by the following metric: $|x - y|$. Show that E_u is a normal and Lindelöf space but $E_u \times E_u$ is not normal and not Lindelöf. (Hence the product of two normal spaces is not necessarily normal.)

9. Show that the product of a locally compact space and a normal space need not be a normal space (use Example 8).

10. Show that the n-dimensional Euclidean space $R^n (n \geq 1)$ is a locally compact space but not compact.

11. Show that the space R^n in Ex. 10 is a complete metric space with the metric

$$d(x, y) = \left[\sum_{i=1}^{n} |x_i - y_i|^2 \right]^{1/2}$$

when $x = (x_1, \ldots, x_n)$, $y = (y_1, \ldots, y_n)$.
    Show that R^n is connected and locally connected.

12. Show that if $I = [0, 1]$, then I^A is compact, where A is any cardinal number.

13. If R is the set of real numbers with the usual metric topology and A an uncountable cardinal, then R^A is not metrizable and not locally compact.

14. If R^2 is the real plane, show that there exists a real-valued function f: $(x, y) \to f(x, y)$ such that for each fixed x or each fixed y, f is continuous but not continuous on R^2.

15. If for each $\alpha \in A$, E_α is a topological space and B, C are disjoint subsets such that $A = B \cup C$, then $\prod_{\alpha \in A} E_\alpha$ is homeomorphic to $\prod_{\alpha \in B} E_\alpha \times \prod_{\alpha \in C} E_\alpha$.

16. Show that a subspace of a connected space need not be connected.

17. Is the product of locally connected spaces locally connected?

18. Let E_α ($\alpha \in A$) and F_α ($\alpha \in A$) be two families of topological spaces and $f_\alpha: E_\alpha \to F_\alpha$ a map for each $\alpha \in A$. Let $E = \prod_\alpha E_\alpha$, $F = \prod_\alpha F_\alpha$ and $f: E \to F$ be defined by $f(x)(\alpha) = f_\alpha(x_\alpha)$ for each $\alpha \in A$, where

$x = \{x_\alpha\} \in E$. Then f is continuous if each f_α is continuous, and open if all but a finite number of f_α's are onto and open.

19. Let F_α be a subset of a topological space E_α for each $\alpha \in A$. Then $\prod_\alpha F_\alpha$ is dense in $\prod_\alpha E_\alpha$ iff each F_α is dense in E_α for each $\alpha \in A$.

20. Let E be a topological space and $E^\sim$ a quotient space of E with the quotient topology where $\sim$ is an equivalence relation.
    (a) Show that $E^\sim$ need not be first countable even though E is.
    (b) The same as (a) for second countable spaces.
    (c) Show that $E^\sim$ is a countably compact T_1-space if E is.

21. Show that the quotient topology on a quotient space $E^\sim$ need not be Hausdorff even though E is Hausdorff. Show that the quotient space is a T_1-space iff the relation $\sim$ is closed.

22. Let $E^\sim$ be the quotient space of a topological space E with the relation $\sim$. Let $\dot{q}$ denote the quotient mapping: $x \rightarrow x^\sim$ of E onto $E^\sim$. Show that $\dot{q}$ is a closed map iff for each subset A of E, $\overline{A^\sim} \subset \bar{A}^\sim$ or $\overline{\dot{q}(A)} \subset \dot{q}(\bar{A})$; $\dot{q}$ is open iff $\dot{q}(A^\circ) \subset [\dot{q}(A)]^\circ$.

23. Let $E = \{0, 2\}$, the two-point set. Let E^N denote the countable product of E. Then each element of E^N is a sequence having 0 or 2 as elements. Let $\{x_i\} \in E^N$, and define

$$\varphi(\{x_i\}) = \sum_{i=1}^\infty \frac{x_i}{3^i}.$$

Then it is clear that $\varphi(\{x_i\}) \in [0, 1]$ and φ is one-to-one. The image of E^N under φ is called the *Cantor set, C*.
    (a) Show that C is a compact subset of $[0, 1]$.
    (b) Show that the map

$$\psi(\{x_i\}) = \sum_{i=1}^\infty \frac{x_i}{2^{i+1}}$$

    of C into $[0, 1]$ is continuous and open.
    (c) Let I^n denote the n-times product of $I = [0, 1]$. Show that there exists a continuous mapping of I onto I^n. (This is called the *Peano curve*.)

24. (a) Show that the inductive limit topology of two locally compact spaces need not be locally compact.
    (b) Show that the inductive limit topology of metric spaces need not be metrizable. (Take E the set of all real sequences with only a finite number of nonzero elements. Then $E = \bigcup_{n=1}^\infty R^n$, where R^n is the n-dimensional Euclidean space. Let f_n denote the embedding of R^n into E for each $n \geq 1$. The inductive limit topology cannot be metrizable. Use results of Chapter VI regarding Baire spaces.)

V

Uniform Spaces

In this chapter, we study a generalization of metric spaces. Some fundamental properties on metric space, e.g., uniform continuity and completeness, can be defined for more general spaces called uniform spaces. The notion of uniform spaces is due to A. Weil [91]. As we shall see in the sequel, there is a close relationship between uniform spaces and completely regular spaces. Some special classes of uniform spaces, viz., topological groups and topological vector spaces, are also treated.

§25. Uniformities and Topologies

Let E be a set. By $E^2 = E \times E$, we mean the set of all ordered pairs (x, y), $x, y \in E$. As we know from Chapter I, E^2 is the product of E by itself. For any subset A of E^2, we define

$$A^{-1} = \{(y, x) \in E^2 \colon (x, y) \in A\}.$$

Let A, B be two subsets of E^2. Then we define $A \circ B$ as a collection of pairs (x, y) in E^2 if for some $z \in E$; $(x, z) \in A$ and $(z, y) \in B$. If $B = A$ in $A \circ B$, we put $A \circ A = A^2$.

A subset A of E^2 is said to be *symmetric* if $A = A^{-1}$. The set $\{(x, x) \colon x \in E\}$ is called the *diagonal* of E^2 and is sometimes denoted by $\Delta(E)$ or simply Δ.

Let A be a subset of E and U a subset of E^2. Then we define

$$U[A] = \{y \in E \colon (x, y) \in U \text{ for some } x \in A\},$$

103

$$U^{-1}[A] = \{y \in E: (y, x) \in U \quad \text{for some } x \in A\}$$
$$= \{y \in E: (x, y) \in U^{-1} \text{ for some } x \in A\}.$$

If A is a singleton $\{x\}$, then

$$U[x] = U[\{x\}] = \{y \in E: (x, y) \in U\},$$
$$U^{-1}[x] = \{y \in E: (y, x) \in U\}.$$

Thus $U = U^{-1}$ if and only if $U[x] = U^{-1}[x]$ for each $x \in E$.

Definition 1. A filter $\mathcal{U} = \{U\}$ in E^2 is said to be a *uniformity* for E or defines a *uniform structure* on E, if the following conditions are satisfied:

(a) Each U in $\mathcal{U}$ contains the diagonal $\varDelta$;
(b) for each U in $\mathcal{U}$, $U^{-1} \in \mathcal{U}$;
(c) for each U in $\mathcal{U}$ there exists a V in $\mathcal{U}$ such that $V \circ V \subset U$.

A set E with a uniformity $\mathcal{U}$ is called a *uniform space* and is denoted by $(E, \mathcal{U})$ or $(E, \{U\})$. A set E with a filter $\mathcal{U}$ satisfying only (a) and (c) is called a *quasiuniform space*.

If we compare properties (a) through (c) with those of metric spaces, we cannot help detecting the close relationship between metric and uniform spaces. The latter generalizes the former. More precisely we have:

Proposition 1. Each metric space can be endowed with a uniform structure.

Proof. Let E be a metric space with the metric d. For each real number $r > 0$, define

$$U_r = \{(x, y) \in E^2: d(x, y) < r\}.$$

It is not hard to check that the family $\{U_r\}_{r>0}$ is a filter base and hence generates a filter $\mathcal{U}$. For each $x \in E$, $d(x, x) = 0 < r$ implies that $\varDelta \subset U_r$ for each $r > 0$. Also since $d(x, y) = d(y, x)$, $U_r^{-1} = U_r$. Moreover, for each $r > 0$, $U_{r/2} \circ U_{r/2} \subset U_r$. Thus all conditions (a) to (c) of Definition 1 above are satisfied by $\mathcal{U}$. Hence E with $\mathcal{U}$ is a uniform space.

Remark. In the above proposition we have not used the fact that $d(x, y) = 0$ implies $x = y$ in a metric space. Thus one can say that each pseudometric space can be endowed with a uniform structure.

Remark. On each set E, there are at least two uniformities: the *largest* uniformity consisting of all subsets of E^2, each of which contains the diagonal, and the *smallest* one, consisting of $E \times E$ only.

As the topologies are generated by their bases, so can one generate uniformities by their bases.

Definition 2. A subfamily $\{B\}$ of a uniformity $\{U\}$ is called a *base* of $\{U\}$ if each U in $\{U\}$ contains a $B \in \{B\}$.

Proposition 2. A filter base $\{B\}$ of subsets of E^2 is a base for some uniformity on E if and only if the following conditions hold:

(i) Each $B \in \{B\}$ contains Δ;

(ii) if $B \in \{B\}$ then B^{-1} contains a member of $\{B\}$;

(iii) if $B \in \{B\}$, then there exists $C \in \{B\}$ such that $C \circ C \subset B$.

Proof. Assume $\{B\}$ is a base of a uniformity $\{U\}$. Then clearly (i) is satisfied because each member of $\{U\}$ contains Δ. Furthermore, each $B \in \{B\}$, being a member of $\{U\}$, implies $B^{-1} \in \{U\}$ and hence there exists a member of $\{B\}$ which is contained in B^{-1}. Also for each $B \in \{B\}$ there is a $V \in \{U\}$ such that $V \circ V \subset B$. But since each $V \in \{U\}$ contains a $C \in \{B\}$ (since B is a base), we have $C \circ C \subset B$ and hence (iii) is also satisfied.

Conversely, assume $\{B\}$ is a subfamily of subsets of E^2 satisfying (i)–(iii). Let $\{U\}$ denote the family of all subsets U of E^2 such that U contains a member of $\{B\}$. Since $\{B\}$ is a filter base, it is clear that $\{U\}$ is a filter. Further, since $\{B\}$ satisfies (i), each $U \in \{U\}$ contains Δ. Also for each $U \in \{U\}$ there exists a $B \in \{B\}$ such that $B \subset U$. Hence $B^{-1} \subset U^{-1}$. Since by (ii) B^{-1} contains a member of $\{B\}$, $U^{-1} \in \{U\}$. Similarly using (iii) we show that for each $U \in \{U\}$ there is a $V \in \{U\}$ such that $V \circ V \subset U$. Thus $\{U\}$ is a uniformity and $\{B\}$ is a base of $\{U\}$.

Definition 3. A subfamily $\mathscr{S}$ of a uniformity $\{U\}$ is said to be a *subbase* of $\{U\}$ if the family of finite intersections of $\mathscr{S}$ is a base for $\{U\}$.

Proposition 3. A family $\mathscr{S}$ of subsets of E^2 is a subbase for some uniformity for E, if the conditions (i) to (iii) of Proposition 2 above are satisfied by the members of $\mathscr{S}$.

Proof. It is straightforward and therefore omitted.

Now we show that a uniformity for a set E defines a topology. The converse problem—What topological spaces have the topologies defined by a uniformity?—will be considered later.

Definition 4.   Let $(E, \{U\})$ be a uniform space. The topology defined by the uniformity $\{U\}$ is the collection of all subsets T of E such that for each $x \in T$ there is a $U \in \{U\}$ with $U[x] = \{y \in E: (x, y) \in U\} \subset T$.

That the collection $\mathscr{C}$ of all subsets T of E satisfying the condition in the above definition does indeed define a topology is a simple matter of verification.

The relevant information concerning the open and closed subsets of the topology defined by a uniformity is given in the following:

Proposition 4.   Let $E, \{U\})$ be a uniform space and let $\mathscr{C}$ be the topology on E defined by $\{U\}$.

(i)   If $\{B\}$ is a base (or subbase) of the uniformity $\{U\}$, then the family $U[x]$, where U runs over $\{B\}$ is a base (or a subbase) of the neighborhood filter of x. Hence each $x \in E$ has a base of neighborhood filters, each member of which is symmetric (i.e., when U in $\{U\}$ is symmetric).

(ii)   If A° is the $\mathscr{C}$-interior of a subset $A \subset E$, then

$$A^\circ = \{x \in E: \text{ for some } U \in \mathscr{U}, U[x] \subset A\}.$$

(iii)   If $\bar{A}$ is the $\mathscr{C}$-closure of $A \subset E$, then

$$\bar{A} = \cap \{U[A]: U \in \mathscr{U}\}.$$

Proof.   (i)   If $\mathscr{B}$ is a base (or subbase) for the uniformity $\mathscr{U}$, then for each $x \in E$, the family $\{U[x]\}$, where U runs over $\mathscr{U}$, forms a base (or subbase) for the neighborhood filter of x. Consequently the symmetric neighborhoods $U[x]$ form a base of the neighborhood filter of x and $U[x] \cap U^{-1}[x]$ is a symmetric neighborhood of x.

(ii)   Put $B = \{x \in E: \text{ for some } U \in \mathscr{U}, U[x] \subset A\}$. Then for each $x \in B$, there exists $U \in \mathscr{U}$ such that $U[x] \subset B$. Also there exists $V \in \mathscr{U}$ such that $V^2 = V \circ V \subset U$. To show that $V[x] \subset B$, let $y \in V[x]$. Then $V[y] \subset V^2[x] \subset U[x] \subset A$. Hence $y \in B$. But since $V[x]$ is $\mathscr{C}$-open by definition and x is arbitrary, it follows that B is $\mathscr{C}$-open. It is now clear that each $\mathscr{C}$-open subset C of A is contained in B since C contains a subset of the type $U[x]$. Hence $B = A^\circ$.

(iii) $x \in \bar{A}$ if and only if, for each $U \in \mathcal{U}$, $U[x] \cap A \neq \emptyset$ if and only if $x \in U^{-1}[A]$. Since each $U \in \mathcal{U}$ contains a symmetric member, it follows that $x \in \bar{A}$ iff $x \in U[A]$ for each $U \in \mathcal{U}$. Therefore, $\bar{A} = \cap \{U[A]: U \in (\mathcal{U})\}$.

Let $(E, \mathcal{U})$ be a topological space. For each pair (U, V) of open sets U and V in E, recall that $U \times V$ is an open set in E^2 in the product topology.

Proposition 5. Let u be the product topology on E^2, where $(E, \mathcal{U})$ is a uniform space and u is defined by the uniform topology on E. Then:

(i) For each U in $\mathcal{U}$, the interior U° of U in the product topology is also in $\mathcal{U}$.

(ii) The family of all symmetric and open (in the product topology) members in $\mathcal{U}$ forms a base for $\mathcal{U}$.

(iii) For each subset $A \subset E \times E$,

$$\bar{A} = \cap \{U \circ A \circ U: U \in \mathcal{U}\},$$

where $\bar{A}$ is the closure in the product topology on E^2.

(iv) The family of all closed and symmetric sets in $\mathcal{U}$ forms a base for $\mathcal{U}$.

Proof. (i) For each subset $G \subset E^2$, the interior G° of G is given by

$$G^{\circ} = \{(x, y): \text{for some } U \text{ and } V \text{ in } \mathcal{U}, U[x] \times V[y] \subset G\},$$

because the collection of sets $U[x] \times V[y]$ forms a base of the neighborhood system of (x, y) in E^2 in the product topology. Since $\mathcal{U}$ is a filter, $U \cap V \in \mathcal{U}$ and so

$$G^{\circ} = \{(x, y): \text{for some } V \in \mathcal{U}, V[x] \times V[y] \subset G\}. \tag{*}$$

For each $U \in \mathcal{U}$, there is a symmetric $V \in \mathcal{U}$ such that $V \circ V \circ V \subset U$. First, we show that for any symmetric set V and any set W in $\mathcal{U}$, we have

$$V \circ W \circ V = \cup \{V[x] \times V[y]: (x, y) \in W\}. \tag{**}$$

Observe that $(a, b) \in V \circ W \circ V$ if and only if $(a, x) \in V$, $(x, y) \in W$, and $(y, b) \in V$ for some x and some y. Since V is symmetric, $(a, b) \in V \circ W \circ V$ if and only if $a \in V[x]$ and $b \in V[y]$ for some $(x, y) \in W$, which proves (**). It is now clear that $a \in V[x]$ and $b \in V[y]$ if and only if $(a, b) \in V[x] \times V[y]$ and hence (*) follows. Replacing W by V in (**), we have

$$V \circ V \circ V = \{(a, b): (a, b) \in V[x] \times V[y] \text{ for some } (x, y) \in V\}$$
$$= \cup \{V[x] \times V[y]: (x, y) \in V\}.$$

But this shows, in view of (*), where we replace G by U, that every point of V is an interior point of U and since the interior U° of U contains V, it follows that $U^\circ \in \mathscr{U}$.

(ii)   Since each $U \in \mathscr{U}$ contains a symmetric set in $\mathscr{U}$, (ii) follows from (i).

(iii)   For each symmetric member U of $\mathscr{U}$, $(U[x] \times U[y]) \cap A \neq \varnothing$ if and only if $(x, y) \in U[a] \times U[b]$ for some $(a, b) \in A$, if and only if $(x, y) \in \cup \{U[a] \times U[b]: (a, b) \in A\} = H$. Since $H = U \circ A \circ U$ [see the proof of (i)], it follows that $(x, y) \in A$ if and only if $(x, y) \in \cap \{U \circ A \circ U: U \in \mathscr{U}\}$. Thus we have shown that $(x, y) \in \bar{A}$ iff $(x, y) \in \cap \{U \circ A \circ U: U \in \mathscr{U}\}$. This proves (iii).

(iv)   For each $U \in \mathscr{U}$, there is a $V \in \mathscr{U}$ such that $V \circ V \circ V \subset U$. But $\bar{V} \subset V \circ V \circ V$ by (iii). Hence U contains a closed member $\bar{V} = W$ of $\mathscr{U}$. But then $W \cap W^{-1}$ is a closed and symmetric member of $\mathscr{U}$, which is contained in U.

§26. Uniformity and Separation Axioms

We have seen in the previous sections how a uniformity on E^2 induces a topology on E and then how the product topology on E^2 is defined by the topology on E. It is natural to expect some relation between the uniformities and the separation axioms defined for the associated topologies.

Proposition 6.   Let E_u be a topological space, where the topology is induced by a uniformity $\mathscr{U}$. Then E_u is a T_3-space. Hence if E_u is a T_1-space, then E_u is regular.

Proof.   Let $x \in E$. For each neighborhood $U[x]$ of x, there exists a $V \in \mathscr{U}$ such that $V \circ V \subset \mathscr{U}$. Then $\bar{V}[x] = \cap \{W \circ V[x]: W \in \mathscr{U}\}$ is a closed neighborhood of x and $\bar{V}[x] \subset U[x]$. By Theorem 3, §18 it follows that E_u is a T_3-space.

We shall show later that E_u is even a T_4-space. But first we have:

Theorem 1.   Let $(E, \mathscr{U})$ be a uniform space and let u be the topology defined by $\mathscr{U}$ on E. The following statements are equivalent:

(i) E_u is a T_1-space;

(ii) E_u is a Hausdorff space;

(iii) $\cap \{U: U \in \mathscr{U}\} = \Delta$, the diagonal set;

(iv) E_u is regular.

Proof. The implications (iv) $\Rightarrow$ (ii) $\Rightarrow$ (i) hold in general by Propositions 19 and 14, Chap. III, and (i) $\Rightarrow$ (iv) by Proposition 6 above. Hence (i) $\Leftrightarrow$ (ii) $\Leftrightarrow$ (iv). Further, (i) $\Leftrightarrow$ (iii) because for each $x \in E$, by Proposition 4, §25, we have:

$$\overline{\{x\}} = \cap \{U[x]: U \in \mathscr{U}\}.$$

Hence $\overline{\{x\}} = \{x\}$ if and only if $\cap \{U: U \in \mathscr{U}\} = \varDelta$.

Now we show that a uniform space endowed with the topology defined by its uniformity is a T_4-space. There is an indirect but simple proof of this fact. We, however, prove this directly.

Theorem 2. Every uniform space $(E, \mathscr{U})$ endowed with the topology defined by $\mathscr{U}$ is a T_4-space and hence completely regular if it is a Hausdorff space. Conversely, every T_4-space can be endowed with a uniform structure.

Proof. Let $x \in E$ and C a closed subset of E such that $x \notin C$. We choose a family of neighborhoods $U_r[x]$ of x for each dyadic rational number $r \in D$, where $D = \{k/2^n: k, n \text{ positive integers}, k \leq 2^n\}$ is the set of all dyadic rationals in $[0, 1]$.

Put $E \backslash C = U_0$, which is an open neighborhood of x. For each integer $n \geq 0$, there exists a $U_n \in \mathscr{U}$ such that $U_{n+1} \circ U_{n+1} \subset U_n$. Hence it follows that

$$U_{n+1}[x] \times U_{n+1}[x] \subset U_n[x].$$

If $r = 1/2^n \in D$, we define $V_{1/2^n}[x]$ by putting $V_{1/2^n}[x] = U_n[x]$ for $n \geq 0$. Suppose by induction $V_r[x]$ has been defined for all $r = k/2^n$, $k \leq 2^n$. Define

$$V_{k'/2^{n+1}}[x] = \begin{cases} U_{k/2^n}[x], & \text{if } k' = 2k; \\ U_{1/2^n}[x] \cap U_{k/2^n}[x], & \text{if } k' = 2k + 1. \end{cases}$$

It is not hard to verify that

$$V_{1/2^n}[x] \cap U_{k/2^n}[x] \subset U_{(k+1)/2^n}[x]$$

for all integers k, with $0 \leq k + 1 \leq 2^n$. By our choice it follows that if r_1, $r_2 \in D$ and $r_1 < r_2$, then $U_{r_1}[x] \subset \bar{U}_{r_1}[x] \subset U_{r_2}[x]$.

Now we define a real-valued function f:

$$f(y) = \begin{cases} \inf r, & \text{if } y \in V_r[x] \text{ for } r \in D, \\ 1, & \text{if } y \notin V_1[x]. \end{cases}$$

Since $x \in V_r[x]$ for all $r \in D$, it is clear that $f(x) = 0$ and also $f(C) = 1$. In order to complete the proof of the first part, we have to show only that f is continuous. But since for $0 \le a, b < 1$, $(b, 1]$ and $[0, a)$ form a subbase of the topology on $[0, 1]$, and since

$$f^{-1}([0, a)) = \bigcup_r \{V_r[x]: r < a\},$$

which is an open set in E, it follows that $f^{-1}([0, a))$ is open. Similarly, for $f^{-1}((b, 1])$. Hence f is continuous. Thus, we have shown that each uniform space is a T_4-space. So if the uniform topology satisfies T_1-axiom or is Hausdorff, then it is completely regular.

For the converse, assume E_u is a T_4-space. Let f be an arbitrary continuous function from E_u to $[0, 1]$. For each $\varepsilon > 0$, define

$$U(f, \varepsilon) = \{(x, y) \in E^2: |f(x) - f(y)| < \varepsilon\}.$$

It is clear that $U(f, \varepsilon_1) \cap U(g, \varepsilon_2) \ne \varnothing$ for any continuous functions f, g on E and ε_1, $\varepsilon_2 > 0$. Hence the family $[U(f, \varepsilon)]$, where ε runs over positive real numbers and f over all continuous functions from E to $[0, 1]$, is a subbase of a filter on $E \times E$. Clearly $(x, x) \in U(f, \varepsilon)$ for each f and ε. Also for each $U(f, \varepsilon)$ there exists $U(f, \varepsilon/2)$ such that $U(f, \varepsilon/2) \circ U(f, \varepsilon/2) \subset U(f, \varepsilon)$ and $U^{-1}(f, \varepsilon) = U(f, \varepsilon)$. Hence the family $\{U(f, \varepsilon): f$ continuous and $\varepsilon > 0\}$ is a subbase for a uniformity $\mathscr{U}$.

Now to complete the proof, we show that the topology defined by $\mathscr{U}$ coincides with the given topology u on E. Let $U(f, \varepsilon)[x]$ be a member of the subbase of the neighborhood filter of x. Since $f: E_u \to [0, 1]$ is continuous, there exists an u-open set P of x in E_u such that for all $y \in P$, $|f(y) - f(x)| < \varepsilon$. Hence $P \subset U(f, \varepsilon)[x]$. Therefore the topology u is finer than the topology defined by $\mathscr{U}$. On the other hand, if P is an open set containing x, then there exists a continuous function $f: E \to [0, 1]$ such that $f(x) = 0$ and $f(E \backslash P) = 1$. Let $U(f, 1)[x]$ be the open neighborhood of x under the topology defined by $\mathscr{U}$. Then for each $y \in U(f, 1)[x]$, $|f(y) - f(x)| < 1$, or $|f(y)| < 1$, because $f(x) = 0$. Hence $y \in P$. This shows that $U(f, 1)[x] \subset P$ or, in other words, the topology defined by $\mathscr{U}$ is finer than u. Hence u is equal to the uniform topology.

Now it is clear from the above theorem that completely regular spaces are precisely the spaces whose topology can be defined by a uniformity. Since there exist lots of nonconstant continuous real-valued functions on a completely regular space, it is possible to describe a uniform space by a family of continuous maps, viz., pseudometrics, as is the case in the section below.

§27. Uniformizable Spaces

Definition 5. Let $\{d_\alpha, \alpha \in A\}$ be a family of pseudometrics on a set E. E is called a *uniformizable space* if E is endowed with the topology whose subbase consists of d_α-balls:

$$B_\varepsilon^\alpha[y] = \{x: d_\alpha(x, y) < \varepsilon, \varepsilon > 0\},$$

where $\alpha \in A$. The family $\{B_\varepsilon^\alpha[y]: \varepsilon > 0\}$ forms a subbasis of the topology on E means that the finite intersections: $\bigcap_{i=1}^{n}\{B_\varepsilon^{\alpha_i}[y]: \varepsilon > 0\}$ form a base. We let D denote the family of pseudometrics: $\{\max_{i \in B} d_{\alpha_i}: B \text{ finite}, B \subset A\}$. It is easy to check that $d = \max_{i \in B} d_{\alpha_i}$ is a pseudometric.

Remark. A uniformizable space is also called a *gauge* space.

Definition 6. A family $\{d_\alpha, \alpha \in A\}$ of pseudometrics on E is said to be *separating* if for each pair (x, y), $x \neq y$, there exists $\alpha \in A$ such that $d_\alpha(x, y) \neq 0$.

Proposition 7. A uniformizable space is Hausdorff if and only if its defining family of pseudometrics, $\{d_\alpha, \alpha \in A\}$, is separating.

Proof. Let $x \neq y$. Then there exists $\alpha \in A$ such that $d_\alpha(x, y) > 0$. Hence if we put $\varepsilon = d_\alpha(x, y)$, then $x \in B_{\varepsilon/2}^\alpha(x) = \{z: d_\alpha(z, x) < \varepsilon/2\}$, $y \in B_{\varepsilon/2}^\alpha(y) = \{z: d_\alpha(z, y) < \varepsilon/2\}$, and $B_{\varepsilon/2}^\alpha(x) \cap B_{\varepsilon/2}^\alpha(y) = \varnothing$. Since $B_{\varepsilon/2}^\alpha(x)$, $B_{\varepsilon/2}^\alpha(y)$ are open neighborhoods of x and y respectively, it follows that E is Hausdorff space.

On the other hand, assume E is a Hausdorff uniformizable space. Let $x \neq y$. Then there are two open neighborhoods U and V of x and y respectively such that $U \cap V = \varnothing$. Since $\{B_\varepsilon^\alpha: \varepsilon > 0\}$ form a subbase, there exists an $\varepsilon > 0$ such that $y \notin \bigcap_{1 \leq i \leq n} B_\varepsilon^{\alpha_i}(x)$, which implies that there exists α_i such that $y \notin B_\varepsilon^{\alpha_i}(x)$, i.e., $d_{\alpha_i}(x, y) \geq \varepsilon > 0$.

Theorem 3. A uniform space is uniformizable and vice versa.

Proof. Let E be a uniformizable space. Then the uniformity generated by each pseudometric d_α on E makes E a uniform space. Thus the largest uniformity generated by the family $\{d_\alpha: \alpha \in A\}$ makes it into a uniform space. It is easy to see that the topology generated by such a uniformity gives the original topology.

Conversely, suppose E is a uniform space, then it is a T_4-space (Theorem 2). Let $C_I(E)$ denote the set of all continuous real-valued functions from E into the unit interval $I = [0, 1]$. $C_I(E) \neq \varnothing$, because E is a nontrivial T_4-space. Hence the product $I^{C_I(E)}$ is a completely regular space. Also, I is clearly a metric space with the metric given by the absolute value $|\cdots|$. Hence by putting

$$d_f(x, y) = |x(f) - y(f)|$$

for $f \in C_I(E)$ we obtain a family $\{d_f : f \in C_I(E)\}$ of pseudometrics on $I^{C_I(E)}$. Each such pseudometric induces a pseudometric on each of its subspaces. Clearly $I^{C_I(E)}$ is uniformizable with the family of metrics, $\{d_f : f \in C_I(E)\}$. Thus the topology so generated coincides with the product topology on $I^{C_I(E)}$. Since a completely regular space E is homeomorphic with a subspace of $I^{C_I(E)}$ (cf. §44, Proposition 51), it follows that E is uniformizable.

Corollary 1. For a topological space E, the following statements are equivalent:

(1) E is a uniform space;

(2) E is a uniformizable space;

(3) E is a T_4-space.

Remark. If we seek a characterization of Hausdorff uniform spaces, then in Corollary 1, we replace "topological" by "Hausdorff" space E.

Proposition 8. Let E be a uniformizable Hausdorff space with its defining family of pseudo-metrics, $\{d_\alpha, \alpha \in A\}$. Then a net $\{x_\beta : \beta \in B\}$ is Cauchy in E if and only if each α, $\{x_\beta : \beta \in B\}$ is d_α-Cauchy. Further $\{x_\beta : \beta \in B\}$ converges to x_0 in E if and only if for each $\alpha \in A$, $d_\alpha(x_\beta, x_0) \to 0$.

Proof. Easy verification.

Proposition 9. Let $(E, d_\alpha, \alpha \in A)$ and $(F, d_\beta', \beta \in B)$ be two uniformizable spaces. Then $f : E \to F$ is continuous if for each d_β' there exist d_α and $M > 0$ such that

$$d_\beta'(f(x), f(y)) \leq M\, d_\alpha(x, y)$$

for all $x, y \in E$.

Proof. It is sufficient to show that if a net $\{x_\gamma, \gamma \in D\}$ converges to x_0 in E then $f(x_\gamma)$ converges to $f(x_0)$ in F. Suppose the condition is satisfied. Clearly from the above definition, $\{x_\gamma\}$ converges to x_0 if for each $\alpha \in A$, $d_\alpha(x_\gamma, x_0) \to 0$, i.e., for a given $\varepsilon > 0$ there exists $\gamma_0 \in D$ such that for all $\gamma \geq \gamma_0$, $d_\alpha(x_\gamma, x_0) < \varepsilon/M$. But then by hypothesis we have that for any d_β', there exist d_α and $M > 0$ such that

$$d_\beta'\big(f(x_\gamma), f(x_0)\big) \leq M\, d_\alpha(x_\gamma, x_0) < M\, \frac{\varepsilon}{M} = \varepsilon$$

for $\gamma \geq \gamma_0$. This shows that $\{f(x_\gamma), \gamma \in D\}$ converges to $f(x_0)$ and hence f is continuous.

Remark. It is clear from the above that a subspace of a uniformizable space is uniformizable in the uniformity induced by pseudometrics.

Let $(E_\alpha, \mathscr{U}_\alpha, \alpha \in A)$ be a family of uniform spaces. Let $E = \prod_{\alpha \in A} E_\alpha$ and let p_α be the projection mapping of the product E onto E_α for each α. Let

$$p_{\alpha 2}^{-1}(\mathscr{U}_\alpha) = \{(x, y) \in E^2 : \big(p_\alpha(x), p_\alpha(y)\big) \in U_\alpha, \ U_\alpha \in \mathscr{U}_\alpha\}$$

be the family of inverse images of $\mathscr{U}_\alpha$ in E^2, where $p_{\alpha 2}(x, y) = \big(p_\alpha(x), p_\alpha(y)\big)$, $x, y \in E$. Let $\mathscr{U}$ denote the filter whose subbase is $\{p_{\alpha 2}^{-1}(\mathscr{U}_\alpha), \alpha \in A\}$ in E^2. Then it is easy to verify that $(E, \mathscr{U})$ is a uniform space. More generally we see that if $(E_\alpha, \mathscr{U}_\alpha)$ is a family of uniform spaces, E a set, and f_α a mapping of E into E_α for each α, then there exists a uniformity of E by the same arguments as before.

Let $(E, \mathscr{U})$ be a uniform space and F a subset of E. Let $\mathscr{U}'$ denote the trace of the filter $\mathscr{U}$ on F^2, i.e., $\mathscr{U}'$ consists of sets U' of F^2 such that $U' = U \cap F^2$, for some $U \in \mathscr{U}$. It is easy to check that $(F, \mathscr{U}')$ is also a uniform space. $(F, \mathscr{U}')$ is called a *subspace* of the uniform space $(E, \mathscr{U})$.

§28. Uniform Continuity and Uniform Spaces

One of the reasons for introducing uniform space is to generalize the usual concept of uniform continuity on the real line or more generally on a metric space. With little effort, one discovers that uniform continuity cannot be defined for arbitrary topological spaces. It was demonstrated by A. Weil [91] that the notion of uniform spaces does appreciably improve this situation and the concept of uniform continuity can be carried over.

Definition 7.   Let $(E, \mathscr{U})$ and $(F, \mathscr{V})$ be two uniform spaces and let f be a mapping of E into F. Then f is said to be *uniformly continuous* if for each V in $\mathscr{V}$ there exists U in $\mathscr{U}$ such that whenever $(x, y) \in U$, it follows that $(f(x), f(y)) \in V$.

The following proposition gives the expected result.

Proposition 10.   Let $(E, \mathscr{U})$, $(F, \mathscr{V})$, and $(G, \mathscr{W})$ be uniform spaces. Then:

(a)   Each uniformly continuous mapping f of E into F is continuous in the topologies defined by the respective uniformities on E and F.

(b)   If $f: E \to F$ and $g: F \to G$ are uniformly continuous, so is $g \circ f$.

Proof.   (a)   Let $V[f(x)]$ be a neighborhood of $f(x) \in F$ in the topology on F. Since f is uniformly continuuos, there exists a U in $\mathscr{U}$ such that $(f(y), f(x)) \in V$ whenever $(y, x) \in U$. In other words, for each $y \in U[x]$, $f(y) \in V[f(x)]$. Since $U[x]$ is a neighborhood of x, it follows that f is continuous. (b) This is straightforward.

In general the converse of (a) in Proposition 10 is not true. For example, $f(x) = x^2$ is a continuous mapping of R into R where the uniformity is defined by the absolute value, but is not uniformly continuous. However, under certain conditions, the converse does hold. First we have:

Lemma 1.   Let $(E, \mathscr{U})$ be a compact uniform space. Then every neighborhood of the diagonal in $E \times E$ is a member of $\mathscr{U}$.

Proof.   Let W be a subset of $E \times E$, which is an open neighborhood of Δ. By Proposition 6, $\mathscr{U}$ has a base $\mathscr{B}$ consisting of closed subsets. Now if $(x, y) \in \cap \{B: B \in \mathscr{B}\}$, then y is in each neighborhood of x (because $\mathscr{B}$ is a base) and hence $(x, y) \in W$, because $W[x]$ is an open neighborhood of x. Therefore, $\cap \{B: B \in \mathscr{B}\} \subset W$. Clearly the family consisting of W and $E^2 \setminus B$, $B \in \mathscr{B}$ forms an open covering of the compact space $E \times E$. By Tychonoffs' theorem (Theorem 3, §35) a finite subcovering covers $E \times E$. But this shows that W must contain a finite intersection of B's from $\mathscr{B}$, which means W is in $\mathscr{U}$.

Theorem 4.   Let $(E, \mathscr{U})$ be a compact uniform space and $(F, \mathscr{V})$ a uniform space. Then each continuous mapping f of E into F is uniformly continuous.

Proof. Define, for $x, y \in E$,

$$f_2(x, y) = (f(x), f(y)),$$

a mapping from $E \times E$ into $F \times F$. Then it is clear by the hypothesis that f_2 is continuous. Since each member V in $\mathscr{V}$ is a neighborhood of the diagonal in $F \times F$ and since f_2 is continuous, it follows that $f_2^{-1}(V)$ is a neighborhood of the diagonal of $E \times E$ and is in the uniformity $\mathscr{U}$, since E is a compact uniform space (Lemma 1). Thus $f_2^{-1}(V) \in \mathscr{U}$ and f is uniformly continuous.

Remark. If the space E is not compact, the above theorem is not necessarily true as pointed out before Lemma 1.

It follows from (b) in Proposition 10 that the collection of all uniform spaces and uniform maps forms a category.

Since each metric space is a uniform space, the following corollaries are immediate consequences of the above.

Corollary 2. A continuous mapping of a metric compact space into a metric space is uniformly continuous.

Corollary 3. A continuous real-valued function on $[a, b]$, $-\infty < a \leq b < \infty$, is uniformly continuous.

Now we establish a connection between the metrics and uniform continuity.

Proposition 11. Let E be a set and $(E_\alpha, \mathscr{U}_\alpha)$ a family of uniform spaces. Let f_α be a mapping of E into E_α for each α. Let E be endowed with the uniformity $\mathscr{U}$ defined above (after Proposition 9). Then $\mathscr{U}$ is the coarsest uniformity with respect to which each $f_\alpha: E \to E_\alpha$ is uniformly continuous.

Proof. Simple verification.
A particular case of the above is:

Corollary 4. Let $E = \prod_{\alpha \in A} E_\alpha$, where $(E_\alpha, \mathscr{U}_\alpha)$ is a uniform space. Then the product uniformity on E is the coarsest such that each projection $p_\alpha: E \to E_\alpha$ is uniformly continuous.

The following proposition relates the uniform continuity of a pseudometric on a uniform space.

Theorem 5.   Let $(E, \mathscr{U})$ be a uniform space and d a pseudometric on E. Then d is uniformly continuous if and only if for each $\varepsilon > 0$,

$$\{(x, y) \in E^2 : d(x, y) < \varepsilon\} \in \mathscr{U}.$$

Proof.   Clearly d maps E^2 into the reals R, where the uniformity on R^2 is defined by the sets $\{(\alpha, \beta) : |\alpha - \beta| < \varepsilon\}$. The uniformity on $E^2 \times E^2$ has a subbase $\{(U, U) : U \in \mathscr{U}\}$. Now first suppose that d is uniformly continuous. Then for each $\varepsilon > 0$ there exists $U \in \mathscr{U}$ such that

$$(d(U), d(U)) \equiv d_2(U, U) \subset \{(\alpha, \beta) : |\alpha - \beta| < \varepsilon\}.$$

Now if $(x_1, y_1) \in U$ and $(x_2, y_2) \in U$, then $|d(x_1, x_2) - d(y_1, y_2)| < \varepsilon$. Putting $x_1 = x$, $x_2 = y_1 = y_2 = y$, we see that $d(x, y) < \varepsilon$, whenever $(x, y) \in U$. Thus $U \subset \{(x, y) : d(x, y) < \varepsilon\}$ for each $\varepsilon > 0$. Hence the latter set is in $\mathscr{U}$.

Conversely, suppose that for each $\varepsilon > 0$, $B_\varepsilon = \{(x, y) : d(x, y) < \varepsilon\}$ is in $\mathscr{U}$. Then $B_{\varepsilon/2} \in \mathscr{U}$. Clearly for $(x_1, y_1), (x_2, y_2)$ in $B_{\varepsilon/2}$, we have

$$|d(x_1, y_1) - d(x_2, y_2)| < \varepsilon,$$

which proves the uniform continuity of d.

Proposition 12.   Let $F = \prod_{\alpha \in A} F_\alpha$, where each F_α is a uniform space. A mapping f of a uniform space E into F is uniformly continuous if and only if for each α, the composition mapping $p_\alpha \circ f$ is uniformly continuous, where $p_\alpha : F \to F_\alpha$ is the projection mapping.

Proof.   By Corollary 4, each p_α is uniformly continuous. Thus it is clear that uniform continuity of f implies that $p_\alpha \circ f$ is uniformly continuous. Conversely, suppose that $p_\alpha \circ f$ is uniformly continuous for each α. To prove that f is uniformly continuous, it is sufficient to show that the inverse image under $f_2 = (f, f)$ of each member of a subbase of the product uniformity on F is a member of the uniformity in E. Since $p_\alpha \circ f$ is uniformly continuous, for each U_α in the uniformity of F_α,

$$A = \{(x, y) \in E^2 : (p_\alpha(f(x)), p_\alpha(f(y))) \in U_\alpha\}$$

is in the uniformity $\mathscr{U}$ of E. But then

$$A = \{(x, y) \in E^2 : (f(x), f(y)) \in p_{2\alpha}^{-1}(U_\alpha)\}$$

implies that $A = f_2^{-1}(p_{2\alpha}^{-1}(U_\alpha)) \in \mathscr{U}$. Since the family $\{p_{2\alpha}^{-1}(U_\alpha)\}$ forms a subbase of the uniformity of F, we have shown that f is uniformly continuous.

§29. Completeness in Uniform Spaces

As remarked before while introducing uniform spaces, one of the main purposes of introducing uniform spaces is to carry over two well-known classical notions of uniform continuity and completeness available for metric spaces. We have already come across uniform continuity. Now we discuss completeness.

Definition 8. Let $(E, \mathscr{U})$ be a uniform space. A net $\{x_\alpha, \alpha \in \Gamma\}$ in E is said to be a *Cauchy net* if for each $U \in \mathscr{U}$, there exists $\alpha_0 \in \Gamma$ such that for all $\alpha, \beta \in \Gamma$, $\alpha, \beta \geq \alpha_0 \Rightarrow (x_\alpha, x_\beta) \in U$.

Similarly, a filter $\mathscr{F} = \{F_\alpha\}$ on E is said to be a *Cauchy filter* if for each $U \in \mathscr{U}$ there exists an F_α such that $(x_\alpha, y_\alpha) \in U$ for all $x_\alpha, y_\alpha \in F_\alpha$. F_α is sometimes called a *U-small* set.

A net $\{x_\alpha, \alpha \in \Gamma\}$ is said to *converge* to a point $x_0 \in E$ if for each $U \in \mathscr{U}$, there exists $\alpha_0 \in \Gamma$ such that for $\alpha > \alpha_0$, $(x_\alpha, x_0) \in U$. In other words, $x_\alpha \in U[x_0]$ for each $\alpha \geq \alpha_0$.

A filter $\mathscr{F} = \{F_\alpha\}$ is said to *converge* to a point $x_0 \in E$ if for each $U \in \mathscr{U}$, there exists F_α such that $(x_\alpha, x_0) \in U$ for all $x_\alpha \in F_\alpha$, or $F_\alpha \subset U[x_0]$.

The following propositions strike a familiar note from metric spaces.

Proposition 13. Every convergent net (or filter) on a uniform space $(E, \mathscr{U})$ is a Cauchy net (or filter). But the converse is not true.

Proof. Let $\{x_\alpha, \alpha \in \Gamma\}$ be a net converging to x_0. Let U be a symmetric member of the uniformity $\mathscr{U}$. There exists $U_1 \in \mathscr{U}$ such that $U_1 \circ U_1 \subset U$. There exists $\alpha_0 \in \Gamma$ such that for all $\alpha \geq \alpha_0$, $(x_\alpha, x_0) \in U_1$. But then for all α, $\beta \geq \alpha_0$,

$$(x_\alpha, x_\beta) \in U_1 \circ U_1 \subset U.$$

Hence $\{x_\alpha, \alpha \in \Gamma\}$ is a Cauchy net. Similar arguments work for filters, too. Further, we see that the set of the rationals with the uniformity induced by the metric on the real line R gives the converse of Proposition 13.

Definition 9. A uniform space is said to be *complete* if each Cauchy net (or filter) in it converges.

Remark. Clearly each Cauchy sequence, being a Cauchy net, converges in a complete uniform space. However, it is not sufficient for completeness. Very precisely, if in a uniform space every Cauchy sequence converges in it, then it need not be a complete uniform space, as mentioned above.

Definition 10.   A uniform space is said to be *sequentially complete* if every Cauchy sequence in it converges.

Remark.   Clearly every complete uniform space is sequentially complete but the converse does not hold. However, the converse does hold for metric spaces, as is shown in the sequel.

Recall that the limit points of convergent filters are unique if and only if the space is Hausdorff (Theorem 2, §18); thus in a complete Hausdorff uniform space Cauchy filters have unique limits. The following also holds:

Proposition 14.   A Cauchy net or filter in a (Hausdorff) uniform space $(E, \mathscr{U})$ converges to its (unique) cluster point, if it has one.

Proof.   Let $\mathscr{F}$ be a Cauchy filter and x_0 a cluster point. Then $x_0 \in \cap \{\bar{F}: F \in \mathscr{F}\}$. Let $U[x_0]$ be an arbitrary neighborhood of x_0 determined by $U \in \mathscr{U}$. There exists $V \in \mathscr{U}$ such that $V \circ V \subset U$. Since $\mathscr{F}$ is Cauchy, there is some $G \in \mathscr{F}$ such that $G \times G \subset V$. Since $x_0 \in \bar{F}$ for all $F \in \mathscr{F}$, there exists a $y \in V[x_0] \cap G$. Hence $(x_0, y) \in V$. Also, $y \in G$ implies that for any $x \in G$, $(y, x) \in G \times G \subset V$ and hence $(x_0, x) \in V \circ V \subset U$. That means that $x \in U[x_0]$ or $G \subset U[x_0]$. This proves that $\mathscr{F}$ converges to x_0. Since the limit of a convergent filter in a Hausdorff space is unique, it proves the other part.

Proposition 15.   (a)   Every closed subset of a complete uniform space is complete.

(b)   Every complete subspace of a Hausdorff uniform space is closed.

(c)   Let $\{E_\alpha, \alpha \in A\}$ be a family of Hausdorff uniform spaces and $E = \prod_\alpha E_\alpha$. Then E is complete iff each E_α is complete.

Proof.   (a)   Let $(E, \mathscr{U})$ be a uniform space and F a closed subset of E. Let $\mathscr{F}$ be a Cauchy filter in F. Then $\mathscr{F}$ converges to $x_0 \in E$. Since x_0 is a cluster point of $\mathscr{F}$ and F is closed, $x_0 \in F$ and thus F is complete.

(b)   Let F be a complete subspace of a Hausdorff uniform space $(E, \mathscr{U})$. Let $x \in \bar{F}$. Then there exists a net $\{x_\alpha, \alpha \in \Gamma\} \subset F$ such that $x_\alpha \to x$. Clearly, by Proposition 13, $\{x_\alpha, \alpha \in \Gamma\}$ is a Cauchy net and hence converges in F, because F is complete. By Proposition 14, then $\{x_\alpha, \alpha \in \Gamma\}$ has the unique limit $x \in F$. This proves that $\bar{F} = F$, i.e., F is closed.

The proof of (b) can be given by using filters instead of nets.

(c)   Since each E_α is a closed subspace of E, the completeness of E implies the completeness of E_α by (a). For the converse, let $\{x_\beta, \beta \in \Gamma\}$ be a Cauchy net in E. Let U_α be a member of the uniformity $\mathscr{U}_\alpha$ of E_α and p_α: $E \to E_\alpha$ the αth projection. Since the inverse images of $\mathscr{U}_\alpha$ under p_α in E form a subbase of the uniformity of E, it is easy to see that $\{p_\alpha(x_\beta): \beta \in \Gamma\}$ is a Cauchy net in E_α for each $\alpha \in A$. Since E_α is complete, $\{p_\alpha(x_\beta): \beta \in \Gamma\}$ converges to $x_\alpha^\circ \in E_\alpha$ for each $\alpha \in A$. Hence (x_α°) is a limit point of $\{x_\beta, \beta \in \Gamma\}$. But $\{x_\beta, \beta \in \Gamma\}$ is Cauchy and so it converges to (x_α°) (Proposition 14) and this proves that E is complete.

Remark.   Recall that a metric space is said to be *complete* if each Cauchy sequence in it converges.

Theorem 6.   A metric space (E, d) is complete if and only if E, as a uniform space, is complete.

Proof.   Assume (E, d) is a complete metric space. Let $\mathscr{F}$ be a Cauchy filter in E. Then, for each n, there exists $F_n \in \mathscr{F}$ such that

$$F_n \times F_n \subset \left\{(x, y): d(x, y) < \frac{1}{n}\right\}.$$

Since by definition of a filter, each F_n is nonempty, choose $x_n \in \bigcap_{k=1}^n F_k$ for all $n \geq 1$. Clearly $\{x_n\}$ is a Cauchy sequence in E and hence converges to x_0 by our assumption. Let $U[x_0]$ be an arbitrary neighborhood of x_0 determined by a member U of the uniformity, determined in turn by the metric d. Then for some n, $x_0 \in B_{1/n}(x_0) \subset U[x_0]$. Since $\{x_n\}$ is a Cauchy sequence, for some $m > 2n$, choose $x_m \in B_{1/2n}(x_0)$. We show that $F_m \subset U[x_0]$. For all $y \in F_m$, $d(x_m, y) < 1/m \leq 1/2n$ because $(x_m, y) \in F_m \circ F_m \subset \{(x, y): d(x, y) < 1/m\}$. Thus

$$d(x_0, y) \leq d(x_0, x_m) + d(x_m, y) < \frac{1}{m} + \frac{1}{m} = \frac{2}{m} < \frac{1}{n}.$$

This shows that $F_m \subset B_{1/n}(x_0) \subset U[x_0]$, and the proof of the "only if" part is completed.

On the other hand, if E is a complete uniform space and if $\{x_n\}$ is a Cauchy sequence, then the filter whose base is $\{F_n\}$, where $F_n = \{x_m: m \geq n\}$, is clearly a Cauchy filter. Hence $\{F_n\}$ converges to $x_0 \in E$, because E is complete. Given $\varepsilon > 0$ there exists n_0 such that $F_{n_0} \subset \{x \in E: d(x, x_0) < \varepsilon\}$. But this means that for all $n \geq n_0$, $d(x_n, x_0) < \varepsilon$, or that $\{x_n\}$ converges to x_0 in the metric topology.

The preceding results demonstrate abundantly that the some well-known theorems proved for metric spaces can be carried over to uniform spaces. Furthermore, the notions of completeness and sequential completeness coincide for metrix spaces.

§30. Completeness, Compactness, and Completions

In this section we establish a relation between complete and compact spaces. Observe that the compactness is defined on arbitrary topological spaces while completeness is defined for uniform spaces only. One can easily see that a complete uniform space need not be compact, as the example of real numbers shows. However, we have the following.

Proposition 16. Every compact uniform space is complete in the topology induced by the uniformity.

Proof. Let $\mathscr{F}$ be a Cauchy filter in a compact uniform space $(E, \mathscr{U})$. By Theorem 1, §35, each filter on a compact space has a cluster point. Hence the Cauchy filter $\mathscr{F}$ converge to its cluster point (Proposition 14, §29) and this proves that $(E, \mathscr{U})$ is complete.

We can improve this proposition, but first we need another notion.

Definition 11. A uniform space $(E, \mathscr{U})$ is said to be *totally bounded* or *precompact* if for each member U of the uniformity $\mathscr{U}$, there exists a finite set x_i $(1 \leq i \leq n)$ (n depending upon U) such that $E \subset \bigcup_{i=1}^{n} U[x_i]$.

Proposition 17. A uniform space $(E, \mathscr{U})$ is precompact iff each filter is contained in a Cauchy filter.

Proof. Suppose $(E, \mathscr{U})$ is precompact and $\mathscr{F}$ is any filter in E. Let $\tilde{\mathscr{F}}$ denote an ultrafilter in E containing $\mathscr{F}$. We show that $\tilde{\mathscr{F}}$ is a Cauchy filter. Let U and V be symmetric members of the uniformity $\mathscr{U}$ such that $V \circ V \subset U$. Since E is precompact, there exists a finite set x_i $(1 \leq i \leq n)$, such that $E \subset \bigcup_{i=1}^{n} V[x_i]$. First we show that for each $\tilde{F} \in \tilde{\mathscr{F}}$, $\tilde{F}_i \cap V[x_i] \neq \emptyset$ for some i. For, if for each $i = 1, \ldots, n$, there is $\tilde{F}_i \in \tilde{\mathscr{F}}$ such that $F_i \cap V[x_i] = \emptyset$, then for $\tilde{F} = \bigcap_{i=1}^{n} \tilde{F}_i \in \tilde{\mathscr{F}}$, we would have:

$$\tilde{F} = \tilde{F} \cap E = \tilde{F} \cap \bigcup_{i=1}^{n} V[x_i] \subset \bigcup_{i=1}^{n} \{\tilde{F}_i \cap V[x_i]\} = \emptyset,$$

which is false because $\mathscr{F}$ is a filter. Hence we have for each $\tilde{F} \in \mathscr{F}$, $\tilde{F} \cap V[x_i] \neq \emptyset$ for the same i. Since $\mathscr{F}$ is an ultrafilter, it follows that $V[x_i] \in \mathscr{F}$, and clearly $V[x_i] \times V[x_i] \subset U$. Hence $\mathscr{F}$ is a Cauchy filter.

Conversely, suppose that each filter in $(E, \mathscr{U})$ is contained in a Cauchy filter. Let $U \in \mathscr{U}$ and assume for each finite set X of E, $U[X] \neq E$ or, equivalently, $(U[X])^c \neq \emptyset$. Thus as X runs over the family of all finite subsets of E, $(U[X])^c$ forms a base of a filter, which, by our assumption, is contained in a Cauchy filter $\mathscr{K}$, say. Thus for some $K \in \mathscr{K}$, $K \times K \subset U$ and so for any $x \in K$, $K \subset U[x]$. Hence $U[x] \in \mathscr{K}$. On the other hand, $\{x\}$ being a finite set, it follows that $(U[x])^c \in \mathscr{K}$. Hence $U[x] \cap (U[x])^c = \emptyset \in \mathscr{K}$, a contradiction. Therefore, $U[X] = E$ for some finite set X and $(E, \mathscr{U})$ is precompact.

Theorem 7. A uniform space $(E, \mathscr{U})$ is compact iff it is precompact and complete.

Proof. Suppose $(E, \mathscr{U})$ is compact. Then, by Proposition 16, it is complete. Moreover, for each $U \in \mathscr{U}$, $\{U[x]\}$ is an open covering of E when x runs over E. Since E is compact, only a finite subcovering of $\{U[x]\}$ covers E and so E is precompact.

Conversely, suppose $(E, \mathscr{U})$ is complete and precompact. Since each filter $\mathscr{H}$ of E is contained in a Cauchy filter $\mathscr{K}$ by Proposition 17, it follows that $\mathscr{K}$ is convergent in E because E is complete. Hence $\mathscr{H}$ has a cluster point and so E is compact (Theorem 1, §35).

Corollary 5. In a metric space a subset is compact iff it is precompact and complete.

Proof. It is a particular case of the above theorem, since each metric space is a uniform space.

Remark. Note that not each precompact uniform space is compact [for example $(0, 1)$], nor is a complete uniform space compact, e.g., the real line.

Now we wish to show that every uniform space can be densely embedded into a complete uniform space. In particular, this would imply that a metric space can also be densely embedded into a complete metric space.

Definition 12. Let $(E, \mathscr{U})$ be a uniform space. A uniform space $(\hat{E}, \hat{\mathscr{U}})$ is said to be a *completion* of $(E, \mathscr{U})$ if:

(i) $(\hat{E}, \hat{\mathscr{U}})$ is a complete uniform space.

(ii) $(E, \mathscr{U})$ with its topology defined by its uniform structure is homeomorphic to a dense subspace of $(E^\sim, \mathscr{U}^\sim)$, when the latter is endowed with the topology defined by its uniform structure $\hat{\mathscr{U}}$. Moreover, the uniform structure $\hat{\mathscr{U}}$ induces $\mathscr{U}$ on E when E is regarded as a subspace of $\hat{E}$.

Remark.   In particular cases, e.g., rational numbers, the reader knows how to obtain completions. For instance, one considers the set of all Cauchy sequences of rational numbers, which includes the rationals as a subset provided we identify a rational number by the constant sequence consisting of that rational number only. The completion so obtained is the set of all real numbers. We apply a similar procedure for uniform spaces as well.

Definition 13.   A Cauchy filter $\mathscr{F} = \{F_\alpha\}$ in a uniform space $(E, \mathscr{U})$ is said to be *minimal* if $\mathscr{F}$ is a minimal element in the partially ordered family of all filters in E.

Proposition 18.   Each Cauchy filter in a uniform space $(E, \mathscr{U})$ contains a unique minimal Cauchy filter.

Proof.   Let $\mathscr{B} = \{B\}$ be a base of a Cauchy filter in E. Let $\mathscr{U} = \{U\}$ be the uniformity. For each pair $B_1, B_2 \in \mathscr{B}$ and $U_1, U_2 \in \mathscr{U}$, there are $B_3 \in \mathscr{B}$ and $U_3 \in \mathscr{U}$ such that $B_3 \subset B_1 \cap B_2$ and $U_3 \subset U_1 \cap U_2$. It is clear that $U_3[B_3] \subset U_1[B_1] \cap U_2[B_2]$, where $U_i[B_i] = \{y \in E: (x, y) \in U_i$ for some $x \in B_i\}$, $i = 1, 2, 3$, as before. Thus it follows that the family $\{U[B]\}$, where U runs over $\mathscr{U}$ and B over $\mathscr{B}$, forms a filter base. If $U \in \mathscr{U}$, is symmetric, there exists $B \in \mathscr{B}$ such that $B \times B \subset U$ (because B is Cauchy) and therefore $U[B] \times U[B] \subset U^3$. This shows that $\{U[B]\}$ forms a Cauchy filter base and it is coarser than $\mathscr{B}$. To show that the filter having $\{U[B]\}$ as base is minimal, let $\mathscr{C} = \{C\}$ be a Cauchy filter coarser than $\mathscr{B}$. Choose $C \in \mathscr{C}$, $B \in \mathscr{B}$ and U, $V \in \mathscr{U}$ such that $V \circ V \subset U$, $C \times C \subset V$ and $B \times B \subset V$. Then $B \times C \subset V \circ V \subset U$. Since $C \in \mathscr{B}$, $C \cap B \neq \varnothing$ and hence $C \subset U[B]$. Therefore $U[B] \in \mathscr{C}$. Clearly the minimal Cauchy filter is unique.

The following is immediate from the above:

Corollary 6.   (i)   Each neighborhood filter of a point in a uniform space $(E, \mathscr{U})$ is a minimal Cauchy filter.

(ii)   Every cluster point of a Cauchy filter is its limit.

(iii)   Every Cauchy filter which is coarser than a convergent filter converges to the same limit.

Corollary 7.   Every minimal Cauchy filter $\mathscr{F}$ in a uniform space $(E, \mathscr{U})$ has a base consisting of open sets.

Proof.   Let U be a member of the uniformity $\mathscr{U}$. Then, as shown before (Proposition 5(i), §25) there is an open set V in $\mathscr{U}$ such that $V \subset U$. Therefore, for any subset $A \subset E$, $V[A] \subset U[A]$. Now from Proposition 18, the corollary follows.

Theorem 8.   To each uniform space $(E, \mathscr{U})$ there corresponds a completion $(\hat{E}, \hat{\mathscr{U}})$. If $(E, \mathscr{U})$ is a Hausdorff uniform space, then the completion $(\hat{E}, \hat{\mathscr{U}})$ is unique up to a homeomorphism, more precisely up to a uniform homeomorphism, i.e., the continuous maps involved are uniformly continuous.

Proof.   Let $\hat{E}$ denote the set of all minimal Cauchy filters on E. First we define a filter which will make $\hat{E}$ a uniform space. For each symmetric $U \in \mathscr{U}$, let $\tilde{U}$ denote the set of all pairs $\tilde{U} = (\mathscr{F}_1, \mathscr{F}_2)$ of elements $\mathscr{F}_1, \mathscr{F}_2$ in $\hat{E}$ which have a common U-small set. To show that the totality $\tilde{\mathscr{U}}$ of such $\tilde{U}$ forms a filter base for a uniformity, we first see that for each symmetric set U in $\mathscr{U}$, and for each $\mathscr{F} \in \hat{E}$, $(\mathscr{F}, \mathscr{F}) \in \tilde{U}$. Since each U is symmetric, we see easily that axioms (a) and (b) of Definition 1, §25 are satisfied by $\tilde{\mathscr{U}}$. To show $\tilde{\mathscr{U}}$ is a filter base, let U_1, U_2 be two symmetric elements in $\mathscr{U}$. Then $U_1 \cap U_2 = U_3$, say, is a symmetric element of $\mathscr{U}$. Moreover, every set which is U_3-small is also U_1-and U_2-small. Therefore, $\tilde{U}_3 \subset \tilde{U}_1 \cap \tilde{U}_2$ and so $\tilde{\mathscr{U}}$ forms a filter base.

Finally, we show that axiom (c) of Definition 1, §25, is also satisfied. Let U and V be symmetric elements in $\mathscr{U}$ such that $V^2 \subset U$. Let $\mathscr{F}_1, \mathscr{F}_2$, $\mathscr{F}_3$ be the elements in $\hat{E}$ such that $(\mathscr{F}_1, \mathscr{F}_2) \in \tilde{V}$ and $(\mathscr{F}_2, \mathscr{F}_3) \in \tilde{V}$. Then there are two V-small sets G_1 and G_2 such that $G_1 \in \mathscr{F}_1$ and $\mathscr{F}_2$ and also $G_2 \in \mathscr{F}_2$ and $\mathscr{F}_3$. Since G_1, $G_2 \in \mathscr{F}_2$, we have $G_1 \cap G_2 \neq \varnothing$ and, therefore, $G_1 \cap G_2$ is V^2-small and hence U-small. Now since $G_1 \cup G_2 \in \mathscr{F}_1$ and $\mathscr{F}_3$, we have $\tilde{V}^2 \subset \tilde{U}$. Hence (c) of Definition 1 is satisfied and $\tilde{\mathscr{U}}$ defines a uniform structure on $\hat{E}$.

To show that the uniform structure defined by $\tilde{\mathscr{U}}$ is Hausdorff, let $\mathscr{F}_1$ and $\mathscr{F}_2$ be two minimal Cauchy filters on $\hat{E}$ such that $(\mathscr{F}_1, \mathscr{F}_2) \in \tilde{U}$ for all symmetric members U of the uniformity $\mathscr{U}$ of E. Clearly $\mathscr{F}_1 \cup \mathscr{F}_2 = \{M_1 \cup M_2, M_1 \in \mathscr{F}_1, M_2 \in \mathscr{F}_2\}$ forms a base of a filter $\mathscr{F}_3$ coarser than $\mathscr{F}_1$ and $\mathscr{F}_2$. Since $\mathscr{F}_1 \cap \mathscr{F}_2$ contains a V-small set for each $V \in \mathscr{U}$, $\mathscr{F}_1 \cup \mathscr{F}_2$ is a Cauchy filter base. Since $\mathscr{F}_1$ and $\mathscr{F}_2$ are minimal Cauchy filters, $\mathscr{F}_1 = \mathscr{F}_2 = \mathscr{F}_3$ and hence $\hat{E}$ is a Hausdorff uniform space. Now if to each $x \in E$, we associate the Cauchy filter of neighborhood $\mathscr{N}(x)$ of x, then $i: x \to \mathscr{N}(x)$ is a mapping of E into $\hat{E}$, since $\mathscr{N}(x)$ is a Cauchy minimal filter. If E is Hausdorff, i is 1 : 1 and we show that the inverse image of $\hat{\mathscr{U}}$ under i induces the uniformity $\mathscr{U}$ on E. For each symmetric U in $\mathscr{U}$, if $i \times i = j$, then we have

$$j^{-1}(\tilde{U}) \subset U \cup j^{-1}(\tilde{U}^3). \tag{$\circledast$}$$

For, if $\big(i(x), i(y)\big) \in \tilde{U}$, there is a U-small set F such that F is a neighborhood of x and y, and hence $(x, y) \in U$. On the other hand, if $(x, y) \in U$ then $U[x] \cup U[y]$ is U^3-small and hence a neighborhood of x and y. Therefore, $\circledast$ is established. Now it is not hard to conclude from $\circledast$ that the inverse image of the uniformity $\mathscr{U}$ on E is the uniformity $\hat{\mathscr{U}}$ on $\hat{E}$.

Finally, we show that $\hat{E}$ is complete and E is homeomorphic with a dense subspace of $\hat{E}$.

To show that $i(E)$ is dense in $\hat{E}$, let $\mathscr{F} = \hat{x}$ be an element of $\hat{E}$, i.e., $\mathscr{F}$ is a minimal Cauchy filter of E. Let $\tilde{V}[\hat{x}]$ be a neighborhood of $\hat{x}$ in $\hat{E}$. Clearly

$$\tilde{V}[\hat{x}] \cap i(E) = \{i(x): \; (\hat{x}, i(x)) \in \tilde{V}\}.$$

But this means that there is a V-small neighborhood of x in $\hat{x}$ or x is an interior point of a V-small set of $\mathscr{F}$. But the union U of the interiors of all V-small sets of $\mathscr{F}$ is in $\mathscr{F}$, i.e., $\tilde{V}[\hat{x}] \cap i(E) = i(U)$. This proves that $\tilde{V}[\hat{x}] \cap i(E) \neq \varnothing$ and hence $i(E)$ is dense in $\hat{E}$. Moreover, $\tilde{V}[\hat{x}] \cap i(E)$ is contained in the filter base $\{\tilde{V}[\hat{x}]\}$ on $\hat{E}$ which converges to $\hat{x}$.

Now to end the proof, we show that $\hat{E}$ is complete. Let $\mathscr{F}$ be a Cauchy filter in $\hat{E}$. Without loss of generality, we can assume that $\mathscr{F}$ is minimal. Then $\mathscr{F} \cap i(E)$ is a Cauchy filter, since $i(E)$ is dense in $\hat{E}$ and $\mathscr{F}$ has a base consisting of open sets. It is clearly sufficient to show that $\mathscr{F} \cap i(E)$ is convergent in $\hat{E}$ or, in other words, each Cauchy filter $\mathscr{K}$ on $i(E)$ is convergent in $\hat{E}$. Clearly $i^{-1}(\mathscr{K})$ is a base of a Cauchy filter on E. Let $\mathscr{G}$ be a minimal Cauchy filter coarser than it. Then $i(\mathscr{G})$ is a Cauchy filter base on $i(E)$ and $i^{-1}(i(\mathscr{G})) = \mathscr{G}$ is finer than the filter whose base is $i^{-1}(\mathscr{K})$. Since $i(\mathscr{G})$ converges in $\hat{E}$, so does $\mathscr{K}$.

The fact that $\hat{E}$ is unique up to a uniform homeomorphism if E is Hausdorff is left for the reader to verify.

Note that one may use nets instead of filters in Theorem 8.

Corollary 8.   For each metric space E there exists a complete metric space E such that E is isometric with a dense subset of $\hat{E}$. $\hat{E}$ is unique up to homeomorphism.

Proof. This is a particular case of the above.

Now we wish to prove some Tietze's-extension-type theorems for uniformly continuous maps on uniform spaces.

Remark. The mapping $i: E \to \hat{E}$ in the completion is called the *canonical embedding* of E into its completion $\hat{E}$. In the sequel, we shall identify E with $i(E)$ and thus E can be regarded as a dense subspace of $\hat{E}$. Clearly the minimal Cauchy filters on E are just the traces of the neighborhood filter of points of $\hat{E}$.

Remark. It is very well known that the real line is a complete metric space. Thus one might ask if each uniformly continuous function on a dense subset of a uniform space can be extended to a uniformly continuous function of the whole space. The example $i: Q \to Q$ (Q = rational numbers) shows that this may not be possible if the range space is incomplete. However if the range space is a complete uniform space, the situation is pleasant. First we prove the following:

Proposition 19. Let E and F be two uniform spaces. The image of each Cauchy filter base (or net) under a uniformly continuous map $f: E \to F$, is a Cauchy filter base (or net).

Proof. Let V be a member of the uniformity of F and let $\mathscr{F}$ be a Cauchy filter on E. By uniform continuity of f, there exists a member U of the uniformity on E such that $g(U) \subset V$, where $g = (f, f)$. Now if there is $K \in \mathscr{F}$ and such that $K \times K \subset U$, then the relations $f(K) \times f(K) \subset g(U) \subset V$ prove the proposition.

Corollary 9. Let E be a set with two uniformities $\mathscr{U}$ and $\mathscr{V}$ such that $\mathscr{V}$ is coarser than $\mathscr{U}$. Then each $\mathscr{U}$-Cauchy filter is a $\mathscr{V}$-Cauchy filter.

Proof. This is a particular case of Proposition 19, since the identity mapping $i: (E, \mathscr{U}) \to (E, \mathscr{V})$ is uniformly continuous. In particular, we have:

Corollary 10. Let f be a uniformly continuous mapping of a metric space E into another metric space F. Then the image of each Cauchy sequence in E is also a Cauchy sequence in F.

Theorem 9. Let A be a dense subset of a uniform space E and f a uniformly continuous mapping of A into a complete uniform Hausdorff space F. Then there exists a uniformly continuous mapping $\tilde{f}$ of E into F such that $\tilde{f}$ coincides with f on A.

Proof. Let $x \in E$, then there is a net $\{x_\alpha\}$ in A converging to x. Clearly $\{x_\alpha\}$ is a Cauchy net. Hence $\{f(x_\alpha)\}$ is a Cauchy net in F by Proposition 19. Since F is complete, $\{f(x_\alpha)\}$ converges to a unique point in F. We put

$$\tilde{f}(x) = \lim f(x_\alpha).$$

Then $\tilde{f}$ is well-defined on E as it can be shown that two different nets converging to x give the same value, and clearly $\tilde{f}(x) = f(x)$ for each $x \in A$. We have to show that $\tilde{f}$ is uniformly continuous. Let V be a closed symmetric member of the uniformity of F. Since $f: A \to F$ is uniformly continuous, there exists a member U of the uniformity of E such that $W = A^2 \cap U$ is a member of the uniformity on A and whenever $(x, y) \in W$, $x, y \in A$ then $(f(x), f(y)) \in V$. Clearly we may assume that V is closed. Thus, by definition of $\tilde{f}$, we have $(\tilde{f}(x), \tilde{f}(y)) \in V$ whenever $(x, y) \in U$, $x, y \in E$ because V is closed, $(\tilde{f}, \tilde{f})$ is continuous, and A is dense in E. This proves the uniform continuity of $\tilde{f}$.

Corollary 11. Let E be a uniform space and F a complete Hausdorff uniform space. Then each uniformly continuous mapping f of E into F can be extended to a unique uniformly continuous mapping of $\hat{E}$ into F.

Proof. This is a particular case of Theorem 9.

Note that one can specialize Corollary 11 for metric spaces.

§31. Topological Groups and Topological Vector Spaces

We have seen that an important subclass of uniform spaces is the class of metric spaces. In this section, we show that with some additional algebraic hypotheses on a topological space we can obtain a uniform space, even though not every topological space is uniformizable.

Definition 14. Let G be a group (§5), the point-set of which is a topological space. G is said to be a *topological group* if the mappings

  (a) $f: (x, y) \to xy$ of $G \times G$ into G;
  (b) $g: x \to x^{-1}$ of G into itself

are continuous or, equivalently, the mapping: $(x, y) \to xy^{-1}$ is continuous.

Remark. It is easy to check that for each fixed $a \in G$, the mappings: $x \to ax$, $x \to xa$, and $x \to a^{-1}xa$ are homeomorphisms (Proposition 5, §65).

Theorem 10. In a topological group, there exists a base $\{U\}$ of neighborhoods of the identity e, satisfying the following conditions:

(i) $U = U^{-1}$ (or $-U$ for an additive group), i.e., U is *symmetric*;

(ii) for each U in $\{U\}$ there is a V in $\{U\}$ such that $V^2 \subset U$;

(iii) for each U in $\{U\}$ and $a \in G$, there is a V in $\{U\}$ such that $aVa^{-1} \subset U$ or $V \subset a^{-1}Ua$.

Conversely, given any filter $\{U\}$ in G satisfying (i)–(iii), there exists a unique topology on G which makes G a topological group.

Proof. Since the mapping: $x \to x^{-1}$ is a homeomorphism, for each neighborhood U of e, U^{-1} is a neighborhood of e.

Now choose $W = U \cap U^{-1}$; then clearly W is a symmetric neighborhood of e. This proves (i). To show (ii), let U be a neighborhood of e. Since the mapping $(x, y) \to xy$ is continuous at (e, e), for each neighborhood U of e there is a neighborhood V of e such that $V^2 \subset U$. Since $x \to a^{-1}xa$ and $x \to axa^{-1}$ are continuous, we obtain (iii).

For the converse, we first see that each U contains e, since by (i) and (ii) each U contains $V \in \{U\}$ such that $VV^{-1} \subset U$ and so $e \in U$. Hence for each $x \in G$, each member of the systems $\{xU\}$ and $\{Ux\}$ contains x. The totality of $\{xU\}$ and $\{Ux\}$ forms a filter base and hence there exists a unique topology (Proposition 12, §10) on G such that (i) implies that the map $x \to x^{-1}$ is continuous at e and (ii) implies that $(x, y) \to xy$ is continuous at (e, e). To show that $(x, y) \to xy^{-1}$ is continuous at any point (a, b) $a, b \in G$, let $x = ar, y = bs, r, s \in G$. Clearly $(ab^{-1})^{-1}(xy^{-1}) = brs^{-1}b^{-1}$. Now let P be an arbitrary neighborhood of e; then there exists a $U \in \{U\}$ such that $U \subset P$. Also if $rs^{-1} \in b^{-1}Ub$ then $brs^{-1}b^{-1} \in U$ and by (iii) there exists $V \in \{U\}$ such that $V \subset b^{-1}Ub$ and by (i) and (ii) there exists $W \in \{W\}$ such that if $r, s \in W$ then $rs^{-1} \in WW^{-1} \subset b^{-1}Ub$. Hence

$$(ab^{-1})^{-1}(xy^{-1}) = brs^{-1}b^{-1} \in U \subset P,$$

for all $r, s \in W$. This proves the continuity of $(x, y) \to xy^{-1}$ at any any point (a, b) of G^2 and this proves that G is a topological group.

Theorem 11. Every topological group is a uniform space.

Proof. Let $\{U\}$ be the system of neighborhoods at e satisfying conditions (i)–(iii) of Theorem 10. For each $U \in \{U\}$, define

$$L(U) =: \{(x, y) \in G \times G : x^{-1}y \in U \text{ or } y \in xU\};$$
$$R(U) = \{(x, y) \in G \times G : yx^{-1} \in U \text{ or } y \in Ux\}.$$

Clearly $\Delta = \{(x, x): x \in G\} \subset L(U) \cap R(U)$, since the identity $e \in U$. Also for each U, $U^{-1} \in \{U\}$ and

$$L(U^{-1}) = \{(x, y) \in G \times G: x^{-1}y \in U^{-1}\}$$
$$= \{(x, y) \in G \times G: y^{-1}x \in U\} = L^{-1}(U).$$

Further, for $U \in \{U\}$ there is a $V \in \{U\}$ such that $V^2 \subset U$. Then $L^2(V) = L(V) \circ L(V) \subset L(U)$. Also $L(U \cap V) = L(U) \circ L(V)$. Similar relations hold for $R(U)$ in place of $L(U)$. Thus $\{L(U)\}$ and $\{R(U)\}$ and $\{L(U) \cap R(U)\}$, where U runs over $\{U\}$, respectively define the uniformities called the left, right, and two-sided uniformities which induce the original topology.

Remark. Observe that each incomplete uniform space can be completed. Similarly a topological group can be completed in the uniformity $\{L(U) \cap R(U)\}$. Thus each abelian group, since $\{L(U)\}$ and $\{R(U)\}$ are the same, can be completed.

Proposition 20. A topological group G with the system $\{U\}$ of neighborhoods of its identity e is Hausdorff if and olny if $\cap U = \{e\}$.

Proof. If G is Hausdorff and if $e \neq x \in \cap U$, then e has a neighborhood U such that $x \notin U$ and so $x \notin \cap U$, a contradiction. For the converse, assume $\cap U = \{e\}$. If $x \neq y$, x, yG, then $y^{-1}x \neq e$. Hence there is a neighborhood U of e such that $y^{-1}x \notin U$. Choose a symmetric neighborhood V of e such that $V^2 \subset U$. But then xV and yV are neighborhoods of x and y, respectively. Moreover, they are disjoint, because if $z \in xV \cap yV$, then there exist $v_0, v_2 \in V$ such that $z = xv_1 = yv_2$ implies that $y^{-1}x = v_2v_1^{-1} \in VV^{-1} = V^2 \subset U$, a contradiction. This proves that G is Hausdorff.

Corollary 12. Every Hausdorff topological group G is completely regular, hence uniformizable.

Proof. Every uniform space is a T_4-space (Theorem 2, §26). But since G is Hausdorff, it follows that G is completely regular or uniformizable.

Definition 15. Let E be a real or complex linear (or vector) (see §5) space, the point-set of which is a topological space. Then E is a *topological vector space* if the following mappings:

$$(x, y) \to x + y \quad \text{of } E \times E \text{ into } E$$
$$(\lambda, x) \to \lambda x \quad \text{of } K \times E \text{ into } E$$

are continuous in both variables together, where $x, y \in E$ and $\lambda \in K$, in which K is the field of real or complex scalars.

Observe that each topological vector space is a topological group since each linear space is an abelian additive group. Thus we have the following as for topological groups:

Proposition 21. Every real or complex topological linear space E is a uniform space and hence completely regular if E is Hausdorff.

Remark. Every topological linear space, if incomplete, can be completed to a complete topological linear space. Furthermore for each $a \in E$, the mappings: $x \to a + x$ and $x \to \lambda x$ ($\lambda \neq 0$) are homeomorphisms.

Definition 16. (a) A metric d on a group E is said to be *translation-invariant* if

$$d(x + a, y + a) = d(x, y) \qquad \text{for } x, y, a \in E.$$

(b) A real or complex topological linear space whose topology is defined by an invariant metric d which makes the space complete is called an *F-space*.

(c) A set A in a linear space E is said to be *convex* if for all $x, y \in A$,

$$\lambda x + (1 - \lambda)y \in A \qquad \text{for } 0 \leq \lambda \leq 1.$$

A set A is called *circled* if for all $x \in A$ and scalars λ, $|\lambda| \leq 1$ implies $\lambda x \in A$. A is *absorbing* if for each $x \in E$ there is $\alpha > 0$ such that for $|\lambda| \geq \alpha$, $x \in \lambda A$.

Theorem 12. In each real or complex topological vector space E, there exists a basis $\{U\}$ of closed neighborhoods of the origin 0 such that

(i) each U is circled and absorbing;

(ii) for each $U \in \{U\}$ there is a $V \in \{U\}$ such that $V + V \subset U$.

Conversely, given a filter $\mathscr{F} = \{U\}$ of subsets of E such that (i) and (ii) are satisfied, then there exists a unique topology $\mathscr{E}$ such that $(E, \mathscr{E})$ is a topological vector space.

Proof. Similar to that of Theorem 10. See [45] for a complete proof.

Definition 17. (a) A real or complex topological linear space in which there exists a fundamental system or base of convex neighborhoods of 0 is called a *locally convex space*.

(b) A locally convex complete metric topological vector space is called a *Fréchet space*.

(c) Let E be a real or complex linear space. Let p be a mapping of E into the nonnegative reals. Then p is called a *seminorm* if (i) $p(\lambda x) = |\lambda| p(x)$ and (ii) $p(x + y) \leq p(x) + p(y)$, where $x, y \in E$ and $\lambda \in K$.

(d) A seminorm p on E is called a *norm* and is denoted by $\| x \|$ instead of $p(x)$, if $p(x) = 0 \Leftrightarrow x = 0$.

(e) Clearly $d(x, y) = \| x - y \|$ defines a metric. If a normed space is complete in the metric induced by the norm, then it is called a *Banach space*.

(f) A normed space is called a *pre-Hilbert* space if the norm, in addition, satisfies the parallelogram law:

$$\| x + y \|^2 + \| x - y \|^2 = 2(\| x \|^2 + \| y \|^2).$$

A complete pre-Hilbert space is called a *Hilbert space*.

It is easy to see that each Hilbert space is a Banach space, which is a Fréchet space, which is a locally convex space. But the converses need not hold in general (see [45]).

Since Banach and Fréchet spaces are metric spaces, they are Hausdorff uniform spaces. We shall give the conditions under which a topological group or topological linear space can be given a metric topology in the sequel. Let F be a subgroup (respectively linear subspace) of a topological group (respectively topological linear space) E. Then in the induced topology F is a topological group (respectively topological linear space).

Let F be a normal subgroup (respectively linear subspace) of a topological group (respectively topological linear space) E. Then the quotient space E/F is a group (respectively linear space). If we endow it with the quotient topology on E/F, viz., the finest topology which makes E/F a topological group (respectively topological linear space) such that the cononical mapping $\varphi: E \to E/F$ is continuous, where $\varphi(x) = x + F$, then it is easy to see that φ is a continuous, open, and onto homomorphism (respectively linear mapping). The quotient topology on E/F is Hausdorff if and only if F is a closed subgroup (respectively closed linear subspace) of E.

It is also easy to check that the arbitrary product $\prod_{\alpha \in A} E_\alpha$ of topological groups (respectively topological vector spaces) E_α's is also a topological group (respectively topological vector space).

For the proof of these facts, see Husain [45], [46] or Köthe [55(b)], etc.

§32. Metrizability

We have seen that metric spaces play a very central role in many branches of topology and analysis. Thus it is natural to determine when a topological space can be metrizable. Indeed, not every topological space can be metrizable, e.g., the space consisting of more than one point with the indiscrete topology cannot be a metric space.

The main idea of metrizability is to be able to embed a given topological space into a metric space such that the induced metric topology coincides with the initial one. For some spaces this can easily be achieved, as we see in the sequel. But first we show the following:

Proposition 22.   Every regular Lindelöf space E is normal.

Proof.   Let A, B be two disjoint closed subsets of E. For each $x \in A$, there exists an open neighborhood U_x of x with $U_x \cap B = \varnothing$, because $E \backslash B$ is open. Clearly the collection $\{U_x : x \in A\}$ forms an open covering of A. Similarly, $\{V_y : y \in B\}$ is an open covering of B. Since $E \backslash (A \cup B)$ is open, $\{U_x : x \in A\} \cup \{V_y : y \in B\} \cup (E \backslash A \cup B)$ is an open covering of E and hence there is a countable subcovering, viz., $\{U_i : i \geq 1\}$ covering A, $\{V_i : i \geq 1\}$ covering B and $(E \backslash A \cup B)$, because E is Lindelöf. Now we make the U_i's and V_i's disjoint as follows: Put

$$P_n = U_n \backslash \bigcup_{m \leq n} \bar{V}_m, \qquad Q_m = V_m \backslash \bigcup_{n \leq m} \bar{U}_n.$$

Since $P_n \cap V_m = \varnothing$ for $m \leq n$, it follows that $P_n \cap Q_m = \varnothing$ for $m \leq n$. Now m and n being symmetric, we see that $P_n \cap Q_m = \varnothing$ for all m, n. Hence $P = \bigcup_{n=1}^{\infty} P_n$, $Q = \bigcup_{m=1}^{\infty} Q_m$ are disjoint open sets. Moreover, $P \supset A$, $Q \supset B$. Hence E is a normal space.

Corollary 13.   Every regular second countable space is normal.

Proof.   Since every second countable space is Lindelöf, Proposition 22 applies.

Definition 18.   A topological space is said to be *metrizable* if there exists a metric which induces a topology equivalent to the given one.

Proposition 23.   Let E be a completely regular space and let X be a set of continuous functions on E with values in $[0, 1] = I$ such that

(i) for $x \neq y$ in E, there exists $f \in X$ such that $f(x) \neq f(y)$, (i.e., X *separates* points of E);

(ii) for any closed subset C and $x \notin C$, there is $f \in X$ such that $f(C) = 1$, $f(x) = 0$ (i.e., X separates points and closed sets).

Then the mapping $x \rightarrow (f(x))_{f \in X}$ of E into I^X is a homeomorphism into.

Proof. We shall see (Proposition 5.1 §44) that the mapping φ: $x \rightarrow (f(x))_{f \in X}$ is a homeomorphism of E into I^X, since (i) ensures that φ is $1 : 1$. Continuity of φ follows because each f is continuous. (ii) implies that φ is open. Hence φ is a homeomorphism into.

Theorem 13. (Urysohn). Every second countable regular space E is metrizable.

Proof. Let $\{B_n\}$ be a countable base of the open sets E. Since E is regular, we can find $B_m \in \{B_n\}$ such that $\bar{B}_m \subset B_n$ for any $n \geq 1$. Let $\mathscr{B} = \{(B_m, B_n)\}$ be the countable family of such pairs. Since every regular Lindelöf space is normal, by Urysohn's lemma (Theorem 7, §19) there exists a continuous function f_{mn} on E with values in $[0, 1]$ such that $f_{mn}(\bar{B}_m) = 0$ and $f_{mn}(E \backslash B_n) = 1$. Certainly $F = \{f_{mn}\}$ is a countable family of continuous functions. We show that F separates points and also points and closed subsets. Let C be a closed subset of E and $x \notin C$. Choose B_n such that $x \in B_n \subset E \backslash C$ and choose B_m in $\{(B_m, B_n)\}$ such that $x \in \bar{B}_m \subset B_n$. Then (B_m, B_n) is a pair in $\mathscr{B}$ and the corresponding f_{mn} satisfies the property: $f_{mn}(\bar{B}_m) = 0$ and $f_{mn}(E \backslash B_n) = 1$. In particular, $f_{mn}(x) = 0$ and $f_{mn}(y) = 1$ for all $y \in E \backslash B_n$ (§22). Now to prove the metrizability, we consider the countable product I^F of the closed unit interval, which is certainly metrizable [Theorem 3(h) §22]. The fact that E is homeomorphic (Proposition 23) with a subspace of I^F shows that E is metrizable.

Now one can characterize those Hausdorff spaces which are metrizable. Observe that a metric space need not be second countable. However, it is if it is separable (Proposition 19, §12). Thus we have:

Theorem 14. Let E be a Hausdorff space or at least a T_1-space. The following statements are equivalent:

(a) E is metrizable and separable;

(b) E is regular and second countable;

(c) E is homeomorphic with a subspace of the countable product of the interval $[0, 1]$.

Proof. (a) $\Rightarrow$ (b) by Proposition 19, §12 and Theorem 6, §18. (b) $\Rightarrow$ (a) by Theorem 13. Finally, (c) $\Leftrightarrow$ (a) by Theorem 3(h), §22 and Proposition 23.

Theorem 15. The uniformity of a Hausdorff uniform space (E, U) has a countable base if and only if E is metrizable.

Proof. Suppose E is a metric space with metric d. Then the sets $U_n = \{(x, y): d(x, y) < 1/n)\}$ certainly form a countable base of its uniformity.

Conversely, assume E has a countable base $\{U_n\}$ of its uniformity $\mathscr{U}$. Clearly, we may choose U_n's so that:

    (i)   each U_n is symmetric;

    (ii) for each $n \geq 1$, $U_n \circ U_n \circ U_n \subset U_{n-1}$.

Now define $f: E \times E \to R$ by

$$f(x, y) = \begin{cases} 2^{-n} & \text{if } (x, y) \in U_{n-1} \backslash U_n; \\ 0 & \text{if } (x, y) \in U_n \text{ for all } n \geq 1. \end{cases}$$

Since the diagonal $\Delta \subset U_n$ for all n, $f(x, x) = 0$. Since E is Hausdorff, $\Delta = \bigcap_{n=1}^{\infty} U_n$ and hence $f(x, y) = 0$ iff $(x, y) \in \Delta$, i.e., $x = y$.

Now we define another function:

$$d(x, y) = \inf\{f(x_i, x_{i+1}): \text{ for a finite sequence } \{x_i\}_{i=1}^{n},$$
$$\text{such that } x_1 = x, \ x_{n+1} = y\}.$$

Clearly, $d(x, y) = 0$ iff $f(x, y) = 0$, which is true if and only if $x = y$. Furthermore, $d(x, y) + d(y, z) \geq d(x, z)$ and by symmetry of U_n's, $d(x, y) = d(y, x)$ because clearly $f(x, y) = f(y, x)$. Thus there exists a metric d on E. We show that this metric induces the initial uniform topology. It is sufficient to show that

$$U_n \subset \{(x, y): d(x, y) < 1/2^n\} \subset U_{n-1}$$

for all $n \geq 1$. But for this it is enough to prove that

$$f(x_1, x_{n+1}) \leq 2 \sum_{i=1}^{n} f(x_i, x_{i+1}).$$

We use induction: For $n = 1$, the inequality is quite clear. Assume it holds up to n. Put

$$a = \sum_{i=1}^{n} f(x_i, x_{i+1}).$$

Let k be the largest integer such that

$$\sum_{i=1}^{k} f(x_i, x_{i+1}) < a/2.$$

Then clearly

$$\sum_{i=k+1}^{n} f(x_i, x_{i+1}) \le a/2$$

and hence by induction each of $f(x_1, x_k)$ and $f(x_{k+1}, x_{n+1})$ is $\le a$. Also clearly $f(x_k, x_{k+1}) \le a$. Choose m to be the smallest integer such that $1/2^m \le a$. Then $(x_1, x_k), (x_k, x_{k+1})$, and $(x_{k+1}, x_{n+1}) \in U_m$ and therefore $(x_1, x_{n+1}) \in U_{m-1}$ by (ii). This shows that

$$f(x_1, x_{n+1}) < 1/2^{m-1} < 2a = 2 \sum_{i=1}^{n} f(x_i, x_{i+1}).$$

This completes the proof.

Corollary 14. (a) A Hausdorff topological group is metrizable if and only if the neighborhood system of its identity has a countable base.

(b) A Hausdorff topological vector space is metrizable if and only if the neighborhood system at its origin has a countable base.

Proof. Both (a) and (b) follow from the above theorem and the fact that the countable base for the neighborhood system at the identity gives a countable base for its uniformity and vice versa, since each topological group as well as each topological vector space is a uniform space (Theorem 11 and Proposition 21, §31).

A more general metrization theorem can be obtained for a more general class of topological spaces. We investigate this in the following:

Theorem 16. A regular space E is metrizable provided its topology has a σ-locally finite base (§10) or, in other words, if it is the countable union of neighborhood-finite families of open sets.

Proof. Let $\mathscr{U} = \bigcup_{n=1}^{\infty} \mathscr{U}_n$ be a base of the topology, where each $\mathscr{U}_n = \{U_{n,\alpha} : \alpha \in A_n\}$ is a neighborhood-finite family of open sets. Given any open covering $\{P_\alpha\}$ of E, each P_α can be replaced by a union of sets from the base. Hence there is a refinement of $\{P_\alpha\}$ which is locally finite. Since E is regular, by Theorem 21, §43, E is paracompact. Now let U be any open set in E and $x \in U$. Then by regularity there is a member $U_{n(x)}$ in $\mathscr{U}_n$ such that

$$x \in U_{n(x)} \subset \bar{U}_{n(x)} \subset U.$$

Now, let $E_k = \cup \{\bar{U}_{n(x)} \colon U_{n(x)} \in \mathscr{U}_k\}$; then E_k is closed because $\{U_n\}$ is neighborhood-finite, E is regular and clearly $\bigcup_{k \geq 1} E_k = U$. In other words, each open set, being the countable union of closed sets, is a F_σ-set.

Let $\mathscr{U}_n = \{P_{n,\alpha} \colon \alpha \in A\}$ for each $n \geq 1$. Then for each (n, α) there is a nonzero continuous real function $f_{n,\alpha} \colon E \to [0, 1]$ such that $f_{n,\alpha}(x) = 1$ and $f_{n,\alpha}(P^c_{n,\alpha}) = 0$. Since each point x belongs to only a finite number of sets in $\mathscr{U}_n$ for each fixed n, the sum $g_n(x) = \sum_\alpha f^2_{n,\alpha}(x)$ is defined for each fixed n and each g_n is continuous.

Now let

$$\varphi_{n,\alpha}(x) = \frac{1}{n} \frac{f_{n,\alpha}(x)}{(1 + g_n(x))^{1/2}};$$

then clearly $\varphi_{n,\alpha}$ is continuous and

$$\sum_{n,\alpha} \varphi^2_{n,\alpha}(x) = \sum_n \sum_\alpha \varphi^2_{n,\alpha}(x) \leq \sum_n \frac{1}{n^2} < \infty.$$

Thus the mapping $\varphi \colon x \to (\varphi_{n,\alpha}(x))$ maps E into a Hilbert space

$$L_2(N \times A) = \{f \colon N \times A \to R \text{ with } \sum_{n,\alpha} f^2(n, \alpha) < \infty\},$$

where all $f(n, \alpha) = 0$, $\alpha \in A = \bigcup_{n=1}^\infty A_n$, except for a countably many (n, α).

We show that φ is an embedding into, i.e., φ is 1 : 1, continuous, and open. If $x \neq y$, then there is some $P_{n,\alpha}$ such that $x \in P_{n,\alpha}$ and $y \notin P_{n,\alpha}$. Then there is $f_{n,\alpha}$ such that $f_{n,\alpha}(x) > 0$ and $f_{n,\alpha}(y) = 0$. Thus $\varphi_{n,\alpha}(x) > 0$ and $\varphi_{n,\alpha}(y) = 0$. Hence φ is 1 : 1.

To show that φ is continuous at $x_0 \in E$, let $\varepsilon > 0$ be given. There exists an n_0 such that $\sum_{n=n_0}^\infty 1/n^2 < 1/4\varepsilon^2$. Let U_0 be a neighborhood of x_0 which intersects at most finitely many $P_{n,\alpha}(n \leq n_0)$ and let A' be the set of α's appearing in this collection. Let M be the number of $n \leq n_0$ for which $\varphi_{n,\alpha}(U_0) \neq 0$. Let V_0 be a neighborhood of x_0 for which

$$|\varphi_{n,\alpha}(x) - \varphi_{n,\alpha}(x_0)| \leq \frac{\varepsilon}{(2M)^{1/2}}$$

for all $x \in V_0$ and for each $\alpha \in A'$, $n \leq n_0$. Thus we have

$$\sum_{\substack{n,\alpha \\ n < n_0 \\ \alpha \in A'}} |\varphi_{n,\alpha}(x) - \varphi_{n,\alpha}(x_0)|^2 \leq \frac{\varepsilon^2}{2}$$

and

$$\sum_{\substack{n,\alpha \\ n \geq n_0}} |\varphi_{n,\alpha}(x) - \varphi_{n,\alpha}(x_0)|^2 \leq 2 \sum_{n \geq n_0} \frac{1}{n^2} < \frac{\varepsilon^2}{2}$$

for all $x \in V_0$. Thus we see that $\varphi_{n,\alpha}$ is continuous at each x_0.

To show that φ is open, let U be an open set in E and $x \in U$. Then for some (n_0, α_0) we have $x \in U_{n_0, \alpha_0} \subset U$. Since

$$f_{n_0, \alpha_0}(E \setminus U_{n_0, \alpha_0}) = 0$$

and $f_{n_0, \alpha_0}(x) > 0$, we see that for $\varphi(y) \in \varphi(U)^c$, $y \in U^c$,

$$\inf\{d_2(\varphi(y), \varphi(x)): y \in U^c\} \geq \frac{1}{n_0} [f_{n_0, \alpha_0}(x)/(1 + g_{n_0}(x))^{1/2}].$$

That means, for $x \in U$, $\varphi(x) \notin \overline{\varphi(E \setminus U)}$. This shows that φ is open and the proof is complete.

Corollary 15. Every compact Hausdorff space E is metrizable iff E is second countable. In particular, every compact second countable Hausdorff space is metrizable.

Proof. This follows from Theorem 16.

§33. Fixed Points

Definition 19. (a) Let E be a uniformizable space, where uniformity is defined by pseudometrics $\{d_\alpha: \alpha \in A\}$. A mapping $f: E \to E$ is said to be a *contraction* if there exists $0 \leq r < 1$, such that

$$d_\alpha(f(x), f(y)) \leq r \, d_\alpha(x, y)$$

for each $\alpha \in A$ and all $x, y \in E$.

(b) $x \in E$ is said to be a *fixed point* of a mapping $f: E \to E$ if $f(x) = x$.

Remark. If (E, d) is a pseudometric space, then a map $f: E \to E$ is a contraction if there exists $0 \leq r < 1$ such that $d(f(x), f(y)) \leq r \, d(x, y)$ for all $x, y \in E$. Clearly a contraction map on a metric space is continuous and hence sequentially continuous.

Theorem 17. Every sequentially continuous contraction mapping on a sequentially complete Hausdorff uniformizable space into itself has a unique fixed point.

Proof. *Uniqueness*: Suppose there are two fixed points x, y, i.e., $f(x) = x$ and $f(y) = y$. If $x \neq y$, then there exists $\alpha \in A$ such that $d_\alpha(x, y) > 0$ (Proposition 7, §27). But then

$$d_\alpha(x, y) = d_\alpha(f(x), f(y)) \leq r \, d_\alpha(x, y) < d_\alpha(x, y)$$

is impossible.

Existence: We show that f has a fixed point. For each $x \in E$, we show that $\{f^n(x)\}$ is a Cauchy sequence, i.e., for each α, $d_\alpha(f^n(x), f^m(x)) \to 0$ as $n, m \to \infty$. For, if $m < n$,

$$
\begin{aligned}
d_\alpha(f^n(x), f^m(x)) &= d_\alpha(f(f^{n-1}(x)), f(f^{m-1}(x))) \\
&\leq r \, d_\alpha(f^{n-1}(x), f^{m-1}(x)) \\
&\leq r^m d_\alpha(f^{n-m}(x), x) \\
&\leq r^m[d_\alpha(f^{n-m}(x), f^{n-m-1}(x)) + d_\alpha(f^{n-m-1}(x), f^{n-m-2}(x)) \\
&\quad + \cdots + d_\alpha(f(x), x)] \\
&\leq r^m[r^{n-m-1} + r^{n-m-2} + \cdots + 1] \, d_\alpha(f(x), x) \\
&\leq \frac{r^m}{1 - r} \, d_\alpha(f(x), x) \to 0
\end{aligned}
$$

as $m \to \infty$ because $r < 1$. Hence $\{f^n(x)\}$ is a Cauchy sequence. Since E is sequentially complete, $f^n(x) \to x_0 \in E$. But, f being sequentially continuous, $f^{n+1}(x) \to f(x_0)$. Since $\{f^{n+1}(x)\}$ is a subsequence of $\{f^n(x)\}$, also $f^{n+1}(x) \to x_0$. Since E is Hausdorff, $f(x_0) = x_0$. This proves that x_0 is a fixed point.

Remark. We deduce the classical fixed-point theorem from the above. But first observe that a contraction mapping in a metric space is necessarily continuous. Also we know that a sequentially continuous function of a metric space is continuous.

Corollary 16. Each contraction mapping on a complete metric space into itself has a unique fixed point.

Proof. This follows immediately from Theorem 17.

Remark 1. In classical analysis, $\{f^n(x)\}$ are usually called "successive approximations" to x_0, no matter what the initial choice of x is.

Remark 2. It is not in general possible to weaken the condition $r < 1$. For example, if $r = 1$, the theorem fails, as is seen by the following example: $f(x) = \log(1 + e^x)$ on the real line, which is a complete metric space in the usual metric topology.

Remark 3. One may obtain other generalizations of the corollary by weakening the hypothesis slightly. For example, the contraction map f may be required to satisfy an inequality which is slightly weaker than a contraction map is supposed to do by definition.

There are fixed-point theorems for continuous functions on certain restricted subsets of linear spaces. But such fixed points need not be unique. The most useful theorems are:

Brouwer fixed-point theorem. Every continuous map of a Euclidean closed disk D in R^n, i.e., $D = \{x, \ldots, x_n) \in R^n : \sum x_i^2 \leq 1\}$ into itself has a fixed point. This can be generalized:

Banach-Schauder fixed-point theorem. Every continuous map of a convex compact subset of a Banach space into itself has a fixed point (see Day [25b]).

By using exactly similar arguments one has:

Tychonoff fixed-point theorem. Every continuous map of a convex compact subset of a locally convex Hausdorff space into itself has a fixed point (cf. Day [25b]).

There are also fixed-point theorems concerning a common fixed point for a family of commuting affine maps on a convex compact subset into itself (cf. Bourbaki [14]).

The fixed-point theorem is very useful in analysis. We give two applications of this theorem—one to the implicit function theorem and the other to the existence of a solution of an ordinary differential equation.

Application I (Implicit function). Let $f(x, y)$ be a continuous real-valued function on the rectangle $|x - x_0| \leq a$, $|y - y_0| \leq b$, where a, $b > 0$ and (x_0, y_0) is a given point in the rectangle such that $f(x_0, y_0) = 0$. Assume f satisfies a Lipschitz condition in y, i.e., there exists α, $0 \leq \alpha < 1$ such that

$$|f(x, y_1) - f(x, y_2)| \leq \alpha |y_1 - y_2|$$

for all x in $|x - x_0| < a$ and all y_1, y_2 such that $|y_0 - y_1| < b$ and $|y_0 - y_2| \leq b$. Then there exists $0 < a_1 < a$ and a continuous map $h(x)$ on the interval $|x - x_0| < a_1$ with values in the interval $|y - y_0| < b$ such that $h(x_0) = y_0$ and $h(x) = y_0 + f(x, h(x))$ for all x in the interval $|x - x_0| < a_1$.

Proof. In order to apply the above fixed-point theorem, first we require a complete metric space and a contraction map. Let R_c denote the rectangle

$$\{(x, y): |x - x_0| \leq c \text{ and } |y - y_0| \leq b\},$$

where $0 < c \leq b$. Consider $C(R_c)$ to be the set of all continuous real-valued functions on R_c. For $f, g \in C(R_c)$, let

$$d(f, g) = \sup\{|f(x, y) - g(x, y)|: (x, y) \in R_c\}.$$

Then d is a metric on the space $C(R_c)$, which is a complete space. Let $C_0 = \{g \in C(I_c): g(x_0) = y_0\}$, where $I_c = \{x: |x - x_0| \leq c\}$. Then C_0 is a closed subspace of $C(I_c)$ and hence a complete metric space.

Define a map φ on C_0 into itself by

$$\varphi(g)(x) = y_0 + f(x, g(x)),$$

for each $g \in C_0$, $x \in I_c$. Then

$$\varphi(g)(x_0) = y_0 + f(x_0, g(x_0)) = y_0 + f(x_0, y_0) = y_0 + 0 = y_0.$$

Furthermore,

$$\begin{aligned}
|\varphi(g)(x) - y_0| &= |f(x, g(x))| \\
&\leq |f(x, g(x)) - f(x, g(x_0))| + |f(x, g(x_0))| \\
&\leq \alpha|g(x) - g(x_0)| + |f(x, g(x_0))| \\
&\leq \alpha b + |f(x, g(x_0))|.
\end{aligned}$$

Since f is continuous on the rectangle R_c, we may choose c sufficiently small, say a_1, such that

$$|f(x, y_0)| \leq b(1 - \alpha) \text{ for all } x \in I_{a_1}.$$

Hence

$$|\varphi(g)(x) - y_0| \leq \alpha b + b(1 - \alpha) = b \text{ for all } x \in I_{a_1}.$$

That means φ maps C_0 into itself. Moreover,

$$| \varphi(g)(x) - \varphi(h)(x) | = | (f(x, g(x)) + y_0) - (f(x, h(x)) + y_0) |$$
$$\leq \alpha | g(x) - h(x) |.$$

This shows that φ is a contraction mapping of C_0 into itself. Hence by the fixed-point theorem (Corollary 16) there exists a unique $h \in C_0$ such that $\varphi(h) = h$. Thus we have shown the existence of $h(x)$ such that $h(x_0) = y_0$ and

$$h(x) = y_0 + f(x, h(x))$$

for all $x \in I_{a_1}$. This completes the proof.

Application II (Differential equations).   Let $f(x, y)$ be a continuous real function on an open connected domain D of the plane satisfying the Lipschitz condition:

$$| f(x, y_2) - f(x, y_1) | \leq \alpha | y_2 - y_1 |$$

for $(y_1, y_2) \in D$ and let $(x_0, y_0) \in D$. Then the differential equation $dy/dx = f(x, y)$ has a unique solution passing through (x_0, y_0), i.e., there is a real function $y = g(x)$ on an open interval I such that $g(x_0) = y_0$ and $g'(x) = dy/dx = f(x, g(x))$ for $x \in I$. (It is easy to see that such a g exists if and only if

$$g(x) = y_0 + \int_{x_0}^{x} f(t, g(t)) \, dt$$

for $x \in I$.)

Since D is locally compact, we find another open connected set $D' \subset D$ such that $(x_0, y_0) \in D'$ and f is bounded on D', i.e., there exists $M > 0$ such that $f(D') \leq M$.

Choose $\sigma > 0$ such that the rectangle

$$[x_0 - \sigma, x_0 + \sigma] \times [y_0 - \sigma M, y_0 + \sigma M] \subset D'$$

and $\alpha \sigma < 1$. Now using the arguments as used in Application I, we obtain the desired function.

Remark.   The existence of the solution of a differential equation $dy/dx = f(x, y)$ in which f need not satisfy a Lipschitz condition can also be established. But it requires the Arzèla–Ascoli Theorem, to be given in a later chapter.

§34. Proximity Spaces

The notion of proximate spaces is due to Efremovic [105]. It is motivated by metric spaces. In a metric space (E, d) we say that a point x is close to a set A if $d(x, A) = 0$, where $d(x, A) = \inf_{y \in A} d(x, y)$, observe that x need not belong to A unless A is closed. This idea is abstracted in the following definition.

Definition 20. Let E be a set and $\mathscr{P}(E)$ the collection of all subsets of E. A *proximity relation* is a binary relation δ on $\mathscr{P}(E)$ and written $A \, \delta \, B$, $A, B \in \mathscr{P}(E)$ with the following properties:

   (i)   If $A \, \delta \, B$, then $B \, \delta \, A$;
   (ii) $A \, \delta \, E$ if and only if $A \neq \varnothing$;
   (iii) $(A \cup B) \, \delta \, C$ if and only if either $A \, \delta \, C$ or $B \, \delta \, C$;
   (iv) if for every $X \in \mathscr{P}(E)$, either $A \, \delta \, X$ or $B \, \delta \, (E \setminus X)$, then $A \, \delta \, B$;
   (v)   if for $x, y \in E$, $\{x\} \, \delta \, \{y\}$ then $x = y$.

The set (E, δ) with a proximity relation δ is called a *proximity space* and the proximity relation "$A \, \delta \, B$" is read as "A is close to B." If $A \, \not\delta \, B$, then we say that A is remote from B. We note that $A, B, C, \ldots$, etc., denote subsets of a set E, i.e., $A, B, C, \ldots \in \mathscr{P}(E)$.

Proposition 24. Let (E, δ) be a proximity space.

   (a) If $A \subset C$, $B \subset D$, and if $A \, \delta \, B$, then $C \, \delta \, D$;
   (b) if $A \cap B \neq \varnothing$, then $A \, \delta \, B$;
   (c) $\varnothing \, \not\delta \, A$ for all $A \in \mathscr{P}(E)$;
   (d) if for some $x \in E$, $\{x\} \, \delta \, A$ and $\{x\} \, \delta \, B$, then $A \, \delta \, B$.

Proof.   (a)   Clearly $C = A \cup C$ and $D = B \cup D$. By axiom (i) of proximity relations, $A \, \delta \, B$ implies $B \, \delta \, A$ and by (iii) $B \, \delta \, C$. Similarly $A \, \delta \, D$ and so $C \, \delta \, D$.

   (b)   Let $x \in A \cap B$. Then $\{x\} \, \delta \, A$ and $\{x\} \, \delta \, B$. Therefore $A \, \delta \, B$ by (a);

   (c)   follows from axiom (ii);
   (d)   follows from axiom (iv).

Theorem 18. Every proximity space (E, δ) can be given a topology with which it is a regular (hence Hausdorff) space.

Proof. For each subset $A \subset E$, we define

$$\bar{A} = \{y \in E: \{y\} \, \delta \, A\}.$$

First we check that A satisfies the closure axioms. Let $A, B \in \mathscr{P}(E)$. For $A \subset \bar{A}$, let $x \in A$, then $\{x\} \subset A$ and hence $\{x\} \, \delta \, A$ by (b) of Proposition 24. Further $\bar{\bar{A}} = \bar{A}$. Clearly $\bar{A} \subset \bar{\bar{A}}$. Let $x \in \bar{\bar{A}}$, then $\{x\} \, \delta \, \bar{A}$. This means that either $x \in \bar{A}$ or $\{x\} \, \delta \, A$, which means $x \in \bar{A}$. Hence $\bar{\bar{A}} \subset \bar{A}$ and so $\bar{\bar{A}} = \bar{A}$.

Let $A, B \in \mathscr{P}(E)$. If $A \subset B$ then $\bar{A} \subset \bar{B}$. Let $x \in \bar{A}$, then $\{x\} \, \delta \, A$ and, since $A \subset B$, it follows that $\{x\} \, \delta \, B$. That means $x \in \bar{B}$. Thus if $A, B \in \mathscr{P}(E)$, then $\bar{B} \cup \bar{A} \subset \overline{A \cup B}$. Now let $x \in \overline{A \cup B}$. Then $\{x\} \, \delta \, (A \cup B)$, which implies either $\{x\} \, \delta \, A$ or $\{x\} \, \delta \, B$ by axiom (iii), i.e., either $x \in \bar{A}$ or $x \in \bar{B}$, i.e., $\overline{A \cup B} \subset \bar{A} \cup \bar{B}$. Thus $A \to \bar{A}$ is a closure operation on $\mathscr{P}(E)$ and hence it defines a topology on E. We show that it is a T_1-space. Let $x \neq y$, $x, y \in E$. Then $\overline{\{x\}}$ is a closed subset. Thus, if $y \in \overline{\{x\}}$ then $\{y\} \, \delta \, \{x\}$ and so $x = y$ by axiom (v). Hence E is a T_1-space. Now let $x \neq y$. Since $\overline{\{x\}}$ and $\overline{\{y\}}$ are closed subsets, $E \backslash \overline{\{x\}}$ is an open neighborhood of y. Let U be a neighborhood of y such that $\bar{U} \subset E \backslash \overline{\{x\}}$. Suppose $\overline{\{x\}} \cap \bar{U} = \varnothing$, then $x \in \bar{U}$, i.e., $\{x\} \, \delta \, U$, which is a contradiction. Hence E is a regular space and hence Hausdorff.

Proposition 25. Let (E, δ) be a proximity space. Then for $A, B \in \mathscr{P}(E)$, $A \, \delta \, B$ iff $\bar{A} \, \delta \, \bar{B}$, where the closure $\bar{A}$ is taken in the topology defined by the proximity.

Proof. Clearly $A \, \delta \, B$ implies $\bar{A} \, \delta \, \bar{B}$. Conversely, assume $A \, \delta\!\!\!/ \, B$. Then by axiom (iv) there exists $C \in \mathscr{P}(E)$ such that $A \, \delta\!\!\!/ \, C$ and $(E \backslash C) \, \delta\!\!\!/ \, B$. By axiom (i) it follows that $\bar{B} \subset C$. But $A \, \delta\!\!\!/ \, C$ implies $A \, \delta\!\!\!/ \, \bar{B}$. Hence by symmetry $\bar{A} \, \delta\!\!\!/ \, \bar{B}$. Thus $A \, \delta \, B$ implies $\bar{A} \, \delta \, \bar{B}$.

Definition 21. A proximity on a topological space is said to be a *compatible proximity* if the topology induced by the proximity coincides with the initial topology.

Proposition 26. Let E be a normal space. Then for $A, B \in \mathscr{P}(E)$, $A \, \delta \, B$ iff $\bar{A} \cap \bar{B} \neq \varnothing$ defines a compatible proximity.

Proof. That δ defines a proximity is easy to check since $\bar{A} \cap \bar{B} = \varnothing$ in a normal space iff there is a continuous function $f: E \to [0, 1]$ such that

$f(\bar{A}) = 0$ and $f(\bar{B}) = 1$. Now we apply the same argument as for regular spaces to show that δ defines a compatible proximity (see Theorem 18).

Now we prove the following result, which is parallel to uniformities on compact spaces:

Proposition 27. On a compact Hausdorff space E there is a unique proximity relation given by $A \delta B$ iff $\bar{A} \cap \bar{B} \neq \varnothing$.

Proof. Clearly E is normal and so δ is a compactible proximity and the result follows immediately since E is compact (Proposition 26).

Proposition 28. Every metric space is a proximity space.

Proof. Let (E, d) be a metric space. Let $A, B \in \mathscr{P}(E)$ and put $d(A, B) = \inf\{d(a, b): a \in A, b \in B\}$. Define $A \delta B$ iff $d(A, B) = 0$. Then axioms (i)–(v) are easily seen to be satisfied. Hence E is a proximity space and the topology induced by the proximity is coincident with the metric topology.

Proposition 29. Every completely regular or Hausdorff T_4-space E is a proximity space.

Proof. Let $A, B \in \mathscr{P}(E)$. Now define $A \not\delta B$ iff there is a continuous function $f: E \to [0, 1]$ such that $f(A) = 0$ and $f(B) = 1$. Then axioms (i)–(iii) and (v) are easy to check To verify (iv) put $C = \{x \in E: 1/2 \leq f(x) \leq 1\}$. Then $A \not\delta C$ and $(E \backslash C) \not\delta B$. For example, there exists a continuous function $g: [0, 1] \to [0, 1]$, $g(x) = 2x$ when $0 \leq x \leq 1/2$ and $=1$ when $1/2 \leq x \leq 1$. Then g is continuous and so is $h = g \circ f$. Further, $h(A) = 0$ and $h(C) = 1$. The rest follows easily.

Proposition 30. Every Hausdorff uniform space is a proximity space.

Proof. Let $(E, \mathscr{U})$ be a uniform space and let $A, B \in \mathscr{P}(E)$. For $U \in \mathscr{U}$, denote $U[A] = \{y: (x, y) \in U$ for some $x \in A\}$. We define $A \delta B$ if $U[A] \cap B \neq \varnothing$ or equivalently $U[B] \cap A \neq \varnothing$ for all $U \in \mathscr{U}$. Then it is easy to check all the axioms of a proximity relation and that it is compatible with the uniformity.

Proposition 31. Every Hausdorff topological group G is a proximity space.

Proof. Let $\{U\}$ denote the system of neighborhoods of identity $e \in G$. There are at least two proximity relations that can be defined: Let A, $B \in \mathscr{P}(G)$. Define

(1) $A \,\delta_1 B$ iff $UA \cap B \neq \varnothing$ for $U \in \{U\}$;

(2) $A \,\delta_2 B$ iff $AU \cap B \neq \varnothing$ for $U \in \{U\}$.

It is clear that δ_1, δ_2 are proximity relations. It is easy to verify that these are compatible proximities with the right and left uniformities of G.

Definition 22. Let (E, δ_1) and (F, δ_2) be two proximity spaces. A mapping $f: E \to F$ is said to be a *proximity mapping* if for any $A, B \in \mathscr{P}(E)$, $A \,\delta_1 B$ implies $f(A) \,\delta_2 f(B)$ or equivalently for any $C, D \in \mathscr{P}(F)$, $C \,\bar{\delta}_2 D$ implies $f^{-1}(C) \,\bar{\delta}_1 f^{-1}(D)$.

Remark. There is no reason to suppose that every continuous map on a proximity space is a proximity map. However, the following statement is pleasant to know:

Proposition 32. Let (E, δ_1), (F, δ_2) be proximity spaces and E a compact space. Then every continuous map $f: E \to F$ is a proximity map.

Proof. Let $A, B \in \mathscr{P}(E)$ such that $A \,\delta_1 B$. Then $\bar{A} \cap \bar{B} \neq \varnothing$ by Proposition 27. But then it follows that $f(\bar{A}) \cap f(\bar{B}) \neq \varnothing$. Hence $f(\bar{A}) \,\delta_2 f(\bar{B})$. Since f is continuous, $f(\bar{A}) \subset \overline{f(A)}$ and $f(\bar{B}) \subset \overline{f(B)}$ and so $\overline{f(A)} \cap \overline{f(B)} \neq \varnothing$. Hence $f(A) \,\delta_2 f(B)$ and so f is a proximity mapping.

Examples and Exercises

1. Let $[0, \Omega)$ be the set of all ordinals less than the first uncountable ordinal Ω, with the order topology. Show that there is a unique uniformity which induces the order topology such that $[0, \Omega)$ is not complete. Show that there is a nonmetrizable uniformity on $[0, \Omega)$ although the topology induced by this uniformity is metrizable.

2. If $(E, \mathscr{U})$ and $(F, \mathscr{V})$ are two uniform spaces and R a closed subset of $E \times F$. For each compact subset A of E, show that

$$R[A] = \{y \in F: (x, y) \in R \text{ for all } x \in A\}$$

is a closed subset of F.

3. Let $(E, \mathscr{U})$ and $(F, \mathscr{V})$ be two uniform spaces and consider the product uniformity on $E \times F$. Show that the closure of a set $R \subset E \times F$ is given by
$$\bar{R} = \cap \{V \circ R \circ U^{-1} : U \in \mathscr{U}, V \in \mathscr{U}\}.$$

4. A topological space $(X, \mathscr{T})$ is called *metrically topologically complete* if there is a metric d for X such that $\mathscr{T}$ is the metric topology and X is complete under d.
    (a) Show that each complete metric space is metrically topologically complete.
    (b) Every G_δ set in a complete metric space is homeomorphic to a topological space which is metrically topologically complete.

5. A topological space $(X, \mathscr{T})$ is called a *topologically complete* space if there is a uniformity $\mathscr{U}$ on X such that $\mathscr{U}$ induces the topology $\mathscr{T}$ and $(X, \mathscr{U})$ is complete.
    (i)   Let $(X, \mathscr{U})$ be a complete uniform space and A an F_σ-set. Show that there is a uniformity $\mathscr{V}$ on an open set B such that the topologies induced by $\mathscr{U}$ and $\mathscr{V}$ on B are identical, $(B, \mathscr{V})$ is complete, and $B \supset A$.
    (ii) Show that the intersection of F_σ-sets in a complete uniform space is topologically complete.
    (iii) Each paracompact (hence a metric space) space is topologically complete.

6. In a topologically complete space, a subset is relatively countably compact iff its closure is compact. Also a closed subset is countably compact iff it is compact.

7. Show that $f(x) = x^2$ from R to R is not uniformly continuous.

8. Show that each quasiuniform space is quasimetrizable iff its quasiuniformity has a countable base.

9. Show that each topological space is quasiuniformizable.

10. Let $(E_\alpha, \mathscr{U}_\alpha)$ be a family of uniform spaces and $E = \prod_\alpha E_\alpha$, the product with the product uniformity. Let $(F, \mathscr{V})$ be a uniform space; then $f: F \to E$ is uniformly continuous iff $p_\alpha \circ f: F \to E_\alpha$ is uniformly continuous.

11. Give examples to show that if $(E, \mathscr{U})$ is a complete uniform space and $\mathscr{V}$ is another uniformity which is strictly coarser or finer than $\mathscr{U}$, then $(E, \mathscr{V})$ need not be complete.

12. Let $(E_\alpha, \mathscr{U}_\alpha)$ be uniform spaces and $E = \prod_\alpha E_\alpha$. Then E is totally bounded iff each E_α is totally bounded.

13. Show that for a topological group G_u the following statements are equivalent:
    (a) G_u is a T_0-space;
    (b) G_u is a T_1-space;
    (c) G_u is a T_2-space or Hausdorff;
    (d) $\{e\} = \cap\{U: U \text{ neighborhoods of } e\}$.

14. In a topological group G_u, for any set A

$$\bar{A} = \cap\{AU: U \text{ neighborhoods of the identity}\}$$
$$= \cap\{UA: U \text{ neighborhoods of the identity}\}.$$

   Show that each topological group has a fundamental system of closed neighborhoods of the identity. Hence G_u is a T_3-space.

15. Let A, B be two subsets in a topological group. Then
    (i) AB is open if at least one of A, B is open;
    (ii) A^{-1} is open iff A is open;
    (iii) AB need not be closed if both A, B are closed;
    (iv) If A is closed and B compact with $A \cap B = \varnothing$, then there exists a neighborhood U of the identity with $AU \cap BU = \varnothing$ (cfr. Proposition 4, §20).
    (v)   If A is compact and B closed, then AB is closed.

16. Let G be a topological group and H a closed normal subgroup of G. Then the canonical map $\varphi: G \to G/H$ is a continuous, open, and surjective homomorphism.

17. Every locally compact Hausdorff group is normal.

18. Show that there always exists a compatible proximity relation on a completely regular space.

19. Show that a continuous map need not be a proximity map. Take the real line $\mathbb{R}$ and define

    (i) $A \, \delta_1 \, B$ iff $\bar{A} \cap \bar{B} \neq \varnothing$ for $A, B \subset R$;
    (ii) $A \, \delta_2 \, B$ iff $d(A, B) = \inf\{|\, a - b \,|: a \in A, b \in B\} = 0$.

   Then $A = \{n: n \geq 1\}$ and $B = \{n - 1/n, \ n \geq 1\}$ are such that $A \, \delta_2 \, B$ but $A \, \not\delta_1 \, B$. Thus $i: (R, \delta_2) \to (R, \delta_1)$ is a continuous but not a proximity map.

20. Show that every proximity space can be embedded densely in a compact proximity space (the compactification thus obtained is usually called the *Smirnov compactification*).

21. Let f be a proximity mapping of a subspace F of a proximity space E to $[0, 1]$. Then f can be extended to a proximity map $\tilde{f}$ from E to $[0, 1]$.

22. Let $M(I)$ denote the set of equivalence classes of measurable (Lebesgue) functions on the closed unit interval $I = [0, 1]$, with the metric

$$d(f, g) = \int_0^1 \frac{|f(t) - g(t)|}{1 + |f(t) - g(t)|} \, dt.$$

Show that $M(I)$ is a complete metric topological vector space but not a locally convex space.

23. Let $C(R)$ denote the set of all continuous real- or complex-valued functions on the set of real numbers R. Show that $C(R)$ is a Fréchet space with the metric defined by

$$d(f, g) = \sum_{i=1}^{\infty} \frac{1}{2^i} \frac{\max_{|x| \le i} |f(x) - g(x)|}{1 + \max_{|x| \le i} |f(x) - g(x)|}.$$

Show that $C(R)$ is not a normed space.

24. Show that a real or complex topological vector space E is compact iff it consists of $\{0\}$.

25. Show that a Hausdorff topological vector space is locally compact iff it is finite-dimensional (see Proposition 19, §36).

VI

Compact Spaces and Various Other Types of Spaces

In this chapter we study various notions of compactness, including local compactness, countable compactness, pseudocompactness, paracompactness, etc. In view of the essential role played by compact spaces throughout topology and analysis, it is interesting to know when a topological space can be compactified; we consider this later in the chapter. We will also introduce other spaces, e.g., Baire spaces, k-spaces, which generalize compact spaces.

§35. Compact Spaces

Recall that a topological space is said to be compact if every open covering has a finite open subcovering (§10).

We begin with a characterization of compact spaces.

Theorem 1. The following statements are equivalent for any topological space E:

(i) E is compact;

(ii) for each family $\{C_\alpha, \alpha \in A\}$ of closed sets in E satisfying $\bigcap_{\alpha \in A} C_\alpha = \varnothing$, there is a subfamily $\{C_{\alpha_i}, 1 \leq i \leq n\}$ such that $\bigcap_{i=1}^{n} C_{\alpha_i} = \varnothing$;

(iii) if $\{C_\alpha\}_{\alpha \in A}$ is a family of closed sets in E such that the intersection of each finite subfamily is nonempty (this is called the *finite-intersection property*), then $\bigcap_{\alpha \in A} C_\alpha \neq \varnothing$;

(iv)   every filter (or net) on E has at least one cluster or limit point in E;

(v)   every ultrafilter on E is convergent in E.

Proof. (i) $\Rightarrow$ (ii)   for each family $\{C_\alpha\}$ of closed sets in E satisfying $\bigcap_\alpha C_\alpha = \varnothing$, it follows by De Morgans' laws that $\bigcup_\alpha C_\alpha^c = E$, where each C_α^c is open. Thus by (i) there is a finite subfamily, say, $\{C_{\alpha_i}^c: 1 \le i \le n\}$ such that $\bigcup_{i=1}^n C_{\alpha_i}^c = E$, which implies that $\bigcap_{i=1}^n C_{\alpha_i} = \varnothing$.

(ii) $\Rightarrow$ (i)   This follows similarly to (i) $\Rightarrow$ (ii) by retracting the arguments.

(ii) $\Leftrightarrow$ (iii)   This equivalence is obvious.

(iii) $\Rightarrow$ (iv)   Let $\{F_\alpha\}$ be a filter on E. Then by definition of a filter, the intersection of each finite subfamily of $\{F_\alpha\}$ is nonempty. The same is true of $\{\bar{F}_\alpha\}$. Hence $\{\bar{F}_\alpha\}$ is a family of closed sets satisfying the finite-intersection property and therefore, by (iii), $\bigcap_\alpha \bar{F}_\alpha \ne \varnothing$. Thus the set of limit or cluster points of $\{F_\alpha\}$ is nonempty.

(iv) $\Rightarrow$ (v)   Let $\{F_\alpha\}$ be an ultrafilter on E. Since by (iv), $\bigcap \bar{F}_\alpha \ne \varnothing$, $\{F_\alpha\}$ is convergent by a general theorem about ultrafilters (cf. §14, Proposition 35).

(v) $\Rightarrow$ (iii)   Let $\{C_\alpha\}$ be a family of closed sets in E such that the intersection of each finite subfamily is nonempty. Then $\{C_\alpha\}$ forms a filter base. Therefore there exists an ultrafilter $\{F_\alpha\}$ which contains $\{C_\alpha\}$. By (v), $\{F_\alpha\}$ converges to $x_0 \in E$. Since $\{C_\alpha\}$ is the base of a filter coarser than $\{F_\alpha\}$, $\{C_\alpha\}$ also converges to x_0 (Proposition 33, §14). Therefore $x_0 \in \bar{C}_\alpha = C_\alpha$ for each α. Hence $\bigcap_\alpha C_\alpha \ne \varnothing$.

Proposition 1.   Every indiscrete space is compact.

Proof.   Let E be an indiscrete space, i.e., E is endowed with the topology consisting of E and $\varnothing$ as open sets. Clearly the only open covering of E is E itself or $(E, \varnothing)$, each of which is a finite open covering.

Corollary 1.   A topological space consisting of a single point is compact.

Proof.   Immediate.

Proposition 2.   A discrete space is compact iff it is finite.

Proof.   Suppose E is a discrete compact space. Let $\{P_\alpha\}$ be an open covering of E such that each P_α consists of a singleton. Then $\{P_\alpha\}$ is reducible to a finite subcovering $\{P_{\alpha_i}\}_{i=1}^n$, i.e., $\bigcup_{i=1}^n P_{\alpha_i} = E$. Since P_{α_i} consists of a singleton, E is finite. Conversely, suppose E is a finite discrete space. Then every open covering of E is a finite subcovering. Hence E is compact.

Proposition 3. Let $\mathscr{B}$ be a base of the topology of a topological space E. Then E is compact iff every open covering consisting of elements from $\mathscr{B}$ is reducible to a finite open covering.

Proof. Suppose the condition of the proposition holds. Let $\{P_\alpha\}$ be an open covering of E. Since $\mathscr{B} = \{B_\beta\}$ is the base of the topology, each P_α contains an element of $\mathscr{B}$. Let $\mathscr{B}' = \{B_{\alpha'}\}$ be those sets in $\mathscr{B}$ the union of which covers E and such that $B_{\alpha'}$ is contained in some P_α. Now $\mathscr{B}'$, being an open covering consisting of elements from $\mathscr{B}$, is reducible to a finite subcovering $\{B_{\alpha_i}, 1 \leq i \leq n\}$. Since $B_{\alpha_i} \subset P_{\alpha_i}$ for each i, it follows that $\{P_{\alpha_i}\}$ is a finite open covering of E. Hence E is compact. The other part is obvious from the definition.

A useful condition for compactness is as follows:

Theorem 2. (Alexander) Let $\mathscr{B}'$ be a subbase of the open sets of a topological space E. If every open covering of E consisting of elements from $\mathscr{B}'$ is reducible to a finite subcovering of E, then E is compact and conversely.

Proof. Assume that each open covering consisting of elements from $\mathscr{B}'$ has a finite open subcovering. To show that E is compact, it is sufficient to show that there exists a family of open sets such that if no finite subfamily of $\mathscr{B}'$ covers E, then it is not a covering. By considering the totality of all families of open sets of which no finite subfamilies cover E, we introduce partial ordering by inclusion. It is clear that the intersection of two such families also has the same property. It is clear that the family is inductive. Thus by Zorn's lemma, there exists a maximal family $\mathscr{F}$ of open sets such that no finite subfamily covers E. We show that $\mathscr{F}$ is not a covering of E. Clearly the family $\mathscr{B}' \cap \mathscr{F} = \{B : B \in \mathscr{B}' \text{ and } B \in \mathscr{F}\}$ has the property that no finite subfamily covers E. Since $\mathscr{B}' \cap \mathscr{F}$ is a subfamily of $\mathscr{B}'$, by assumption $\mathscr{B}' \cap \mathscr{F}$ is not a covering of E. We shall end the proof by showing that if $x \in \cup \{A : A \in \mathscr{F}\}$ then there exists some $B \in \mathscr{B}'$ such that $x \in B$ and $B \in \mathscr{F}$. But this shows that $\mathscr{F}$ covers E iff $\mathscr{F} \cap \mathscr{B}'$ covers E. Let $x \in A \in \mathscr{F}$. Then there exists a finite subfamily $\{B_i\}_{i=1}^n$ of sets in $\mathscr{B}'$ such that $x \in \bigcap_{i=1}^n B_i \subset A$, because A is open and $\mathscr{B}'$ forms a subbase of the topology. First we show that for some i, $B_i \in \mathscr{F}$. If not, then $B_i \notin \mathscr{F}$, $i = 1, \ldots, n$.

We show that then $A \notin \mathscr{F}$. We argue for $n = 2$ and the same argument can be repeated by finite induction. Since $B_1 \notin \mathscr{F}$ and B_1 is open, there is a finite family F_i $(1 \leq i \leq k)$ of $\mathscr{F}$ such that $B_1 \cup (\bigcup_{i=1}^{k} F_i) = E$ because of the maximality of $\mathscr{F}$. Similarly there is a finite subfamily F_i' $(1 \leq i \leq m)$ of $\mathscr{F}$ such that $B_2 \cup (\bigcup_{i=1}^{m} F_i') = E$. But then

$$(B_1 \cap B_2) \cup \left(\bigcup_{i=1}^{k} F_k \cap \bigcup_{i=1}^{m} F_k' \right) = E$$

implies that no open set containing $B_1 \cap B_2$ belongs to $\mathscr{F}$. In particular $A \notin \mathscr{F}$, a contradiction. Therefore there is some i such that $B_i \in \mathscr{F}$. Hence $A \in \mathscr{F}$, because $\mathscr{F}$ is maximal (§14, Theorem 7). The converse is obvious. The following proposition contains some of the permanence properties.

Proposition 4. (a)   A closed subspace of a compact space is compact.

(b)   The continuous image of a compact space is compact.

(c)   Let E be a compact space and $\sim$ an equivalence relation. Then the quotient space $E^\sim$ with the quotient topology is compact.

(d)   A finite topological sum of compact spaces is compact.

Proof. (a)   If F is a closed subspace of a compact space E and $\{x_\alpha\}$ is a net in F, then $\{x_\alpha\}$ has a limit point x in E because E is compact (Theorem 1). But then, F being closed, it follows that $x \in F$ and hence F is compact.

(b)   Suppose E is compact, F any topological space, and f a continuous mapping of E onto F. If $\{Q_\alpha, \alpha \in A\}$ is an open covering of F, then for each α, $f^{-1}(Q_\alpha)$ is open and $\{f^{-1}(Q_\alpha), \alpha \in A\}$ is an open covering of E which reduces to a finite subcovering $\{f^{-1}(Q_{\alpha_i})\}_{i=1}^{n}$, say. But then $E = \bigcup_{i=1}^{n} f^{-1}(Q_{\alpha_i})$ implies that $F = \bigcup_{i=1}^{n} Q_{\alpha_i}$ and so F is compact.

(c)   If F is a quotient space of a compact space E, endowed with the quotient topology, then the quotient mapping $f : E \to F$ is continuous and onto. Since E is compact so is F by (b).

(d)   Let $E = \sum_{i=1}^{n} E_i$ be a finite topological sum of compact spaces E_i, $1 \leq i \leq n$. Let $\{P_\alpha, \alpha \in A\}$ be an open covering of E. Then for each i, $\{P_\alpha \cap E_i, \alpha \in A\}$ is an open covering of E_i and hence reducible to a finite subcovering $\{P_{\alpha_{ij}}, 1 \leq j \leq m_i\}$ for each $i = 1, \ldots, n$. Clearly $\{P_{\alpha_{ij}}\}$, $j = 1, \ldots, m_i$, $i = 1, \ldots, n$ is an open finite covering of E and so E is compact.

Remarks. (1)   Part (a) of Proposition 4 is not true if the subspace is not closed. For example, $[0, 1]$ is compact but $(0, 1)$ is not.

(2)   Part (d) is not true for countable topological sums. For example, $\{n\}$ is compact for each integer $n \geq 0$ but the topological sum, being homeomorphic to N, is not compact.

One of the most useful and powerful results in topology is that which pertains to the product of compact spaces and which is due to Tychonoff. It is known (e.g., see [53]) that the following result is equivalent to the Axiom of Choice or Hausdorff's maximal principle.

Theorem 3. (Tychonoff)   An arbitrary product $E = \prod_{\alpha \in A} E_\alpha$ of topological spaces E_α, endowed with the product topology, is compact iff each E_α is compact.

Proof.   Let $\{C_\beta\}$ be a family of closed sets in E, having the finite-intersection property. We show that $\cap\, C_\beta \neq \varnothing$. Consider the totality of all families $\{P\}$ having the following properties:

(a)   each family $\{P\}$ contains $\{C_\beta\}$;

(b)   each family $\{P\}$ has the finite-intersection property.

Clearly $\{C_\beta\}$ generates a filter $\mathscr{G}$ and this filter is contained in an ultrafilter $\overline{\mathscr{F}}$. For each $\alpha \in A$, consider $\mathscr{F}_\alpha = \overline{\{p_\alpha(F)\colon F \in \mathscr{F}\}}$, where p_α is the αth projection mapping: $E \to E_\alpha$. It is clear that $\mathscr{F}_\alpha$ has the finite-intersection property. Hence $\cap\, \overline{\{p_\alpha(F)\colon F \in \mathscr{F}\}} \neq \varnothing$ for each α because E_α is compact (Theorem 1). Let $x = (x_\alpha)$, where $x_\alpha \in \cap\, \overline{\{p_\alpha(F)\colon F \in \mathscr{F}\}}$. Let U_α be a neighborhood of x_α, then $U_\alpha \cap p_\alpha(F) \neq \varnothing$ for each $F \in \mathscr{F}$. Clearly $p_\alpha^{-1}(U_\alpha)$ is a member of the subbase of the product topology on E, and $U_\alpha \cap p_\alpha(F) \neq \varnothing$ for each $F \in \mathscr{F}$ implies that $p_\alpha^{-1}(U_\alpha) \in \mathscr{F}$, because $\mathscr{F}$ is an ultrafilter (Theorem 7, §14). Hence for any finite index subset B of A, $\cap\{p_\alpha^{-1}(U_\alpha)\colon \alpha \in B\} \in \mathscr{F}$. Since any neighborhood U of x contains a member of the base of the type $\cap\{p_\alpha^{-1}(U_\alpha)\colon \alpha \in B\}$, it is clear that $U \cap F \neq \varnothing$ for each $F \in \mathscr{F}$, which means that $U \in \mathscr{F}$, since $\mathscr{F}$ is maximal. Hence $\mathscr{F}$ converges to x, i.e., $x \in \bar{F}$ for all $F \in \mathscr{F}$. In particular, $x \in \bar{C}_\beta = C_\beta \in \mathscr{G}$. Hence $\cap\, C_\beta \neq \varnothing$ and we have shown that E is compact (Theorem 1). The converse follows from Proposition 4(b), because the projection mapping $p_\alpha \colon E \to E_\alpha$ is continuous.

We saw earlier that the separation axioms in an arbitrary topological space are telescopic, that is, the stronger implies the weaker but not conversely. However, in restricted cases, the converses may hold. First we exhibit that in some separation axioms, points can be replaced by compact sets.

Definition 1. A subset of a topological space is said to be compact (relatively compact) if it (its closure) is a compact space in the induced relative topology.

Proposition 5. A subset F of a topological space E is compact iff each open covering (in E) of F has a finite open subcovering.

Proof. If $\{P_\alpha\}$ is an open (in E) covering of F, then $\{P_\alpha \cap F\}$ is an open (in F) covering of F. Hence only a finite subcovering $\{P_{\alpha_i} \cap F\}_{i=1}^n$ covers F, because F is compact. But then clearly $\{P_{\alpha_i}\}_{i=1}^n$ covers F. Conversely, if $\{P_\alpha\}$ is an open (in F) covering of F, then for each α, there exists an open (in E) set Q_α such that $Q_\alpha \cap F = P_\alpha$ and clearly $\{Q_\alpha\}$ is an open (in E) covering of F and thus the assumption implies that F is compact.

Proposition 6. Let E be a T_1-space, K any subset of E, and $x_0 \in E \backslash K$. Then there exists an open set U such that $K \subset U$ and $x_0 \notin U$.

Proof. Since E is a T_1-space, for each point $x \in K$ there is an open set U_x containing x and $x_0 \notin U_x$. Clearly $\{U_x\}$ is an open covering of K. Then $K \subset U = \cup\, U_x$ and $x_0 \notin U$.

Proposition 7. Every compact subset of a Hausdorff space is closed.

Proof. Let K be a compact subset of a Hausdorff space E. Let $x \in E \backslash K$. For each $y \in K$, there exist disjoint open neighborhoods U_y and V_x of y and x, respectively. Let y run over K; then $\{U_y\}$ is an open covering of K which reduces to a finite subcovering $\{U_{y_i}\}_{i=1}^n$, say, because K is compact. Let V_{x_i} be the corresponding open neighborhoods of x for $i = 1$, $\ldots, n$. Then $V = \bigcap_{i=1}^n V_{x_i}$ is clearly an open neighborhood of x and $V \cap U_{y_i} = \varnothing$ for each $i = 1, \ldots, n$ implies that $x \in V \subset E \backslash K$. This shows that $E \backslash K$ is open and so K is closed.

Remark. In a non-Hausdorff space, the above proposition is false. (For example, in an indiscrete space each singleton is a compact nonclosed set.) Even in the T_1-spaces, the above proposition might fail.

Proposition 8. Let E be a Hausdorff space. Then for any pair of disjoint compact subsets K_1, K_2 of E, there exist disjoint open subsets U_1, U_2 such that $K_i \subset U_i$, $i = 1, 2$.

Proof. For $x \in K_1$ and $y \in K_2$, there exist disjoint open sets U_x, V_y such that $x \in U_x$, $y \in V_y$. For a fixed $y \in K_2$, with x running over K_1,

$\{U_x\}$ is an open covering of K_1 and such that $y \notin U_x$. Since K_1 is compact, we find a finite open subcovering $\{U_{x,i}, 1 \leq i \leq n\}$ of K_1 and such that $y \notin U_{x,i}, 1 \leq i \leq n$. Clearly, then, $\bigcup_{i=1}^{n} U_{x,i} = U(y)$, say, is an open set containing K_1. Let $V(y) = \bigcap_{i=1}^{n} V_{y,i}$, where $V_{y,i}$ corresponds to $U_{x,i}$ and which is an open set containing y and not meeting $U_{x,i}$. Then $V(y)$ is an open set containing y and disjoint from $U(y)$. Let now y run over K_2; then we obtain an open covering $\{V(y)\}$ of K_2. Compactness of K_2 reduces $\{V(y)\}$ to a finite subcovering, $\{V_i\}_{i=1}^{m}$, say. Let $V = \bigcup_{i=1}^{m} V_i$; then V is an open set containing K_2. Put $U = \bigcup_{i=1}^{m}(U_i(y)$, where $U_i(y)$ is the open set corresponding to V_i and such that $K_1 \subset U_i(y)$, for each i, $1 \leq i \leq m$, as determined above. Thus U and V are the required disjoint open sets.

Corollary 2. Every compact Hausdorff space is normal.

Proof. Let C_1, C_2 be two disjoint closed subsets of a compact space E. Then C_i ($i = 1, 2$) is compact and hence by the above proposition there exist disjoint open sets U_1, U_2 such that $C_i \subset U_i$, $i = 1, 2$, and so E is normal.

Proposition 9. Let E be a T_3-space and K a compact subset of E. Then for each open set $U \supset K$ there exists an open set V such that $K \subset V \subset \bar{V} \subset U$.

Proof. For each $x \in K$ there is an open set V_x such that $x \in V_x \subset \bar{V}_x \subset U$ because E is a T_3-space (Theorem 3, §18). Clearly $\{V_x\}$ is an open covering of K and hence reduces to a finite open subcovering $\{V_{x_i}, 1 \leq i \leq n\}$ say, such that

$$K \subset \bigcup_{i=1}^{n} V_{x_i} \subset \bigcup_{i=1}^{n} \bar{V}_{x_i} \subset U$$

because K is compact. By putting $V = \bigcup_{i=1}^{n} V_{x_i}$, we get the result, since

$$U \supset \bar{V} = \overline{\bigcup_{i=1}^{n} V_{x_i}} = \bigcup_{i=1}^{n} \bar{V}_{x_i} \supset \bigcup_{i=1}^{n} V_{x_i} \supset K.$$

Proposition 10. Every compact T_3-space (respectively compact regular space) is a T_5-space (respectively normal).

Proof. Combine Proposition 9 and Corollary 2.

A general theorem that includes the above proposition and similar results is due to A. Wallace [90]:

Theorem 4.   Let E, F be topological spaces and A, B compact subsets of E and F, respectively. Let W be an open set of the product space $E \times F$, which contains $A \times B$. Then there exist open sets U and V in E and F, respectively, with $A \subset U$, $B \subset V$, and $U \times V \subset W$.

Proof.   For each $(x, y) \in A \times B$, there are open sets U_x, V_x in E and F with $U_x \times V_y \subset W$. For a fixed y, $\{U_x\}$ is an open covering of A. Since A is compact, only a finite subfamily $\{U_i\}_{i=1}^{n}$ of $\{U_x\}$ covers A. Let $Q_y = \bigcup_{i=1}^{n} U_i$; then Q_y is open and $A \subset Q_y$. (Q_y is denoted in this form so as to suggest that it depends upon a fixed choice of y.) Let V_i be an open neighborhood of y corresponding to U_i which was chosen to satisfy: $U_i \times V_i \subset W$. Put $P_y = \bigcap_{i=1}^{n} V_i$; then P_y is an open neighborhood of y and clearly $Q_y \times P_y \subset W$. Now if y runs over B, $\{P_y\}$ is an open covering of B. Compactness of B implies that only a finite subcovering, say $\{P_i\}_{i=1}^{m}$ covers B. Since each P_i corresponds to some Q_i, by setting $U = \bigcap_{i=1}^{m} Q_i$ and $V = \bigcup_{i=1}^{m} P_i$, we obtain the desired result.

Corollary 3.   In a compact Hausdorff space containing more than one point there exist nonconstant continuous functions.

Proof.   Each compact Hausdorff space is normal and hence the corollary follows from Theorem 7, §19.

Proposition 11.   Each continuous mapping of a compact space into a Hausdorff topological space is closed.

Proof.   If C is a closed subset of a compact space E then C is compact [Proposition 4(a)]. If $f : E \to E$ is continuous, then $f(C)$ is compact [Proposition 4(b)] and hence closed since F is Hausdorff (Proposition 7).

Proposition 12.   Each one-to-one continuous mapping of a compact space onto a Hausdorff space is a homeomorphism.

Proof.   Let E be a compact space, F a Hausdorff space, and f a continuous $1 : 1$ mapping of E onto F. For each open set P of E, $E \backslash P$ is closed and so $f(E \backslash P) = f(E) \backslash f(P) = F \backslash f(P)$ (because f is $1 : 1$ and onto) is closed (Proposition 11). Hence $f(P)$ is open and therefore f is a homeomorphism.

§36. Countable Compactness and Sequential Compactness

In this section we study countable compactness and sequential compactness. In general, neither of these two notions is equivalent to compactness. However, in special cases, for example in metric spaces, these two notions are coincident.

Definition 2. (a) A topological space is said to be *countably compact* if every countable open covering reduces to a finite subcovering.

(b) A topological space is said to be *sequentially compact* if every sequence in it contains a convergent subsequence.

Proposition 13. Every compact space is countably compact. But the converse is not true.

Proof. The first part is obvious. For the converse, we see that the space $[0, \Omega)$ (Exercise 5, Chapter III) of all ordinals less than the first uncountable ordinal Ω with the order topology is countably compact but not compact.

A characterization of countable compactness is:

Theorem 5. The following statements are equivalent for T_1-spaces:

(1) E is countably compact;

(2) for each countable family of closed sets $\{C_n\}_{n=1}^{\infty}$ with the finite-intersection property, it follows that $\bigcap_{n=1}^{\infty} C_n \neq \varnothing$;

(3) every countably infinite set of E has at least one cluster point;

(4) every sequence in E has a limit point.

Proof. Clearly $(1) \Leftrightarrow (2)$, as in the proof of (Theorem 1, §35).

$(1) \Rightarrow (3)$ Let $A = \{a_i\}_{i=1}^{\infty}$ be a countably infinite subset of E. Assume A has no cluster point. Then A is a discrete closed subset because E is a T_1-space and hence each singleton is closed. Therefore for each $a_i \in A$, there is a neighborhood U_i of a_i such that $U_i \cap A = \{a_i\}$. Clearly $\{U_i\}_{i=1}^{\infty} \cup A^c$ is a countable open covering of E which contains no finite subcovering. This contradicts (1).

$(3) \Rightarrow (4)$ Let $\{a_i\}_{i=1}^{\infty}$ be a sequence in E. Then either $a_i = a_j$ for all $j \geq i_0$ or by taking a subsequence we may assume that $a_i \neq a_j$ for all i, $j \geq 1$, $i \neq j$. In the first case, $\{a_i\}$ converges to a_{i_0} and hence a_{i_0} is a limit point. In the second case, $\{a_i\}$ is a countably infinite set and has a cluster point by (3).

$(4) \Rightarrow (1)$ If E is not countably compact, there is a countable open covering $\{P_i\}$ of E which does not reduce to a finite subcovering. Thus for each n, there is some $x_n \in E \setminus \bigcup_{i=1}^{n} P_i$. Clearly $\{x_n\}$ is a sequence in E which does not have a limit point because each $x \in E$ belongs to some P_i and $x_k \notin P_i$ for $k \geq i + 1$.

Remark.   In general, the following relations hold between the statements of Theorem 5: $(1) \Leftrightarrow (2)$, $(3) \Rightarrow (4) \Rightarrow (1)$. The T_1-axiom for the space is needed only to show that $(1) \Rightarrow (3)$. It is not hard to show that $(4) \Rightarrow (1)$ and thus $(1) \Leftrightarrow (4)$ in general.

The following theorem deals with a few permanence properties of countably compact spaces.

Theorem 6.   (i)   A closed subspace of a countably (respectively sequentially) compact space is countably (respectively sequentially) compact.

(ii)   The continuous image of a countably (respectively sequentially) compact space is countably (respectively sequentially) compact.

(iii)   The product of a finite family of countably compact spaces need not be countably compact.

(iv)   A countably (respectively sequentially) compact subspace of a Hausdorff first countable topological space is closed.

Proof.   (We prove these statements only for countably compact spaces. The proofs for sequentially compact spaces are similar.)   (i)   Let E be a countably compact space and F a closed subspace of E. If $\{P_n\}$ is a countable open covering of F, then there exist open sets Q_n in E such that $Q_n \cap F = P_n$.

Clearly $\{Q_n\} \cup (E \setminus F)$ is a countable open covering of E and hence it reduces to a finite subcovering $\{Q_n,\ 1 \leq n \leq k\} \cup (E \setminus F)$. Then clearly $\{P_n\}_{n=1}^{k}$ covers F and F is countably compact.

(ii)   Let $f\colon E \to F$ be a continuous mapping of a countably compact space E onto a topological space F. Let $\{Q_n\}$ be a countable open covering of F. Since f is continuous, $f^{-1}(Q_n)$ is open in E. Hence $\{f^{-1}(Q_n)\}$ is a countable open covering of E and therefore reduces to a finite subcovering $\{f^{-1}(Q_n)\}_{n=1}^{m}$, say, because E is countably compact. But then $\{Q_n\}_{n=1}^{m}$ covers F and it proves that F is countably compact.

(iii)   See Exercise 31.

(iv)   Let A be a countably compact subspace of a first countable Hausdorff topological space E. Let $x \in \bar{A}$; then there is a sequence $\{x_n\} \subset A$ converging to x because E is first countable. Since $\{x_n\}$ is a sequence in a countably compact space A, it has a limit point $y \in A$. Since E is Hausdorff and $\{x_n\}$ is convergent to x and y, $x = y \in A$. Therefore, A is closed.

Remark. Part (iii) holds for sequentially compact spaces but it is not true for a countable product of such spaces.

§37. Compactness in Metric Spaces

Definition 3. A subset B of a metric space (E, d) is said to be *bounded* if there exists a real number $\alpha > 0$ such that $d(x, y) \leq \alpha$ for all $x, y \in B$. The quantity

$$\delta(B) = \sup \{d(x, y): x, y \in B\}$$

is called the *diameter* of B.

Remark. Clearly a subset B of a metric space is bounded iff $\delta(B) < \infty$.

Proposition 14. A compact subset of a metric space (E, d) is closed and bounded. But the converse is not true.

Proof. Since a compact subset B of a Hausdorff (in particular, metric) space is closed (Proposition 7, §35), it suffices to prove that B is bounded. Suppose $\delta(B) = \infty$. Then by induction we choose a sequence $\{x_n\} \subset B$ such that $d(x_n, x_{n+1}) \geq n$ for $n \geq 1$. Clearly $\{x_n\}$ has no limit points. But since B is compact, $\{x_n\}$ has a limit point, which is a contradiction. For the converse, we take the real line R with metric:

$$d'(x, y) = \frac{|x - y|}{1 + |x - y|}.$$

Then (R, d') is closed and bounded but not compact.

The converse of the above proposition, however, holds in Euclidean spaces and this fact goes under the name of the Heine–Borel Theorem. But first we prove a special case for the intervals:

Lemma 1. $[a, b]$, $-\infty < a \leq b < \infty$, is a compact space.

Proof. Let $\{P\}$ be an open covering of $[a, b]$. Let $\alpha = \sup\{x \in [a, b]:$ some finite subfamily of $\{P\}$ covers $[a, x]\}$. Observe that $\{a\}$ is always covered by some member of $\{P\}$ and hence $a \leq \alpha$. We show that $\alpha = b$. Suppose P_0 is a member of $\{P\}$ such that $\alpha \in P_0$. Since P_0 is open, choose a number $\beta \in (a, \alpha)$ such that $[\beta, \alpha) \subset P_0$. Clearly $[a, \beta]$ is covered by a finite subfamily

$\{P_i\}_{i=1}^n$ of $\{P\}$. Thus $\{P_i\}_{i=0}^n$ covers $[a, \alpha]$. But $\alpha \in P_0$ implies that there is $\varepsilon > 0$ such that $(\alpha - \varepsilon, \alpha + \varepsilon) \subset P_0$ and hence $[a, \alpha + \varepsilon]$ is covered by $\{P_i: 0 \leq i \leq n\}$. But this contradicts the definition of α unless $\alpha = b$. This completes the proof. Hence each closed bounded subset of R is compact.

Theorem 7. (Heine–Borel) A subset M of the n-dimensional Euclidean space R^n is compact if and only if it is closed and bounded.

Proof. In view of Proposition 14, we have to show only that a closed bounded set M in R^n is compact, since R^n is a metric space. Let p_i be the ith projection map of R^n onto R, viz., $p_i(x_1, \ldots, x_n) = x_i$. Clearly $\overline{p_i(M)}$ is closed and bounded and hence compact by the above lemma. By the Tychonoff's theorem (Theorem 3, §35), $\prod_{i=1}^n \overline{p_i(M)}$ is compact. Since M is closed and clearly $M \subset \prod_{i=1}^n \overline{p_i(M)}$, it follows that M is compact [Proposition 4(a), §35].

Theorem 8. (a) Every continuous real-valued function f on a compact space E is bounded and attains its maximum and minimum values.

(b) Every continuous real-valued function on a countably compact space is bounded.

Proof. (a) By Proposition 4(b), §35, $f(E)$ is a compact subset of the real line, hence bounded by Theorem 7. Let $t_0 = \sup\{f(x): x \in E\}$ and $s_0 = \inf\{f(x): x \in E\}$. Let $\{x_\alpha\}$ be a net in E such that $f(x_\alpha) \to t_0$. Since E is compact, there is a subnet (x_β) which converges to $x_0 \in E$. By continuity of f, $f(x_\beta) \to f(x_0)$. Since $\{f(x_\beta)\}$ is a subnet of the convergent net $\{f(x_\alpha)\}$, we have $f(x_0) = t_0 = \sup\{f(x): x \in E\}$. Similarly there is $y_0 \in E$ such that $s_0 = f(y_0)$.

(b) Suppose f is a continuous real-valued function on a countably compact space E. For each positive integer n, $P_n = \{x \in E: |f(x)| < n\}$ is an open set because f is continuous. Clearly $\{P_n: n \geq 1\}$ is an open covering of E and hence only a finite number of P_n's, say, $\{P_n: 1 \leq n \leq n_0\}$ covers E, i.e., for each $x \in E$, $x \in P_n$ for some n, $1 \leq n \leq n_0$. Hence $|f(x)| \leq n_0$ for all $x \in E$, i.e., f is bounded. The remainder is easy to see.

Corollary 4. The cube I^X, where I is a bounded closed interval and X is any set, is compact.

Proof. This follows by combining Lemma 1 with Tychonoff's theorem (Theorem 3, §35).

Proposition 15. Every compact metric space is separable and hence second countable.

Proof. Let $\{P_{1/n}\}$ be an open covering of a compact metric space E by open balls of radius $1/n$, $n \geq 1$. Since E is compact, only a finite number of balls, say $\{P_{1/n}(x_i), i = 1, \ldots, k_n\}$, cover E for each $n \geq 1$. Clearly $\bigcup_{n=1}^{\infty}\{x_i\}_{i=1}^{k_n} = A$ is countable. Now for each $x \in E$ and a given $\varepsilon > 0$, we choose n with $1/n < \varepsilon$ and x_i with $d(x_i, x) < 1/n < \varepsilon$, because x_i is in $\bigcup_{i=1}^{k_n}P_{1/n}(x_i)$. This shows that A is dense in E and hence E is separable. Since a separable metric space is second countable, the proof is complete.

The following theorem asserts that the three different notions of compactness, countable compactness, and sequential compactness coincide in a metric space.

Theorem 9. Let E be a metric space; the following statements for a subset $K \subset E$ are equivalent:

   (i) K is compact;

  (ii) K is countably compact;

 (iii) K is sequentially compact;

 (iv) K is totally bounded (or precompact) and complete;

  (v)   if $\{B_{\delta_n}\}$ is a strictly decreasing sequence of closed balls in K with diameter δ_n such that $\delta_n \to 0$ as $n \to \infty$, then $\bigcap_n B_{\delta_n} = \{x\} \subset K$ and K is totally bounded.

Proof. (i) $\Rightarrow$ (ii)   This is generally true (Proposition 13, §36).

(ii) $\Rightarrow$ (iii)   Let $\{x_n\}$ be a sequence in K. We may assume that $\{x_n\}$ is a countable set. Then $\{x_n\}$ has a cluster point x in K because K is countably compact. Thus there is a subsequence $\{x_{n_k}\}$ of $\{x_n\}$ which converges to x.

(iii) $\Rightarrow$ (iv)   Let K be sequentially compact. Let $\{x_n\}$ be a Cauchy sequence in K. Then there is a subsequence $\{x_{n_k}\}$ of $\{x_n\}$ which converges to $x \in K$. Since $\{x_n\}$ is Cauchy, it follows that $\{x_n\}$ converges to x. Hence K is complete. Assume K is not totally bounded. There exists an $\varepsilon > 0$ and $x_n \in K$ such that $d(x_n, x_{n-1}) > \varepsilon$, for $n \geq 1$ by induction, because otherwise K will be covered by a finite number of balls of radius ε. Clearly $\{x_n\}$ has no convergent subsequence. Hence K is not sequentially compact.

(iv) $\Rightarrow$ (i)   Assume K is totally bounded and complete. Then the completion $\hat{K}$ of K coincides with K. Thus $\hat{K} = K$ is a compact subset because K is precompact.

(iv) $\Rightarrow$ (v)   Let $\{B_{\delta_n}\}$ be a strictly decreasing sequence of closed balls in K with diameter $\delta_n \to 0$ as $n \to \infty$. We show that $\bigcap_{n=1}^{\infty} B_{\delta_n} = \{x\}$. Let $\{x_n\}$

be a sequence of points chosen such that $x_n \in B_{\delta_n} \setminus B_{\delta_{n+1}}$. (This is possible because δ_n is a monotonically decreasing sequence of positive real numbers and hence $B_{\delta_{n-1}} \supset B_{\delta_n}$.) $\{x_n\}$ is a Cauchy sequence. For, given $\varepsilon > 0$, choose n_0 such that $\delta_n < \varepsilon$ for $n \geq n_0$. Then $x_m, x_n \in B_{\delta_n}$ for all $m, n \geq n_0$. Hence $d(x_m, x_n) \leq \delta_n \leq \varepsilon$ for all $m, n \geq n_0$. Thus by (iv), $\{x_n\}$ converges to $x \in K$. Clearly $x_m \in B_{\delta_n}$ for all $m \geq n$. Since B_{δ_n} is closed, $x \in B_{\delta_n}$ for all $n \geq 1$ and so $x \in \bigcap_{n=1}^{\infty} B_{\delta_n}$. Suppose $x, y \in \bigcap_{n=1}^{\infty} B_{\delta_n}$ and $x \neq y$. Then $d(x, y) = \varepsilon > 0$. Choose n_0 such that $x_{n_0} \varepsilon B_{\delta_{n_0}}(x) \subset B_{\varepsilon/2}(x)$. But then $y \notin B_{\delta_{n_0}}(x)$ because if $y \in B_{\delta_{n_0}}(x)$ then $\varepsilon = d(x, y) \leq \delta_{n_0} < \varepsilon/2$, which gives a contradiction. This proves (v).

(v) $\Rightarrow$ (iv)   Let $\{x_n\}$ be a Cauchy sequence in K. For each positive integer k, there are $x_{n_k}, x_{n_{k+1}} \in \{x_n\}$ with

$$d(x_{n_k}, x_{n_{k+1}}) \leq \frac{1}{2^{k+1}}$$

for all $k \geq 1$. Consider the strictly decreasing sequence of closed balls, $\{B_{1/2^k}(x_{n_k})\}_{k \geq 1}$. By (v), $\bigcap_{k \geq 1} B_{1/2^k}(x_{n_k}) = \{x\}$. Hence $\{x_{n_k}\}$ converges to x. Since $\{x_n\}$ is Cauchy, $x_n \to x$, and so K is complete and (iv) is established.

Definition 4.   Let A, B be two subsets of a metric space. We define the *distance* between A, B to be $d(A, B) = \inf\{d(x, y) : x \in A, y \in B\}$.

Remark.   If we take $A = \{n\}$ and $B = \{n - 1/n\}$, $n \geq 1$, the two subsets of the real line, then A, B are closed, $A \cap B = \varnothing$, and $d(A, B) = 0$. Thus two disjoint closed subsets may not be at a positive distance. The following proposition, however, shows that the situation is different if one of the sets is compact.

Proposition 16.   Let A, B be two disjoint subsets of a metric space (E, d), where A is closed and B compact. Then $d(A, B) > 0$.

Proof.   Let $f(x) = d(x, A) = \inf\{d(x, y) : y \in A\}$. Then as shown before (Examples and Exercises 19, Chapter III), f is a continuous real-valued function on E. Hence its restriction on the compact set B assumes (Theorem 8) its minimum value at $x_0 \in B$. Hence

$$f(x_0) = d(x_0, A) = \inf\{d(x, A) : x \in B\}$$
$$= d(B, A).$$

Since $x_0 \notin \bar{A} = A$, $d(B, A) > 0$.

The following theorem deals with the well-known classical notion of the *Lebesgue number* of a covering.

Theorem 10 (Lebesgue covering number).   Let (E, d) be a compact metric space and $\{P_\alpha\}$ an open covering of E. Then there exists a positive real number λ, depending upon the covering $\{P_\alpha\}$, such that each ball B_λ is contained in at least one P_α. In other words, one may say that each open covering of a compact metric space may be refined to a "uniform covering" by balls of a constant radius λ. (λ is called a *Lebesgue number of the covering* $\{P_\alpha\}$.)

Proof.   Let $x \in E$. Since $\{P_\alpha\}$ is a covering, $x \in P$ for some α. But then there is an $r = r(x) > 0$ (depending upon x) such that $B_r(x) \subset P_\alpha$. Clearly $\{B_{r/2}(x) : r = r(x), x \in E\}$ is an open covering of the compact space E. Hence only a finite subcovering $\{B_{r_i/2}(x_i) : r_i = r_i(x_i), i = 1, \ldots, n\}$ covers E. Let $\lambda = \min_{1 \le i \le n} r_i/2$. Then $\lambda > 0$. We show that for any $x \in E$, $B_\lambda(x) \subset P_\alpha$ for some α. If $y \in B_\lambda(x)$, then clearly $y \in B_{r_i/2}(x_i)$ for some i and therefore

$$d(x, x_i) \le d(x, y) + d(y, x_i) < \lambda + r_i/2 \le r_i.$$

Hence $x \in B_{r_i}(x_i) \subset P_\alpha$ for some α and the proof is completed.

The last result of this section deals with the uniform continuity of a map which can be derived from the above theorem.

Proposition 17.   Let (E, d) and (F, d') be two metric spaces, where E is compact. Then each continuous mapping $f : E \to F$ is uniformly continuous.

Proof.   Let $\varepsilon > 0$ be given. Clearly $\{B_{\varepsilon/2}(y) : y \in F\}$ is an open covering of F and since f is continuous, $\{f^{-1}(B_{\varepsilon/2}(y)) : y \in F\}$ is an open covering of E. Let δ be the Lebesgue number of the latter covering of E. Observe that δ depends only upon ε, since the covering depends upon ε. Thus each $B_\delta(x)$ is contained in some $f^{-1}(B_{\varepsilon/2}(y))$, i.e., $f(B_\delta(x)) \subset B_{\varepsilon/2}(y)$ for some $y \in F$. We show that $f(B_\delta(x)) \subset B_\varepsilon(f(x))$. Let $z \in B_\delta(x)$. Then

$$d'(f(z), f(x)) \le d'(f(z), y) + d'(y, f(x)) < \varepsilon/2 + \varepsilon/2 = \varepsilon$$

and thus $f(z) \in B_\varepsilon(f(x))$ for all $z \in B_\delta(x)$.

§38. Locally Compact Spaces

The notion of local compactness generalizes that of the compactness. Many spaces that one comes across in analysis and geometry are locally compact.

Definition 5. A topological space is said to be *locally compact* if each point has a compact neighborhood.

Proposition 18. A Hausdorff topological space is locally compact iff each neighborhood of each point contains a neighborhood whose closure is compact.

Proof. Let E be a Hausdorff locally compact space and let $x \in E$. Then there is a compact neighborhood C of x. If U is any neighborhood of x, then $U \cap C$ is a neighborhood of x. Clearly $\overline{U \cap C} \subset \bar{C} = C$ since E is Hausdorff (Proposition 7, §35). Thus $U \cap C$ is the required neighborhood Proposition 4(a), §35. The converse is obvious.

Remark. The above proposition says that E is locally compact if and only if each point has a neighborhood basis consisting of compact neighborhoods.

Proposition 19. Any finite-dimensional Euclidean space R^n $(n \geq 0)$ is locally compact. In particular, the real line R is locally compact.

Proof. Recall the topology on R^n is defined by the sets

$$B_\varepsilon(x_0) = \{(x_1, \ldots, x_n) \in R^n : \sum_{i=1}^{n} |x_i - x_i^0| \leq \varepsilon\},$$

where $x^0 = (x_1^0, \ldots, x_n^0)$. Obviously such sets (called parallelepiped) contain $\prod_{i=1}^{n} I_i$ (i.e., n times product of closed bounded intervals, which are compact sets by the Heine–Borel theorem (Theorem 7, §37), which is compact by the Tychonoff theorem (Theorem 3, §35) and these sets form a basis of the topology. Hence R^n is locally compact. In particular, R is locally compact.

Proposition 20. Every discrete space is locally compact.

Proof. Since each singleton $\{x\}$ is a compact open set and therefore a compact neighborhood of x, the proposition is immediate.

Proposition 21. Every Hausdorff locally compact space is completely regular.

Proof. Let E be a Hausdorff locally compact space, C a closed subset of E, and $x \notin C$. Since $E \backslash C$ is an open neighborhood of x, there exists a neighborhood U of x such that $\bar{U}$ is compact and $U \subset E \setminus C$ (Proposition 18). Choose an open neighborhood V of x such that $x \in V \subset \bar{V} \subset U \subset \bar{U}$ which is possible because $\bar{U}$, being a compact Hausdorff space, is normal. Now there exists a continuous real function f by Urysohn's lemma (Theorem 7, §19) such that $f(\bar{V}) = 1$, $f(\bar{U} \setminus U) = 0$, and $0 \leq f(x) \leq 1$. Put $f(E \setminus \bar{U}) = 0$. It is easy to check that f is continuous on E, $f(x) = 1$, and $f(C) = 0$, Hence E is completely regular.

Remark. Even though each Hausdorff locally compact space is completely regular, it is not true that such a space is normal (see Exercise 34). However, if there is an additional structure on the locally compact space, e.g., if it is a topological group, then it is normal (see [46]).

Proposition 22. An open subset of a locally compact (in particular, compact) space is always locally compact.

Proof. Let E be a locally compact space and F an open subset of E. If U is an open neighborhood of $x \in F$, then U is also open in E. Hence there is a compact neighborhood V of x in E such that $x \in V \subset U$. But then V is a compact neighborhood of x in F. This proves that F is locally compact.

Theorem 11 (Permanence properties).

   (i)   A closed subspace of a locally compact space is locally compact.
   (ii) A topological sum of locally compact spaces is locally compact.
   (iii) The continuous and open image of a locally compact space is locally compact.
   (iv) A finite product of locally compact spaces is locally compact.
   (v) $\prod_{\alpha \in A} E_\alpha$ is locally compact iff each E_α is compact except for a finite number of them, which are locally compact.
   (vi) Let f be a continuous closed map of a Hausdorff space E onto a Hausdorff space F such that for each $y \in F$, $f^{-1}(y)$ is compact. Then F is locally compact if E is.

Proof. (i) Let F be a closed subspace of a locally compact space E. Let $x \in F$ and let C be a compact neighborhood of x in E. Then $C \cap F$ is a compact neighborhood of x in F. Hence F is locally compact.

   (ii) Let $\{E_\alpha : \alpha \in A\}$ be a family of pairwise disjoint locally compact spaces. Let $E = \sum_{\alpha \in A} E_\alpha$ be their topological sum. Clearly for each $x \in E$, $x \in E_\alpha$ for some unique α and there exists a compact neighborhood C of x

in E_α. But then C is compact in E and is a neighborhood of x. Hence E is locally compact.

(iii)   Let f be a continuous and open mapping of a locally compact space E onto a topological space F. Let $y \in F$; then there is an $x \in E$ such that $f(x) = y$. Since E is locally compact, there is a compact neighborhood C of x in E. Since f is continuous, $f(C)$ is compact, and since f is open, $f(C)$ is a neighborhood of $f(x) = y$. Therefore F is locally compact.

(iv)   Let E_i $(1 \le i \le n)$ be a finite number of locally compact spaces. Let $E = \prod_{i=1}^n E_i$ and $x \in E$, then $x = (x_1, \ldots, x_n)$, where $x_i \in E_i$. Since each E_i is locally compact, there exists a compact neighborhood C_i of x_i in E_i. Then by Tychonoff's theorem (Theorem 3, §35) $C = \prod_{i=1}^n C_i$ is a compact neighborhood of x in the product topology of E.

(v)   First suppose $\prod_{\alpha \in A} E_\alpha$ is locally compact. Then since each projection mapping $p_\alpha \colon \prod_{\alpha \in A} E_\alpha \to E_\alpha$ is continuous, open and onto, by part (iii), E_α is locally compact. Further, if V is an open relatively compact subset of $\prod_\alpha E_\alpha$, then each $p_\alpha(\bar{V})$ is compact and since $p_\alpha(\bar{V}) = E_\alpha$ for all α except for a finite number, the result follows. Conversely, suppose all the E_α are compact except a finite number of them, which are locally compact, viz., E_{α_i} $(1 \le i \le n)$ are locally compact. Then

$$F = \prod \{ E_\alpha, \alpha \in A \setminus \{\alpha_i\}_{i=1}^n \}$$

is compact by Tychonoff's theorem (Theorem 3, §35) and hence locally compact. But then E, being homeomorphic with $F \times \prod_{i=1}^n E_{\alpha_i}$, which is locally compact by (iv), is also locally compact.

(vi)   Suppose E is locally compact. Let $y \in F$; then $B = f^{-1}(y)$ is compact by hypothesis. Since E is locally compact, there is a relatively compact open neighborhood U_x of each x in B so that the open covering $\{U_x \colon x \in B\}$ of the compact set B is reducible to a finite subcovering $\{U_{x_i} \colon i = 1, \ldots, n\}$. Let $U = \bigcup_{i=1}^n U_{x_i}$. Then U is a relatively compact open neighborhood of B. By assumption there is a neighborhood V of y such that $f^{-1}(V) \subset U$. Hence $y \in V \subset f(U) = f(\bar{U})$. Since the latter set is compact, we have shown that F is locally compact.

Definition 6.   A locally compact space which can be written as the union of a countable number of compact spaces is called a σ-compact space.

Remark.   Sometimes one defines σ-compact spaces without local compactness, but we use it because of Theorem 12 and some applications elsewhere.

Proposition 23. Every countable set with the discrete topology is σ-compact.

Proof. Obvious, since a discrete space is always locally compact and countable points give a countable union of compact sets.

Remark. It is not true that an uncountable set with the discrete topology is σ-compact. For example, the real line with the discrete topology cannot be written as a countable union of compact spaces.

The set of rationals Q with the metric topology induced from the reals can be written as the countable union of compact spaces, viz., singletons, but fails to be σ-compact since Q is not locally compact.

Indeed, the real line R itself with the usual metric topology is σ-compact.

It is possible to represent a σ-compact space by an increasing sequence of relatively compact spaces, as is shown by:

Proposition 24. If E is a Hausdorff σ-compact space, then $E = \bigcup_{n \geq 1} E_n$, where each E_n is a relatively compact open set and $\bar{E}_n \subset E_{n+1}$ for all $n \geq 1$.

Proof. Since E is σ-compact by definition, E is locally compact and $E = \bigcup_{n \geq 1} C_n$, where each C_n is compact. Hence there exists a relatively compact open set E_1 such that $C_1 \subset E_1$. This can be done as follows: Let $x \in C_1, y \in E \backslash C_1$. Since E is Hausdorff, there exist disjoint neighborhoods U_x and V_y of x and y, respectively. We may pick U_x to be such that $\bar{U}_x$ is compact because E is locally compact. Clearly $\{U_x\}$ is an open covering of the compact set C_1 and hence only a finite number of members, $\{U_{x_i}, 1 \leq i \leq n\}$, cover C_1. Clearly $E_1 = \bigcup_{i=1}^n U_{x_i}$ is an open set such that $C_1 \subset E_1$ and $\bar{E}_1 = \bigcup_{i=1}^n \bar{U}_{x_i}$ is compact. Now, by induction, we complete the proof.

Indeed, this property can be seen to be equivalent to Hausdorff σ-compact spaces. Precisely, we have:

Theorem 12. Let E be a Hausdorff topological space. The following statements are equivalent:

(a) E is σ-compact;
(b) E satisfies the condition stated in Proposition 24;
(c) E is a locally compact Lindelöf space.

Proof. (a) $\Rightarrow$ (b)   Proved in Proposition 24.

(b) $\Rightarrow$ (c)   Suppose E satisfies the property mentioned in Proposition 24, viz., $E = \bigcup_{n=1}^{\infty} E_n$ such that each E_n is relatively compact and $\bar{E}_n \subset E_{n+1}$. Let $\{P_\alpha: \alpha \in A\}$ be an open covering of E. Since each $\bar{E}_n$ is compact, only a finite number of P_α's covers $\bar{E}_n$. Since E is a countable union of $\bar{E}_n$'s, only countably many members of P_α's cover E and so E is a Lindelöf space. Furthermore, for each $x \in E$, $x \in E_n$ for some n such that $\bar{E}_n$ is compact and E_n is open. Hence E is locally compact.

(c) $\Rightarrow$ (a)   Assume E is a locally compact Lindelöf space. For each $x \in E$, there is an open neighborhood U_x of x such that $\bar{U}_x$ is compact. Clearly $\{U_x: x \in E\}$ is an open covering of E and hence there exists a countable subcovering, $\{U_i: i \geq 1\}$, such that each U_i is open and $\bar{U}_i$ is compact, and clearly $E = \bigcup_{i=1}^{\infty} \bar{U}_i$. Hence E is σ-compact.

§39. *MB*-Spaces

Definition 7.   A metric space is said to be an *MB-space* if every closed bounded set in it is compact.

Example.   The n-dimensional Euclidean space R^n (in particular, the real line R) with the usual metric is an MB-space by the Heine–Borel theorem (Theorem 7, §37).

Proposition 25.   A metric compact space is an MB-space.

Proof.   This is obvious since each closed subset of a compact space is compact and therefore bounded.

Proposition 26.   Every MB-space is complete.

Proof.   Let $\{x_n\}$ be a Cauchy sequence in an MB-space E. Then clearly $\{x_n\}$ is bounded. That means there exists $a > 0$ such that the diameter of $\{x_n\} = \delta(\{x_n\}) \leq a$. Thus $d(x_1, x_n) \leq a$ for all $n \geq 1$. Since $\{y: d(x_1, y) \leq a\}$ is a closed bounded set and hence compact because E is an MB-space, let $x_0 \in \overline{\{x_n\}} \subset E$. Then there is a subsequence $\{x_{n_k}\}$ converging to x_0. Since $\{x_n\}$ is Cauchy it follows that $x_n \to x_0$ and E is complete.

Proposition 27.   Let E be a metric space and let M be a bounded subset of E. Then $\bar{M}$ is bounded.

Proof. Since M is bounded, there exists $a > 0$ such that for any $x \in M$, $M \subset \{y: d(x, y) \leq a\}$. Since the latter is a closed set, therefore $\bar{M} \subset \{y: d(x, y) \leq a\}$. This shows that $\bar{M}$ is bounded.

Proposition 28. Let E be an MB-space and f a continuous mapping of E into a metric space F. If M is a bounded subset of E, then $f(M)$ is a bounded subset of F.

Proof. Since M is bounded, $\bar{M}$ is a closed and bounded subset of E and hence compact because E is an MB-space. Thus $f(\bar{M})$ is compact and hence closed and bounded in F. But then $f(M)$, being a subset of the bounded set $f(\bar{M})$, is also bounded.

Theorem 13 (Permanence properties).

    (a)   Every closed subspace of an MB-space is an MB-space;

    (b)   a countable product of compact metric spaces is an MB-space.

Proof. (a) Let F be a closed subspace of an MB-space. If C is a closed bounded subset of F, then C is closed and bounded in E and hence compact. Since a subset of F which is compact in E also remains compact in F, (a) follows.

    (b) Let $\{E_i\}$ be a sequence of compact metric spaces. Clearly $E = \prod_{i=1}^{\infty} E_i$ is a metric compact space and hence an MB-space by Proposition 25.

Proposition 29. Let E be an MB-space and F a metric space. Let f be a continuous mapping of E onto F such that for each bounded subset M of F, $f^{-1}(M)$ is bounded in E. Then F is an MB-space.

Proof. Let M be a closed bounded subset of F. Then $f^{-1}(M)$ is a closed and bounded subset of E, hence compact. Continuity of f implies then that $M = f(f^{-1}(M))$ is compact.

Corollary 5. Let E, F be metric spaces such that E is an MB-space of finite diameter. If f is a continuous mapping of E onto F, then F is an MB-space.

Proof. Since E is of finite diameter, for each bounded subset M of F, $f^{-1}(M)$ is bounded and Proposition 29 applies.

Corollary 6. A continuous image of a compact space in a metric space is an MB-space.

Proposition 30. Every locally compact real or complex topological vector Hausdorff space is an MB-space.

Proof. Since every locally compact Hausdorff topological vector space is homeomorphic with an n-dimensional real or complex Euclidean space R^n or C^n (Exercise 25, Chap. V), the proposition follows from Proposition 19, §38.

Proposition 31. Every MB-space E is locally compact.

Proof. Let $x \in E$. The closed ball $B_\varepsilon(x) = \{y : d(x, y) \leq \varepsilon\}$ is a closed bounded, hence a compact neighborhood of x.

The converse of Proposition 31 is not necessarily true (Exercise 30).

Proposition 32. Every separable metric locally compact space can be given a metric which makes it an MB-space.

Proof. Let (E, d) be a separable metric locally compact space and $E^\sim = E \cup \{\infty\}$, its one-point compactification (Proposition 49, §44 below). Then $E^\sim$ is a compact metric (hence second countable and normal) space. Hence there exists a homeomorphism h of $E^\sim$ into a subspace of the Hilbert space $H = l_2$ by Theorem 9, §19. For each $x \in E$, define $f(x) = x - h(\infty)$, where the translation is defined because H is a linear space. Clearly f is a homeomorphism of H into itself and f maps $h(\infty)$ into 0. Now the composition $f \circ h : E^\sim \to H$ is a homeomorphism (into) such that $f \circ h(\infty) = 0$. For each $x = \{x_i\} \in H$, let

$$\| x \|^2 = \sum_{i=1}^{\infty} | x_i |^2,$$

then

$$\sum_{i=1}^{\infty} \left(\frac{x_i}{\| x \|} \right)^2 = 1,$$

if $\| x \| \neq 0$. Now define

$$\varphi : H \backslash \{0\} \to H$$

by

$$\varphi(x) = \left\{ \frac{x_i}{\| x \|^2} \right\},$$

where $x = \{x_i\}$, $i \geq 1$. Then φ is a homeomorphism of $H \backslash \{0\}$ onto itself. The restriction of the composition of φ with $f \circ g$ is a homeomorphism of E into H and so the closed and bounded subsets of E are mapped into closed bounded subsets of $H \backslash \{0\}$. Since closed bounded subsets of any locally compact subset of H are compact, so are those of E. Hence E is an MB-space.

Theorem 14. Let E be an MB-space and $K \subset E$ a subset. The following statements are equivalent:

(a) K is closed and bounded;

(b) K is compact;

(c) K is sequentially compact;

(d) K is countably compact;

(e) K is totally bounded (precompact) and complete.

Proof. In any metric space (b)–(e) are equivalent (Theorem 9, §37), whereas (a) ⇔ (b) in an MB-space by definition.

Proposition 33. Every MB-space E is separable, second countable, and Lindelöf.

Proof. Let $x \in E$ and $\bar{B}_n(x) = \{y: d(x, y) \leq n\}$. Then clearly each $\bar{B}_n(x)$ is compact because E is an MB-space and, hence, being a subset of a metric space, it is separable. But $E = \bigcup_{n \geq 1} \bar{B}_n(x)$ and therefore E is separable. Since E is a metric space, it is second countable. Hence E is also Lindelöf.

Corollary 7. Every MB-space is σ-compact.

Proof. Follows from Propositions 31 and 33.

§40. *k*-Spaces and *k_r*-Spaces

Definition 8. A topological space E is said to be a *k-space* if each subset U is open in E if and only if $U \cap C$ is relatively open for each compact subset C of E.

Remark. Alternatively, one defines that E is a *k*-space if for each compact set C, $U \cap C$ is closed implies U is closed in E.

Proposition 34. Every Hausdorff locally compact (in particular, compact) space is a k-space.

Proof. Let U be a subset of a Hausdorff locally compact space E. Suppose $U \cap C$ is relatively open for each compact subset C of E and let $x \in U$. Since E is locally compact, there exists an open neighborhood such that $\bar{V}$ is compact. By assumption $U \cap \bar{V}$ is open in $\bar{V}$ and so $U \cap V$ is open in V. Since V is open in E, $U \cap V$ is an open neighborhood of x in E. Hence each point x of U has an open neighborhood which is contained in U. Hence U is open in E and so E is a k-space.

Remark. There do exist k-spaces which are not locally compact. For example l_2 is a k-space (this follows from the following proposition) but not locally compact.

Proposition 35. Every first countable topological space is a k-space.

Proof. Let E be a first countable topological space. Let U be a subset of E such that $U \cap C$ is closed for each compact set C. Let $x_n \in U$ such that $x_n \to x \in \bar{U}$. Clearly $C' = \{x_n\} \cup \{x\}$ is compact and hence $C' \cap U$ is closed by assumption. Since $x_n \in C' \cap U$, it follows that $x \in C' \cap U$, which shows that $x \in U$. This proves that U is closed and hence E is a k-space.

Corollary 8. Every metric space is a k-space.

Theorem 15 (Permanence properties).

(a) A closed subspace of a Hausdorff k-space is a k-space.

(b) A topological sum of k-spaces is a k-space.

(c) Let f be a continuous and open mapping of a k-space E onto a topological space F such that each compact subset of E is the inverse image of a compact subset in F. Then F is also a k-space.

(d) A Hausdorff space E is a k-space iff it is a quotient space of a locally compact space.

Proof. (a) Let F be a closed subspace of a k-space E. Let U be a subset of F such that $U \cap C$ is closed in F for each compact subset C of F. Since F is closed in E, $U \cap C$ is closed in E. Now let K be a compact subset of E; then $K \cap F$ is compact in F because F is closed and E is Hausdorff. Hence

$U \cap (K \cap F) = U \cap K$ (since $U \subset F$) is closed in E for each compact subset K of E. Since E is a *k*-space, U is closed in E and hence closed in F.

(b) Let $E = \sum E_\alpha$ ($\alpha \in A$) be the topological sum of *k*-spaces. It is clear that each E_α is a closed and open subset of E. Furthermore, if for each subset U of E such that $U \cap C$ is closed for each compact subset C of E, then $U \cap C_\alpha$ is closed for each compact subset C_α of E_α for each α, because each compact subset C_α is also compact in E. Hence $U \cap C_\alpha$, being closed, implies that $U \cap E_\alpha$ is closed in E_α for each α. But then by the definition of topological sum U is closed in E and E is a *k*-space.

(c) Let f be a map of a *k*-space E onto a topological space F as described in the hypothesis. Let V be a subset of F such that $V \cap C$ is open in C for each compact set C in F. But then $f^{-1}(V \cap C) = f^{-1}(V) \cap f^{-1}(C)$ is open in $f^{-1}(C)$ because f is continuous. Since every compact subset of E is the inverse image of some compact subset of F, we conclude that $f^{-1}(V)$ is an open subset of E because E is a *k*-space. But since f is open, $f(f^{-1}(V)) = V$ is open in F and thus F is a *k*-space.

(d) Let $= \{C_\alpha : \alpha \in \Gamma\}$ denote the family of all compact subsets of E. For each $\alpha \in \Gamma$, let $C_\alpha{}'$ denote the space $\{\alpha\} \times C_\alpha$ (as a product of the discrete space $\{\alpha\}$ and C_α). Then it is clear that $C_\alpha{}'$ is homeomorphic to C_α and the spaces $\{C_\alpha{}' : \alpha \in \Gamma\}$ are pairwise disjoint. Let $F = \sum \{C_\alpha{}' : \alpha \in \Gamma\}$ (called the free union of $\{C_\alpha : \alpha \in \Gamma\}$) be endowed with the topological direct sum topology (cf., §21). Let $h_\alpha : C_\alpha{}' \to C_\alpha$, defined by $h_\alpha(\alpha, x) = x \in C_\alpha$, be the homeomorphism. Then the map $h : F \to E$ defined by $h \mid C_\alpha{}' = h_\alpha$ is continuous (because E is a *k*-space) and so $F/\ker h$ [where $\ker h = \{(x, y) \in F \times F : h_\alpha(x) = h_\alpha(y)$ for $\alpha \in \Gamma]$ is homeomorphic with E. Since each C_α is compact, so is $C_\alpha{}'$. But then F is locally compact and so the "only if" part follows.

For the "if" part, let q be the quotient map of F onto E, where F is locally compact. Let U be a subset of E such that $U \cap C$ is open in C for each compact set $C \subset E$. To show that E is a *k*-space, we wish to show that U is open. Clearly for each relatively compact open subset $D \subset F$, $U \cap q(\bar{D})$ is open in the compact set $q(\bar{D})$ and so $U \cap q(\bar{D}) = V \cap q(\bar{D})$, where V is open in E. But

$$q^{-1}(U) \cap q^{-1}(q(\bar{D})) = q^{-1}(V) \cap q^{-1}(q(\bar{D}))$$

implies

$$q^{-1}(U) \cap q^{-1}(q(\bar{D})) \cap D = q^{-1}(V) \cap q^{-1}(q(\bar{D})) \cap D$$

and so $q^{-1}(U) \cap D = q^{-1}(V) \cap D$, because $D \subset q^{-1}(q(\bar{D}))$. Hence $q^{-1}(U) \cap D$ is open in D. Since F can be written as an union of relatively compact open subsets, viz., $F = \bigcup_\alpha D_\alpha$, where D_α is relatively compact open, because F is locally compact, we have $q^{-1}(U) = \bigcup_\alpha q^{-1}(U) \cap D_\alpha$, which is open and so $q(q^{-1}(U)) = U$ is open in E because q is open. In other words, E is a k-space.

Remark. The product of two k-spaces need not be a k-space (cf. Exercise 20).

Definition 9. Let E be a topological space and let $\mathscr{C}$ denote a family of compact subsets of E. The family of subsets U such that $U \cap C$ is open in C for each $C \in \mathscr{C}$ defines a topology called the k-*topology* and is denoted by $k(u, \mathscr{C})$, where u is the original topology on E.

Remark. If $\mathscr{C}$ is the family of *all* compact subsets of E_u, then $k(u, \mathscr{C})$ is called the k-*extension* of u. (See Definition 3, §15.)

Proposition 36. Let E_u be a topological space. Then:

(a) $k(u, \mathscr{C}) \supset u$, i.e., $k(u, \mathscr{C})$ is finer than u;

(b)   if $\mathscr{C}_1 \supset \mathscr{C}_2$, then $k(u, \mathscr{C}_1) \subset k(u, \mathscr{C}_2)$;

(c)   if $\mathscr{C}$ denotes the set of all u-compact subsets of E, then E_u is a k-space if and only if $u = k(u, \mathscr{C})$.

Proof.   (a)   For each u-open subset U of E, clearly $U \cap C$ is open in C and hence U is $k(u, \mathscr{C})$-open.

(b) $\mathscr{C}_1 \supset \mathscr{C}_2$ implies that each compact subset of $\mathscr{C}_2$ is also in $\mathscr{C}_1$. Hence for each set U, $U \cap C$ being open in C for all C in $\mathscr{C}_1$ implies that $U \cap C_2$ is open in C_2 for all C_2 in $\mathscr{C}_2$. Thus $k(u, \mathscr{C}_1) \subset k(u, \mathscr{C}_2)$.

(c)   By (a), we have $u \subset k(u, \mathscr{C})$. Now if E is a k-space, then each $k(u, \mathscr{C})$-open set is also u-open and hence $u = k(u, \mathscr{C})$. The fact that $u = k(u, \mathscr{C})$ implies that E_u is a k-space follows from the definition.

Remark.   It is clear that if $u \supset v$ and if $\mathscr{C}_u$, $\mathscr{C}_v$ denote the family of u-compact and v-compact subsets, respectively, then $\mathscr{C}_u \subset \mathscr{C}_v$ and hence we obtain

$$k(u, \mathscr{C}_u) \supset k(v, \mathscr{C}_u) \supset k(v, \mathscr{C}_v) \supset v.$$

Theorem 16. Let E_u be a k-space and f a real-valued function on E_u. Then f is continuous iff its restriction $f \mid C$ for each compact subset C in E_u is continuous.

Proof. Obviously if f is continuous then its restriction $f \mid C$ on each compact subset C is continuous. Conversely, suppose the restriction $f \mid C$ is continuous for each compact subset C of E. Let V be an open subset of the real numbers and C, a compact subset of E. Then $f^{-1}(V) \cap C$ is open in C. Hence $f^{-1}(V)$ is $k(u, \mathscr{C})$-open. Since E is a k-space, we see that $f^{-1}(V)$ is u-open.

Definition 10. A topological space E_u is said to be a k_r-*space* if each real-valued function on E_u whose restriction on each compact subset of E is continuous is actually continuous.

Clearly it follows from Theorem 16 that each k-space is a k_r-space. But the converse need not be valid. [Example 37(c)].

§41. Baire Spaces

Definition 11. (1) Let E be a topological space. A subset F of E is said to be a *nondense* (or *nowhere dense* or a *siéve*) if $(\bar{E})^{\circ} = \varnothing$.

(2) A countable union of nondense sets in a space is called a set of the *first category* or a *meager* set. The complement of a set of the first category is called a *residual* set.

(3) A set which is not of the first category is said to be of the *second category* or a *nonmeager set*.

(4) A topological space E is said to be a *Baire space* if whenever $E = \bigcup_{n=1}^{\infty} E_n$ such that E_n is closed, then for at least one n, $E_n^{\circ} \neq \varnothing$.

Remark. Clearly each singleton in each Hausdorff topological space is nondense and thus a countable Hausdorff space is always of the first category.

Theorem 17. A topological space E is a Baire space if and only if for any countable family $\{U_n\}$ of open dense subsets in E, $\bigcap_{n \geq 1} U_n \neq \varnothing$.

Proof. Let $\{U_n\}$ be a countable family of subsets of E. Put $U_n^{\circ} = E \backslash U_n$. Clearly each U_n is open and dense if each U_n° is closed and nondense. Hence $E = \bigcup_{n=1}^{\infty} U_n^{\circ}$ iff $\bigcap_{n=1}^{\infty} U_n = \varnothing$. Thus E is a Baire space iff for any sequence $\{U_n\}$ of open and dense subsets of E, $\bigcap_{n=1}^{\infty} U_n \neq \varnothing$.

Proposition 37. Every locally compact Hausdorff space E is a Baire space.

Proof. Let $\{E_i\}_{i \geq 1}$ be a sequence of open dense subsets of E. We wish to show that $U \cap (\cap_{i=1}^{\infty} E_i) \neq \varnothing$ for each open subset U of E. Since E_1 is dense, $U \cap E_1 \neq \varnothing$ and is open. Thus there exists an open set V_1 such that $\bar{V}_1$ is compact and $\bar{V}_1 \subset U \cap E_1$, since E is locally compact. By induction, we find open subsets V_n such that $\bar{V}_n$ is compact and $\bar{V}_n \subset V_{n-1} \cap E_n$ for all $n \geq 1$ where $V_0 = U$. Clearly $\{\bar{V}_n\}_{n \geq 1}$ is a decreasing sequence of compact and closed subsets of a compact space $\bar{V}_1$. Hence $\cap_{n \geq 1} \bar{V}_n \neq \varnothing$. Since $\cap_{n=1}^{\infty} \bar{V}_n \subset (\cap_{n=1}^{\infty} E_n) \cap U$, it proves the theorem.

Theorem 18 (Baire category). Every complete metric space E is a Baire space.

Proof. Suppose $E = \cup_{n=1}^{\infty} E_n$, if possible, where each E_n is a closed nondense subset. Since E is nondense, there exist $\varepsilon_1 > 0$ and $x_1 \in E$ such that $B_{\varepsilon_1}(x_1) \cap E_1 = \varnothing$. Again, E_2 being a nondense closed subset, there exists $\varepsilon_2 < \varepsilon_1/2$, $x_2 \in E$ with $B_{\varepsilon_2}(x_2) \subset B_{\varepsilon_1}(x_1)$ and $B_{\varepsilon_2}(x_2) \cap (E_2 \cup E_1) = \varnothing$. By induction, there exists a sequence $\{x_n\} \subset E$ and a sequence of positive reals $\{\varepsilon_n\}$, $\varepsilon_n \leq \varepsilon_1/2^{n-1}$ for $n \geq 1$ such that $B_{\varepsilon_n}(x_n) \subset B_{\varepsilon_{n-1}}(x_{n-1})$ and $B_{\varepsilon_n}(x_n) \cap E_i = \varnothing$ for all $1 \leq i \leq n$. Since $d(x_{n-1}, x_n) < \varepsilon_{n-1} \leq \varepsilon_1/2^{n-2}$, it follows that for $m > n$,

$$d(x_m, x_n) \leq d(x_m, x_{m-1}) + d(x_{m-1}, x_{m-2}) + \cdots$$
$$+ d(x_{n+1}, x_n) < \varepsilon_{m-1} + \varepsilon_{m-2} + \cdots + \varepsilon_n \leq \frac{\varepsilon_1}{2^{m-2}} + \cdots$$
$$+ \frac{\varepsilon_1}{2^{n-1}} \leq \varepsilon_1 \left(\sum_{i=n-1}^{m-2} \frac{1}{2^i} \right) \to 0$$

as $m, n \to \infty$. Hence $\{x_n\}$ is a Cauchy sequence. Since E is complete, there exists $x_0 \in E$ such that $\{x_n\}$ converges to x_0. By construction $x_n \notin E_n$ for all $n \geq 1$. Suppose $x_0 \in E_{n_0}$ for some n_0. Clearly $x_0 \in \bar{B}_{\varepsilon_n}(x_n)$ for all $n \geq 1$ and $\bar{B}_{n+1}(x_{n+1}) \subset B_{\varepsilon_n}(x_n)$. This means that for all $n \geq 1$, $B_{\varepsilon_n}(x_n) \cap E_{n_0} \neq \varnothing$, which is a contradiction for $n \geq n_0$. Hence $x_0 \notin E_n$ for all $n \geq 1$, i.e., $x_0 \notin E$, which is impossible. Hence E cannot be written as a countable union of closed nondense subsets and so E is a Baire space.

Corollary 9. Every Fréchet space (in particular, every Banach space or Hilbert space) (cf. Definition 17, §31) is a Baire space.

Remark. (1) There do exist nonmetric Baire spaces. For example, $R^{\aleph_1}$, where $\aleph_1$ is an uncountable cardinal number.

(2) From the above theorem, it follows that R^2 (more generally R^n, $n \geq 2$) cannot be written as a countable union of circles (respectively spheres).

Proposition 38 (Permanence properties).

(a)   An open subset of a Baire space E is a Baire space;

(b)   every Baire space is of the second category;

(c)   if E is a Baire space, F any Hausdorff space, and if f is a continuous open mapping of E onto F, then F is a Baire space.

Proof.   (a) The proof is contained in Theorem 17. Let U be an open subset of E and if $\{E_i\}$ is a sequence of open dense subsets of U, then we showed that $U \cap (\bigcap_{i=1}^{\infty} E_i) \neq \emptyset$.

(b)   Suppose a Baire space E is of the first category; then $E = \bigcup_{i=1}^{\infty} E_i$, where E_i $(i \geq 1)$ is closed and nondense. This violates the definition of a Baire space.

(c)   Let $\{F_i\}$ be a sequence of open dense subsets of F. Then by continuity of f, $f^{-1}(F_i) = E_i$ is open and by openness of f, E_i is dense. Thus $\bigcap_{i=1}^{\infty} E_i$ is dense in E, because E is a Baire space. Clearly

$$f\left(\bigcap_{i=1}^{\infty} E_i \right) \subset \bigcap_{i=1}^{\infty} f(E_i) = \bigcap_{i=1}^{\infty} F_i.$$

Since f is continuous, $f(\bigcap_{i=1}^{\infty} E_i)$ is dense in E and hence $\bigcap_{i=1}^{\infty} F_i$ is dense. Thus F is a Baire space.

Remark.   A k-space need not be a Baire space. Consider the noncomplete metric space $E = \bigcup_{n=1}^{\infty} E_n$, where $E_n \subsetneq E_{n+1}$ and E_n is closed in E. Then E is of the first category and hence E is not a Baire space. But since E is a metric space, it is a k-space. For example, consider $E = R^{(N)}$, the space of all finite sequences, then $E = \bigcup_{n=1}^{\infty} R^n$, where R^n is the n-dimensional Euclidean space.

§42. Pseudocompact Spaces

It has been noted before [Theorem 8, §37] that each continuous real function on a compact space is bounded. It is worth noting that the converse may not hold, i.e., there exists a completely regular space on which each continuous real function is bounded but the space is not compact, e.g., the space of all ordinals less than the first uncountable ordinal with the order topology. We want to study this phenomenon in this section.

Definition 12.   A Hausdorff topological space is said to be *pseudocompact* if every continuous real function on it is bounded.

Proposition 39.

(a) Every Hausdorff compact space is pseudocompact;

(b) every Hausdorff countably compact space is pseudocompact.

Proof.

(a) This follows from Theorem 8, §37;

(b) this follows also from Theorem 8, §37.

Remark. Let ω denote a countable ordinal and Ω the first uncountable ordinal. Let $X = [0, \Omega] \times [0, \omega] \setminus (\Omega, \omega)$ be the space called the Tychonoff plank. Then X is not countably compact but every continuous real function can be extended to the compact space $[0, \Omega] \times [0, \omega]$ and hence f is bounded, i.e., X is pseudocompact.

Even though the classes of countably compact and pseudocompact spaces need not coincide as the above example indicates, the following holds:

Theorem 19. Let E be a completely regular space. The following statements are equivalent:

(a) E is pseudocompact;

(b) let $\{U_i\}_{i \geq 1}$ be a strictly decreasing sequence of nonempty open sets; then $\bigcap_{i=1}^{\infty} \bar{U}_i \neq \varnothing$;

(c) let $\{P_n\}$ be an open countable covering of E; then there is a finite subfamily $\{P_{n_i}\}_{i=1}^{k}$ such that $\bigcup_{i=1}^{k} \bar{P}_{n_i} = E$.

Proof. (a) $\Rightarrow$ (b) Let $\{U_i\}$ be a strictly decreasing sequence of nonempty open sets and suppose $\bigcap_{i=1}^{\infty} \bar{U}_i = \varnothing$. Let $x \in E$; then each neighborhood U of x meets only a finite number of $\bar{U}_i$'s. For if $U \cap \bar{U}_i \neq \varnothing$ for all i, then $x \in \bar{U}_i$ for all $i \geq 1$, a contradiction. Since E is completely regular, there is a continuous function $f_n : E \to \mathbb{R}$, $n \geq 1$, such that $f_n(y_n) = 1$ for some $y_n \in U_n$ for all $n \geq 1$ and $f_n(E \setminus U_n) = 0$. Since each x can belong to only a finite number of $\bar{U}_n$'s, the function $f(x) = \sum_{i=1}^{\infty} i f_i(x)$ is well-defined and continuous but not bounded. Hence E is not pseudocompact.

(b) $\Rightarrow$ (c) Let $\{P_n\}$ be an open countable covering of E. Let $U_n = E \setminus \bigcup_{i=1}^{n} \bar{P}_i$ for $n \geq 1$. Clearly

$$\bar{U}_n = \overline{\left(\bigcup_{i=1}^{n} \bar{P}_i \right)^c} = \overline{\bigcap_{i=1}^{n} \bar{P}_i^{\,c}} \subset \bigcap_{n=1}^{n} \overline{\bar{P}_i^{\,c}} \subset \bigcap_{i=1}^{n} P_i^{\,c}.$$

If no U_n's are empty, (b) implies

$$\varnothing \neq \bigcap_{n=1}^{\infty} \bar{U}_n \subset \bigcap_{i=1}^{\infty} P_i^{c}.$$

This implies $E \neq \bigcup_{n=1}^{\infty} P_n$, contrary to assumption. Hence there is a finite subfamily of P_n's whose closures cover E.

(c) $\Rightarrow$ (a) Let f be a continuous real function on E. Putting $P_n = \{x \in E: |f(x)| < n\}$ for $n \geq 1$, we obtain an open covering of E. By (c), then there is an integer k such that $E = \bigcup_{n=1}^{k} \bar{P}_n$. But this implies that f is bounded and hence E is pseudocompact.

Theorem 20. Let E be a completely regular space which is weakly normal (i.e., any two disjoint closed subsets such that one of them is countable can be separated by disjoint open neighborhoods). Then E is countably compact if and only if E is pseudocompact.

Proof. We have shown [Theorem 8(b), §37] that each countably compact space is pseudocompact. To show the converse, assume E is pseudocompact but not countably compact. But then there is a countable infinite set $\{x_n\}$ which has no limit point or, in other words, $A = \{x_n\}$ is discrete and closed. Since E is completely regular and therefore regular, we construct by induction a sequence $\{U_n\}$ of open sets such that $\bar{U}_n \cap \bar{U}_m = \varnothing$ for $n \neq m$ and $A \cap U_n \neq \varnothing$ for each $n \geq 1$. We choose $U_0 = \varnothing$. Suppose $\{U_i\}_{i=0}^{n}$ have been defined such that $\{\bar{U}_i\}_{i=1}^{n}$ are pairwise disjoint and $A \cap U_i \neq \varnothing$ for $i = 1, \ldots, n$. Put $A_n = A \backslash \bigcup_{i=1}^{n} U_i$. If A_n is finite, the proofs is clear. If A_n is infinite, let $a, b \in A_n$ and using regularity, we obtain open sets V and W such that

$$a \in V \subset \bar{V} \subset E \backslash \left(\bigcup_{i=0}^{n} \bar{U}_i \cup \{b\} \right)$$

and

$$b \in W \subset \bar{W} \subset E \backslash \left(\bigcup_{i=1}^{n} \bar{U}_i \cup \bar{V} \right)$$

Put $U_{n+1} = V$ or W if $\bar{V} \backslash A$ is finite or otherwise. Now take $B = (\bigcup_{i=1}^{\infty} U_i)^c$. Then by assumption there are disjoint open sets P and Q such that $A \subset P$ and $B \subset Q$. Now it is clear that the open covering $\{U_n\} \cup Q$ has no finite subfamily whose closures cover E, contradicting (c) of Theorem 19.

§43. Paracompact Spaces

In order to introduce paracompact spaces, we recall some notions about coverings on which the definition of a paracompact space depends.

Definition 13. Let $\mathcal{P} = \{P_\alpha: \alpha \in A\}$ and $\mathcal{Q} = \{Q_\beta: \beta \in B\}$ be two coverings of a space E. $\mathcal{P}$ is said to *refine* $\mathcal{Q}$ if for each P_α there is a Q_β such that $P_\alpha \subset Q_\beta$. We write $\mathcal{P} < \mathcal{Q}$ if $\mathcal{P}$ is a refinement of $\mathcal{Q}$. (See §10.)

Clearly every subcovering of a covering is a refinement.

Proposition 40. The relation $<$ on the set of coverings of a set E defines a preorder.

Proof. Easy.

In general, $<$ is not a partial ordering since the symmetric law $P < Q$ and $Q < P \Leftrightarrow P = Q$ may fail to hold. For example, $P_n = \{(-\infty, n): n \geq 1\}$ and $Q_n = \{(-\infty, n + \frac{1}{3}): n \geq 1\}$ refine each other without being equal.

Definition 14. A covering $P = \{P_\alpha: \alpha \in A\}$ of a topological space E is said to be *neighborhood-finite* if each point has an open neighborhood which meets only a finite number of the P_α (this is the same as the definition of locally finite given in §10).

Remark. Clearly every finite covering is neighborhood-finite. If $\{P_\alpha: \alpha \in A\}$ is neighborhood-finite, so is $\{\bar{P}_\alpha: \alpha \in A\}$ and for any $B \subset A$, $\cup \{\bar{P}_\alpha: \alpha \in B\}$ is closed.

Definition 15. A topological space is said to be *paracompact* if each open covering of it has a refinement which is a neighborhood-finite covering.

Proposition 41.

(a) Every compact space is paracompact.
(b) Every discrete space is paracompact.

Proof. (a) This follows from the fact that in a compact space each open covering is reducible to a finite open subcovering and hence it is neighborhood-finite.

(b) In a discrete space E each open covering has the refinement in which each singleton is an open set and therefore the neighborhood consisting of the point itself meets itself and therefore E is a paracompact space.

Proposition 42. Every Hausdorff paracompact space E is normal and hence regular.

Proof. First, to show that E is regular, let C be a closed subset of E and $x \notin C$. Since E is Hausdorff, each $y \in C$ and x have open neighborhoods U_y and V_y such that $U_y \cap V_y = \varnothing$. Clearly $\{U_y : y \in C\} \cup (E \backslash C)$ is an open covering of E. Since E is paracompact, there is a refinement by a neighborhood-finite open covering, $\{P_\alpha : \alpha \in A\} \cup Q$, where $Q \supseteq E \backslash C$. Then $U = \bigcup_{\alpha \in A} P_\alpha$ is open and contains C. Since each neighborhood of $z \in \bar{U}$ meets only a finite number of P_α's, we see that $z \in \cup \bar{P}_\alpha$ and hence $\bar{U} = \bigcup_{\alpha \in A} \bar{P}_\alpha$. Since $x \notin \bar{U}_y \supset \bar{P}_\alpha$ for each $\alpha \in A$, we see that $y \notin \bar{U}$. Thus $E \backslash \bar{U}$ and U are disjoint open neighborhoods of x and C, respectively, and hence E is regular.

Now to show that E is normal, let A, B be two disjoint closed subsets. By regularity, there are disjoint open neighborhoods of $a \in A$ (for each $a \in A$) and B. Let U_a be a neighborhood of $a \in A$ such that $\bar{U}_a \subset E \backslash B$. Thus by the same argument as before we obtain disjoint open neighborhoods of A and B. This completes the proof.

Sometimes the following characterization of paracompact spaces is very useful.

Theorem 21. Let E be a regular space. The following statements are equivalent:

(a) E is paracompact.

(b) Each open covering of E has an open refinement that can be written as at most a countable collection of neighborhood-finite families of open sets (i.e., *countable at infinity*).

(c) Each open covering of E has a neighborhood-finite refinement. (Each set in the new covering is necessarily neither open nor closed.)

(d) Each open covering of E has a neighborhood-finite refinement in which each set is closed;

Proof. (a) $\Rightarrow$ (b) Obvious.

(b) $\Rightarrow$ (c) Let $\{P_\alpha : \alpha \in A\}$ be an open covering of E. Then by assumption (b) there is an open refinement $\{Q_\alpha^{(n)} : \alpha \in A, n \geq 1\}$ where each $\{Q_\alpha^{(n)} : \alpha \in A\}$ for a fixed n is a neighborhood-finite family. Put $U_n = \bigcup_{\alpha \in A} Q_\alpha^{(n)}$; then $\{U_n : n \geq 1\}$ is an open covering of E. Put $Q_n = U_n \backslash \bigcup_{i<n} U_i$. Clearly $\{Q_n : n \geq 1\}$ is a refinement of $\{U_n : n \geq 1\}$, and it is a covering since each $x \in Q_{n(x)}$, where $n(x)$ is the first n for which

$x \in U_i$. Also $\{Q_n: n \geq 1\}$ is neighborhood-finite. But then $\{Q_n \cap Q_\alpha^{(n)}\}$ is the desired neighborhood-finite refinement of $\{P_\alpha: \alpha \in A\}$.

(c) $\Rightarrow$ (d) Let $\{P_\alpha: \alpha \in A\}$ be an open covering of E. For each $x \in E$, choose P_α such that $x \in P_\alpha$, and then find an open set V_x such that $x \in V_x \subset \bar{V}_x \subset P_\alpha$, because E is regular. Then $\{V_x: x \in E\}$ is an open covering and so it has a refinement $\{Q_x: x \in E\}$ which is neighborhood-finite. Since $\bar{Q}_x \subset \bar{V}_x \subset P_\alpha$ for each $x \in E$, it follows that $\{\bar{Q}_x: x \in E\}$ is the desired neighborhood-finite refinement.

(d) $\Rightarrow$ (a) Let $\mathscr{P} = \{P_\alpha: \alpha \in A\}$ be an open covering of E. By (d), there is a neighborhood-finite refinement $\{Q_\beta: \beta \in B\}$ such that each Q_β is closed. Then for each $x \in E$ there is an open neighborhood U_x which meets only a finite number of sets in $\{Q_\beta\}$. Clearly $\{U_x: x \in E\}$ is an open covering of E and hence by (d) there is a refinement by a closed covering $\mathscr{Q}$ which is neighborhood-finite. For $Q \in \mathscr{Q}$, put $Q' = \{\cup G: G \in \mathscr{Q}$ and $G \cap Q = \varnothing\}^c$. Then $Q \subset Q'$ and Q' is open because $\mathscr{Q}$ is closed neighborhood-finite. Clearly a set in $\mathscr{Q}$ intersects Q' if it does Q; moreover, $\mathscr{Q} < \mathscr{P}$. For each $Q \in \mathscr{Q}$ choose $P_\alpha' \in \mathscr{P}$ such that $Q \subset P_\alpha'$. Then $\mathscr{E} = \{Q' \cap P_\alpha': Q \in \mathscr{Q}\}$ is a neighborhood-finite open refinement of $\mathscr{P}$. Hence by definition E is paracompact.

Theorem 22. Each metric space (E, d) is paracompact.

Proof. Let $\mathscr{P} = \{P_\alpha: \alpha \in A\}$ be an open covering of E. Since each set can be well-ordered, we assume that the indexing set A is well-ordered. By transfinite induction, for each positive integer n and each $\alpha \in A$, we define

$$U_\alpha^n = E \backslash B_n\Big(P_\alpha \backslash \bigcup_{\beta < \alpha} U_\beta^n\Big),$$

where $B_n(T) = \{y \in E: d(y, T) < 1/2^n\}$ for any nonempty subset T of E.

Let $x \in E$. There exists a minimum $\lambda \in A$ such that $x \in P_\lambda$. Since P_λ is open, there is an integer $n > 0$ such that $B_n(x) \subset P_\lambda$. If $x \notin U_\lambda^n$, then $B_n(x) \not\subset P_\lambda \backslash \bigcup_{\beta < \lambda} U_\beta^n$ by the definition of U_λ^n. Since $B_n(x) \subset P_\lambda$, we have $B_n(x) \cap \bigcup_{\beta < \lambda} U_\beta^n \neq \varnothing$ and so $B_n(x) \cap U_\alpha^n \neq \varnothing$ for some $\alpha < \lambda$. But then for this $\alpha < \lambda$,

$$x \in B_n(U_\alpha^n) = B_n\Big(E \backslash B_n(P_\alpha \backslash \bigcup_{\beta < \alpha} U_\beta^n)\Big) \subset P_\alpha \backslash \bigcup_{\beta < \alpha} U_\beta^n \subset P_\alpha,$$

which contradicts the definition of λ. Hence $x \in U_\lambda^n$ and so $\{U_\alpha^n\}$ is a covering of E. Let $V_\alpha^n = E \backslash B_{n+3}(U_\alpha^n)$ and $W_\alpha^n = B_{n+2}(U_\alpha^n)$ for $n > 1$ and $\alpha \in A$. Then $V_\alpha^n \subset W_\alpha^n$. Also, for $\alpha < \beta$ and for all β

$$B_n(U_\beta^n) = B_n\Big(E \backslash B_n(P_\beta \backslash \bigcup_{\alpha < \beta} U_\alpha^n)\Big) \subset P_\alpha \backslash \bigcup_{\alpha < \beta} U_\alpha^n \subset E \backslash U_\alpha^n.$$

This shows that for β and $\alpha < \beta$, $B_n(U_\beta{}^n) \cap U_\alpha{}^n = \varnothing$. Hence for all $\alpha \neq \beta$, $d(U_\alpha{}^n, U_\beta{}^n) \geq 1/2^n$; whence we have $d(W_\alpha{}^n, W_\beta{}^n) \geq 1/2^{n+1}$ for $\alpha \neq \beta$. Since each $V_\alpha{}^n$ is closed and for different α's the distance between the $V_\alpha{}^n$ is uniform and positive, $V^n = \bigcup_\alpha V_\alpha{}^n$ is closed.

Now to define the elements of our desired covering, we put $Q_\alpha{}^n = W_\alpha{}^n \setminus \bigcup_{k<n} V^k$ for each $n \geq 1$ and $\alpha \in A$. Clearly each $Q_\alpha{}^n$ is open. Let $x \in E$. Since $\{V_\alpha{}^n\}$ is a covering, $x \in V_\alpha{}^n$ for some α and least integer n. Whence

$$x \in V_\alpha{}^n \setminus \bigcup_{k<n} \bigcup_\beta V_\beta{}^k = V_\alpha{}^n \setminus \bigcup_{k<n} V^k \subset W_x{}^k \setminus \bigcup_{k<n} V^k = Q_\alpha{}^n.$$

This shows that $\{Q_\alpha{}^n\}$ is a covering of E and

$$Q_\alpha{}^n \subset W_\alpha{}^n = B_{n+2}(U_\alpha{}^n) \subset B_n(U_\alpha{}^n)$$
$$= B_n\big(E \setminus B_n(P_\alpha \setminus \bigcup_{\beta<\alpha} U_\beta{}^n)\big) \subset P_\alpha \setminus \bigcup_{\beta<\alpha} U_\beta{}^n \subset P_\alpha.$$

Thus $\{Q_\alpha{}^n\}$ is a refinement of $\{P_\alpha\}$. To end the proof, we must show that $\{Q_\alpha{}^n\}$ is locally finite. Let $x \in E$ and suppose $x \in U_\alpha{}^n$. Then

$$B_{n+3}(x) \subset B_{n+3}(U_\alpha{}^n) \subset E \setminus B_{n+3}(U_\alpha{}^n) = V_\alpha{}^n \subset V^n,$$

and so $B_{n+3}(x) \cap Q_\alpha{}^k = \varnothing$ for all $k > n$ and $\alpha \in A$. Since $\delta\big(B_{n+3}(x)\big) < 1/2^{n+1}$ and $d(W_\alpha{}^n, W_\beta{}^n) \leq 1/2^{n+1}$, it follows that for a fixed $k \leq n$, $B_{n+3}(x)$ can intersect at most one $V_\alpha{}^k$ and hence for a fixed $k \leq n$, $B_{n+3}(x)$ intersects at most one $Q_\alpha{}^n$. This proves that $\{Q_\alpha{}^n\}$ is locally finite; hence E is paracompact.

Remark. There exists a paracompact space which is not metric (e.g., Exercise 36).

Definition 16. A topological space is called a *countably paracompact* space if each countable open covering has a countable open neighborhood-finite refinement.

Proposition 43. Every paracompact space is countably paracompact.

Proof. Simple.

Proposition 44. A topological space E is countably paracompact if and only if for every decreasing sequence of nonempty closed sets $\{P_n\}$,

$n \geq 1$, such that $\bigcap_{n=1}^{\infty} P_n = \emptyset$ there exist open sets $\{Q_n\}$ such that $Q_n \supset P_n$ for $n \geq 1$ and $\bigcap_{n=1}^{\infty} Q_n = \emptyset$.

Proof. For the "only if" part, observe that $\mathscr{P} = \{E \backslash P_n, n \geq 1\}$ is an open covering of E, and let there be an open neighborhood-finite refinement $\{R_n : n \geq 1\}$ of $\mathscr{P}$. Put $\mathscr{Q} = \{Q_n = \bigcup_{i=n}^{\infty} R_i, n \geq 1\}$. It is easy to check that the family $\mathscr{Q}$ satisfies the desired conditions.

For the "if" part, suppose $\{G_n : n \geq 1\}$ is an open covering. Let $P_n = E \backslash \bigcup_{i=1}^{n} G_i$. Now there is a family $\{Q_n : n \geq 1\}$ such that $Q_n \supset P_n$ and $\cap \bar{Q}_n = \emptyset$ (cf. Theorem 19). Then $\{Q_n \cap G_{n+1}, n \geq 1\} \cup G_1$ is the required refinement and hence E is countably paracompact.

Proposition 45. A closed subspace of a paracompact (respectively countably paracompact) is paracompact (respectively countably paracompact).

Proof. Let F be a closed subspace of a paracompact space E. Let $\{P_\alpha : \alpha \in A\}$ be an open covering of F, where $P_\alpha = V_\alpha \cap F$, in which V_α is open in E. Clearly $\{P_\alpha : \alpha \in A\} \cup (E \backslash F)$ is an open covering of E. Hence there is a neighborhood-finite open covering $\{Q_\beta : \beta \in B\} \cup (E \backslash F)$ which refines it. But then $\{Q_\beta : \beta \in B\}$ is a neighborhood-finite open covering which refines $\{P_\alpha : \alpha \in A\}$ and the first part is proved. The proof of the other part is similar.

Remark. It is not difficult to show that even an F_σ-set (i.e., $F = \bigcup_{n=1}^{\infty} F_n$ with F_n closed) of a paracompact space is paracompact.

Definition 17. Let E be a Hausdorff space. A family $\{f_\alpha : \alpha \in A\}$ of continuous real functions is called a *partition of unity* on E if

(i) $U_\alpha = \mathrm{cl}\{x \in E : f_\alpha(x) \neq 0\}$ form a neighborhood-finite closed covering of E.

(ii) For each $x \in E$, $\sum_\alpha f_\alpha(x) = 1$.

Remark. Observe that the sum in (ii) makes sense because x lies only in finite number of U_α's and hence all $f_\alpha(x) = 0$ except for only a finite number of α's.

Definition 18. A partition of unity $\{f_\alpha : \alpha \in A\}$ is said to be subordinated to an open covering $\{P_\alpha : \alpha \in A\}$ of a topological space E if $\mathrm{cl}\{x \in E : f_\alpha(x) \neq 0\} \subset P_\alpha$ for each α.

Theorem 23. (a) If E is a Hausdorff paracompact space, then for each open covering $\mathscr{P} = \{P_\alpha \colon \alpha \in A\}$ of E, there is a partition of unity to which $\mathscr{P}$ is subordinated.

(b) If E is a Hausdorff countably paracompact space, then for each countable open covering $\{P_n \colon n \geq 1\}$, there is a countable partition of unity to which $\{P_n\}$ is subordinated.

Proof. The proof of (b) being similar, we only prove (a). Since E is paracompact, there exists a neighborhood-finite refinement $\{Q_\alpha\}$ of $\mathscr{P}$ which is an open covering of E and such that $\bar{Q}_\alpha \subset P_\beta$ for some β. Now again shrink the Q_β's to find a neighborhood-finite open covering $\{G_\alpha\}$ such that $\bar{G}_\alpha \subset Q_\alpha$. Since E is normal (Proposition 42), there exists a continuous function $f_\alpha \colon E \to [0, 1]$ such that $f_\alpha(\bar{G}_\alpha) = 1$ and $f_\alpha(E \setminus Q) = 0$ for each α. Clearly $\mathrm{cl}\{x \in E \colon f_\alpha(x) \neq 0\} \subset P_\alpha$. Since $\{\bar{G}_\alpha\}$ is a neighborhood-finite covering, it follows that $f = \sum_\alpha f_\alpha(x)$ is well-defined on E, continuous and $\neq 0$ on E. Now let us define $g_\alpha(x) = f_\alpha(x)/f(x)$. Then $\{g_\alpha\}$ is the required partition of unity, since $\sum_\alpha g_\alpha(x) = 1$ on E.

Proposition 46. A Lindelöf space is paracompact if and only if it is regular.

Proof. Since every paracompact space is normal, hence regular, the "only if" part is immediate. Now suppose E is a Lindelöf regular space. Let $\{P_\alpha\}$ be any open covering of E. The Lindelöf property says that there exists a countable open subcovering $\{P_{\alpha_i} \colon i \geq 1\}$ of E. Then $\{P_{\alpha_i}\} < \{P_\alpha\}$ and $\{P_{\alpha_i}\}$ decomposes into countably many neighborhood-finite families, each consisting of a single set P_{α_i}. Hence by Theorem 21, the "if" part follows.

Proposition 47. Each separable paracompact space E is Lindelöf.

Proof. Let $\{P_\alpha \colon \alpha \in A\}$ be an open covering of E and let $\{Q_\beta \colon \beta \in B\}$ be an open neighborhood-finite refinement. We may ignore Q_β if $Q_\beta = \varnothing$ and thus $\{Q_\beta\}$ is at most countable since E is separable. Now for each Q_β choose a P_β containing it; then we have a countable open covering of E and hence E is Lindelöf.

Remark. A paracompact Lindelöf space may fail to be separable, e.g., $[0, \Omega]$, where Ω is the first uncountable ordinal, is a Lindelöf paracompact space which is not separable.

§44. Compactifications

We have seen before [Theorem 2, §26] that certain topological spaces can be uniformized and also certain topological spaces can be metrized. Indeed, it is not true that every topological space is uniformizable and every topological space is metrizable.

Since compact spaces play an important role in topology as well as in analysis, it is of great interest to know whether or not a certain class of topological spaces can be compactified.

There are several kinds of compactifications: one-point compactification, Stone–Čech compactification, Wallman's compactification, etc. We shall mainly deal with the first two kinds of compactifications in the sequel, since these are the most useful and widely used compactifications. Furthermore, the one-point compactification is the smallest whereas the Stone–Čech compactification is the largest.

The motivation for a one-point compactification is provided from the real line. We often add "$\pm \infty$" to the real line to make it compact, or just one point to the complex plane to make it homeomorphic with the sphere by stereographic projection. In these examples, the addition of only one or two points makes the space compact. For such compactifications only locally compact spaces are suitable. But in some other cases one has to add many more points to get a compactification. For such compactifications, completely regular spaces are suitable.

Definition 19. A topological space $\tilde{E}$ is said to be a *compactification* of a topological space E if:

(i) $\tilde{E}$ is a compact Hausdorff space;

(ii) E is homeomorphic to a dense subset of $\tilde{E}$.

Since E is homeomorphic to a dense subset of $\tilde{E}$, it is often considered that E is a dense subset of $\tilde{E}$.

Theorem 24 (Alexandroff). Each Hausdorff locally compact space E can be compactified to a unique (up to a homeomorphism) compact Hausdorff space $\tilde{E}$ such that $\tilde{E} \setminus E$ is a single point.

Proof. Let "∞" be an element which is not in E. Set $\tilde{E} = E \cup \{\infty\}$. We put a topology on E as follows: Each open set of E is open in E and the complement of each compact subset in E is also open in $\tilde{E}$. It is easy to see that this family defines a topology on $\tilde{E}$ and that E is a subspace of $\tilde{E}$. Further, this topology is Hausdorff: For any pair $x \neq y$ in E, indeed there are disjoint open neighborhoods in $\tilde{E}$ and hence in E. Let $x \neq \infty$. Since E is locally compact, there is an open neighborhood U of x such that $\bar{U}$ is compact. Thus $\tilde{E} \backslash \bar{U}$ is an open neighborhood of ∞ which is disjoint from U. Now to show that $\tilde{E}$ is compact, let $\{P_\alpha : \alpha \in A\}$ be an open covering of $\tilde{E}$. Then there is an open set P_{α_0} containing ∞. Now clearly $\{P_\alpha : \alpha \in A, \alpha \neq \alpha_0\}$ is an open covering of $\tilde{E} \backslash P_{\alpha_0}$ which is compact. Hence there is a finite subcovering $\{P_{\alpha_i} : i = 1, \ldots, n, \alpha \neq \alpha_0\}$ of $\tilde{E} \backslash P_{\alpha_0}$. But then $\{P_{\alpha_i} : i = 0, \ldots, n\}$ covers $\tilde{E}$ and hence $\tilde{E}$ is compact. By definition $\tilde{E} \backslash E = \{\infty\}$.

Now to show the uniqueness (up to homeomorphism), assume there is another one-point compactification $\tilde{F}$ of E, and $\tilde{F} \backslash E = \{w\}$ is a singleton. Let f be a map of $\tilde{E}$ onto $\tilde{F}$ as follows: $f(x) = x$ for all $x \in E$ and $f(\infty) = w$. Clearly f is $1 : 1$, onto, and a homeomorphism on E. Let U_∞ be a neighborhood of ∞ in $\tilde{E}$. Then $\tilde{E} \backslash U_\infty$ is a compact subset of E and hence by continuity of f on E, $f(\tilde{E} \backslash U_\infty)$ is compact in $\tilde{F}$. But f being $1 : 1$ and onto, we have $f(\tilde{E} \backslash U_\infty) = f(\tilde{E}) \backslash f(U_\infty) = \tilde{F} \backslash f(U_\infty)$, which implies that $f(U_\infty)$ is an open neighborhood of w. Thus f is an open mapping. Since f and f^{-1} are symmetric, it follows that f^{-1} is also open or, in other words, f is continuous. Thus f is a homeomorphism of $\tilde{E}$ onto $\tilde{F}$.

Now we show that for one-point compactification only locally compact Hausdorff spaces are suitable.

Proposition 48. Let E be a Hausdorff space. Then E is locally compact if and only if E is an open subset of any given compactification.

Proof. Let $\tilde{E}$ be a compactification of E. If E is open in $\tilde{E}$, then $\bar{E}$ in $\tilde{E}$ is compact. Since every open subset of a Hausdorff compact space is locally compact, E is locally compact. Conversely, assume E is locally compact. We have to show that the mapping of E into $\tilde{E}$ by $f(x) = x$ for $x \in E$ is an open mapping. But this has been shown in Theorem 24 and the proof is complete.

Proposition 49. Let $\tilde{E}$ be the one-point compactification of a Hausdorff locally compact space E. Then $\tilde{E}$ is metrizable if and only if E is second countable.

Proof. *"Only if" part*: If $\tilde{E}$ is metrizable then $\tilde{E}$ is separable (Proposition 15, §37) (because it is compact) and hence second countable. But since a subspace of a second countable space is second countable we have the "only if" part.

"If" part: Assume now that E is second countable. Let $\{U_n\}$ be a countable base of open sets in E. We may assume that $\bar{U}_n$ is compact, because E is locally compact. But the sequence $\{\tilde{E} \setminus \bar{U}_n\}$ forms a base of open neighborhoods of ∞ in $\tilde{E}$. Thus $\tilde{E}$ is second countable, because the sequence $\{U_n, \tilde{E} \setminus \bar{U}_n : n \geq 1\}$ forms a base of open sets in $\tilde{E}$. Now $\tilde{E}$, being a compact Hausdorff space, is normal, hence by a metrization theorem (§32, Theorem 13), it is metrizable.

Very often we require the extension of a continuous map from E to $\tilde{E}$. The following proposition deals with such a situation:

Proposition 50. Let F be a Hausdorff space, E a Hausdorff locally compact space, and $\tilde{E}$ its one-point compactification. Let f be a continuous mapping of E into F. Then f can be extended to a continuous mapping $\tilde{f}$: $\tilde{E} \to F$ if and only if the filter base of the images under f of open neighborhoods of ∞ converges.

Proof. If $\tilde{f}$ is the continuous extension of f, then clearly $\{\tilde{E} \setminus C, C \text{ compact in } E\}$ is a neighborhood filter at ∞ which converges to ∞ and hence its image under the continuous map $\tilde{f}$ is also convergent.

Conversely, if the filter base of neighborhoods at ∞ is mapped into a convergent filter base in F, then $\tilde{f}$ can be defined as follows: $\tilde{f}(x) = f(x)$ for $x \in E$ and $\tilde{f}(\infty) = \lim f(\mathcal{U}_\infty)$, where $\mathcal{U}_\infty$ is the neighborhood filter at ∞. Clearly $\tilde{f}$ is continuous at ∞ and so it is continuous everywhere and $\tilde{f}$ coincides with f on E. The proof in now complete.

Now we take up the Stone–Čech compactification, which was declared by fiat to be the largest compactification. In order to demonstrate this, we require a space where lots of continuous functions are available.

Let E be a completely regular space. Then for each closed subset C and $x \notin C$, there is a continuous function $f : E \to [0, 1]$ such that $f(x) = 0$ and $f(C) = 1$. Let $I = [0, 1]$ and consider I^X, the product space. X is a set of functions on E. By Tychonoff's theorem (§35, Theorem 3), I^X endowed with the product topology is a compact Hausdorff space. Now for each $x \in E$, consider $x \to (f(x)_f \in I^X$, where $f \in X$ is a continuous functions on E with values in I. For the embedding of E into I^X, we require X to satisfy the conditions of Proposition 23, §32. Here we take $X = C(E, I) = \{f : E \to [0, 1],$ continuous$\}$.

Proposition 51. Let E be a completely regular space. Then the mapping $\varphi : x \to (f(x))_f$ of E into $I^{C(E,I)}$ is a homeomorphism into.

Proof. φ is $1:1$, for if $\varphi(x) = \varphi(y)$, then $f(x) = f(y)$ for all $f \in C(E, I)$. This shows $x = y$. For if $x \neq y$ then, E being a completely regular space, there is a continuous function $f \in C(E, I)$ such that $f(x) \neq f(y)$, a contradiction. φ is continuous, because $p_f(\varphi(x)) = f(x)$ is continuous on E in view of Proposition 5, §22. To show that φ is open onto $\varphi(E)$, let U be an open set in E. Since E is completely regular (and hence regular) for each $x \in U$, there exists an $f \in C(E, [0, 1])$ such that $f(x) = 1$ and $f(E \setminus U) = 0$. Hence $x \in f^{-1}(0, 1] \subset U$. Thus the family $\{f^{-1}(0, 1]: f \in C(E, I)\}$ is a base of the topology of E. Clearly

$$\varphi(f_0^{-1}(0, 1]) = \{\{f(x)\}_f \in I^{C(E, I)} \cap \varphi(E): f_0(x) > 0\}$$

shows that φ is open.

Definition 20. Let E be a completely regular space and $\varphi: E \to I^{C(E, I)}$ be its embedding in the compact Hausdorff space, as shown above. Then $\overline{\varphi(E)} = \beta E$ is called the *Stone–Čech compactification* of E.

Remark. If E is a compact Hausdorff space, then clearly $E \approx \varphi(E)$ $= \overline{\varphi(E)} = \beta E$ because $\varphi(E)$, being compact, is closed.

Theorem 25 (Stone–Čech). Let E be a completely regular space and βE its Stone–Čech compactification. Let F be a compact Hausdorff space. Then:

(i)   Each continuous mapping $f: E \to F$ has a unique extension to a continuous mapping $\tilde{f}: \beta E \to F$ such that $\tilde{f} \circ \varphi = f$.

(ii)   Any compactification $\tilde{E}$ of E having the property (i) is homeomorphic with βE.

(iii) βE is the largest compactification in the sense that any other compactification of E is a quotient space of βE.

Proof. (i)   Let $\varphi: E \to \beta E$. Obviously $\psi: F \to \beta F \approx F$ is a homeomorphism. If $f: E \to F$ is a continuous map, then φ induces a mapping $\tilde{\varphi}$ from $C(F) \to C(E)$, where $\tilde{\varphi}(g)(x) = g(f(x))$. It is easy to check that $\tilde{\varphi}$ is a continuous mapping of $C(F)$ into $C(E)$ with the topologies induced from I^F and I^E respectively. This in turn induces a mapping from βE to βF, which we denote by $\varphi_0: \beta E \to \beta F$. Thus the desired extension $\tilde{f}$ is given by $\tilde{f} = \psi^{-1} \circ \varphi_0$, which, being the composition of continuous maps, is continuous. Since E is dense in βE, $\tilde{f}$ is clearly unique.

(ii)   Let $\tilde{E}$ be any compactification of E. Then there is a homeomorphism (into) $g: E \to \tilde{E}$ such that $g(E)$ is dense in $\tilde{E}$. We can thus regard E as a subspace of $\tilde{E}$ and βE. If i is the identity map: $E \to E \subset \tilde{E}$, then by (i) there is a unique continuous map $\tilde{f}: \beta E \to \tilde{E}$ such that $\tilde{f} \mid E = i$. Similarly there is a unique continuous map $\tilde{g}: \tilde{E} \to \beta E$ such that $\tilde{g} \mid E = i^{-1} = i$. Since E is dense in both $\tilde{E}$ and βE, from $\tilde{f} \circ (\tilde{g} \mid E) = i$ and $(\tilde{g} \mid E) \circ \tilde{f} = i^{-1} = i$, we obtain $\tilde{f} \circ \tilde{g} = i_{\tilde{E}}$, $\tilde{g} \circ \tilde{f} = i_{\beta E}$. Thus both $\tilde{f}$ and $\tilde{g}$ are homeomorphisms.

(iii)   Let $\tilde{E}$ be another compactification of E. By (i), there is a continuous extension mapping $\tilde{f}: \beta E \to \tilde{E}$, extending the identity $i: E \to E \subset \tilde{E}$. Since βE is compact, $\tilde{f}(\beta E)$ is closed and contains the dense subset $E \subset \tilde{E}$. Thus $\tilde{f}$ is onto. Since $\tilde{f}$ is closed, we have $\tilde{E}$ homeomorphic to $\beta E / K(f)$, where $K(f)$ is the set of closed equivalence relations (Proposition 8, §23).

Corollary 10.   If E, F are homeomorphic completely regular spaces, then their Stone–Čech compactifications βE and βF are also homeomorphic.

Examples and Exercises

1. Show that on a metric space pseudocompactness coincides with compactness.

2. Show that a precompact metric space is separable.

3. If A is a nonempty compact subset of a metric space (X, d), then $\delta(A) = d(x, y)$ for some points $x, y \in A$.

4. If (X, d) is a metric space and A, B nonempty compact subsets, then there exist points $x \in A$ and $y \in B$ such that

$$d(A, B) = d(x, y).$$

5. If (X, d) is a metric space, A a compact and B a closed subset of X, then $d(A, B) > 0$.

6. Let $H = \{\{x_n\}$ real sequences: $0 \le x_n \le 1/n\}$. Then H is a metric space with the metric

$$d(\{x_n\}, \{y_n\}) = \left[\sum_{n=1}^{\infty} (x_n - y_n)^2 \right]^{1/2}$$

H is called the *Hilbert cube*. Show that H is a compact separable, hence second countable metric space. Show that H is isometric with a nowhere dense subset of itself.

7. Show that a metric space is compact if and only if it is bounded in every equivalent metric.

8. Show that every MB-space is separable.

9. Show that every complete metric space is of the second category.

10. Show that every countable complete metric space contains an infinite number of isolated points.

11. Show that every subset of a first category set is a first category set and the countable union of sets of the first category is a set of the first category.

12. Show that $D(I)$, the space of all differentiable real functions on $I = [0, 1]$, is a set of the first category in $C(I)$, the space of all continuous functions. (Hence conclude the existence of a continuous function which is not differentiable.)

13. Show that $C(I)$ is not locally compact.

14. Show that $[0, \Omega)$, the set of all ordinals less than the first uncountable ordinal Ω, has a unique compactification. (See Exercise 17, Chapter III.)

15. Describe the Stone–Čech compactification βN of the positive integers N. What is the cardinality of $\beta N \setminus N$? (See Exercise 32 below.)

16. A Hausdorff space is compact iff it is both countably compact and each open covering has a point finite open refinement. (This statement defines a *metacompact* space).

17. A paracompact space is countably compact iff it is compact.

18. A Lindelöf space is countably compact iff it is compact.

19. If X is a compact Hausdorff space, Y a Hausdorff paracompact space, then $X \times Y$ is paracompact.

20. Show that the product of two k-spaces is not necessarily a k-space. It is, however, true if both of them are first countable or one of them is locally compact.

21. Show that a regular space which is the countable union of compact spaces is paracompact. And so is a locally compact Hausdorff space which is σ-compact or second countable. Also show that a locally compact second countable space is a topological sum of locally compact spaces which is countable at infinity. If $f: X \to X$ is continuous, where X is compact, then there is a closed nonempty set $A \subset X$ such that $f(A) = A$.

22. Show that an isometry φ of a metric compact space X into itself is onto.

23. Let E be a discrete space. Then in βE, the closure of each open set is open.

24. For a completely regular space E, βE is connected iff E is connected.

25. If $f: E \to F$ is continuous, open, and onto and E is a Baire space, then so is F.

26. Show that a completely regular pseudocompact space is a Baire space.

27. There exists a completely regular space which is not a k-space.

28. Let A be an uncountable set of indices and R, the real line. Show that R^A is not a k-space.

29. Discuss some properties of UB-spaces as defined below: A Hausdorff
     uniform (metric) space is defined here to be an UB-space (respectively
     PB-space) in which each closed totally bounded or precompact subset
     is compact. Show that each complete uniform (in particular, complete
     metric) space is an UB-space (respectively PB-space). Every R^n or C^n
     is a PB-space.

30. Show that the open interval $(0, 1)$ with the induced metric topology
     from R, the real line, is a locally compact separable metric space but
     not an MB-space.

31. Let N denote the set of all positive integers with the order topology.
     N is a completely regular space and βN is its Stone–Čech compactifi-
     cation. Let E and O denote the set of all even and odd integers
     respectively. Clearly $N = E \cup O$, $E \cap O = \emptyset$. Let $\bar{E}$, $\bar{O}$ be closures
     of E and O in βN; then $\bar{E} \cap \bar{O} = \emptyset$ and $\beta N = \bar{E} \cup \bar{O}$. (Indeed,
     $\bar{E} \approx \bar{N} \approx \beta N$ and the continuous map $f(x) = 0$ or 1 according to
     whether $x \in E$ or O, extends to $\tilde{f}: \beta N \to [0, 1]$ such that $\bar{E} \subset \tilde{f}^{-1}(0)$
     and $\bar{O} \subset \tilde{f}^{-1}(1)$. Hence show that $\beta N = \bar{E} \cup \bar{O}$.)

     The map $\varphi: E \to O$ defined by $\varphi(2n) = 2n - 1$ extends to a homeo-
     morphism $\tilde{\varphi}: \bar{E} \to \bar{O}$ because $\bar{E}$ and $\bar{O}$ are compact. The map $h(x)$
     $= \tilde{\varphi}(x)$ or $\tilde{\varphi}^{-1}(x)$ according as $x \in \bar{E}$ or $\bar{O}$ is a homeomorphism of βN
     onto βN such that $h \circ h = i$ (identity) but $h(x) \neq x$. Show that car
     $\beta N = 2^c$ [c is the cardinal number of the continuum]. Let $\mathscr{F}$ denote
     the family of all countably infinite subsets of βN. Clearly car $\mathscr{F} = 2^c$.
     By induction show that there exists a subset $A \subset \beta N$ such that A con-
     tains a cluster point of each member of $\mathscr{F}$ and $h(A) \cap A = \emptyset$.

     Put $X = A \cup N \subset \beta N$. Show that N is dense in X and X is countably
     compact and hence pseudo-compact but X^2 is not countably compact
     and not even pseudo-compact.

32. Show that there exists a separable completely regular and countably
     compact space which is not compact (hint: Take X of Example 31).

33. Show that $[0, \Omega]$ is not paracompact.

34. Show that a Hausdorff locally compact space need not be normal.

35. Show that a Hausdorff locally compact topological group is normal.

36. Show that a paracompact space need not be metrizable.

37. (a) Show that in the plane R^2 there exists a countable dense subset
         S such that $S \cap L$ is finite (possibly empty) for each vertical or
         horizontal straight line L. [*Hint*: For each positive integer $n \in N$,
         let S_n denote the set of points $(i/n, j/n^2)$, where $|i| \leq n^2$,
         $|j| \leq n^4$, and (i, n) and (j, n) are relatively prime. Put $S = \bigcup_{n=1}^{\infty} S_n$.]

(b) Show that there exists a topology $\mathscr{C}$ on R^2 such that $(R^2, \mathscr{C})$ is not a k-space. [*Hint*: Take the coarsest topology $\mathscr{C}$ on R^2 with respect to which each real-valued function on R^2 which is continuous on each vertical or horizontal line is continuous. See Proposition 13, §17.]

(c) Show that $(R^2, \mathscr{C})$ is a k_r-space but not a k-space. (This example is due to Katetov; see [97b].)

38. Show that $[0, \Omega]$ is the Stone–Čech compactification of $[0, \Omega)$ and $[0, \Omega] \times [0, \Omega]$ is the Stone–Čech compactification of $[0, \Omega) \times (0, \Omega)$.

39. Show that the Stone–Čech compactification of a discrete space is *extremally disconnected* (i.e., the closure of each open set is open.)

40. Show that in the category of completely regular spaces and continuous functions, the Stone–Čech compactification has the following universal (i.e., characterizing) property: If f is a continuous map of X into $[0, 1]$ then there exists a unique continuous extension $f: \beta X \rightarrow [0, 1]$.

41. Show that an open subset of a Baire space is also a Baire space.

42. Show that

$$l_2 = \left\{ \{x_i\}: \sum_{i=1}^{\infty} |x_i|^2 < \infty \right\}$$

with the metric

$$d(\{x_i\}, \{y_i\}) = \left[\sum_{i=1}^{\infty} |x_i - y_i|^2 \right]^{1/2}$$

is a Baire space which is not locally compact.

43. Let $f: E \times F \rightarrow G$ be a map, where E, F, G are metric spaces such that f is separately continuous. If E is a Baire space then f is continuous at some point of $E \times F$.

VII

Generalizations
of Continuous Maps

In this chapter we study different maps which generalize continuous maps. It will be shown by examples that in general all notions are not identical with continuous maps. However, under certain conditions some of them are coincident.

We shall particularly deal with almost continuous, nearly continuous, graphically continuous, approximately continuous, and semicontinuous maps, not necessarily in this order.

It is worth noting that this chapter considers many different concepts which have been labeled by the same name by different authors, thus causing confusion. Depending upon their usefulness in the wider sense, these notions have been renamed. Our arbitrary action here is solely designed to clear the confusion, and it is hoped that it will not fall short of this goal.

We have included the work of a number of mathematicians, although not each theorem has been labeled by the proper name, e.g., Husain [45-49], Smith [80], Stallings [82], Singal and Singal [77], and Long and McGehee [61], etc.

§45. Almost Continuous Maps

The notion of almost continuous maps probably originated with the work of Blumburg [96], although he did not call it by that name. It has also been called nearly continuous by some authors in the study of locally convex spaces, e.g., Pták [97]. Anyway, for topological spaces we prefer this name (see Husain [47]) for the notion defined below. Many other authors since then have used this notion, e.g., Smith [80], McGehee [61], etc., to mention just a few. Some authors call it almost continuous in the sense of Husain (e.g., [61, 80]).

Definition 1. (a) A mapping f of a topological space E into a topological space F is called *almost continuous* at $x_0 \in E$ if for each neighborhood V of $f(x_0)$, $\overline{f^{-1}(V)}$ is a neighborhood of x_0.

(b) $f: E \to F$ is called *almost continuous* on E if it is so at each point of E.

Proposition 1. Every continuous function is almost continuous but not conversely.

Proof. The first part is immediate. For the converse take $f(x) = 1$ or -1 according to whether x is rational or irrational, which is clearly discontinuous everywhere and almost continuous everywhere.

An useful characterization of almost continuous maps is given as follows:

Proposition 2. $f: E \to F$ is almost continuous at $x_0 \in E$ if and only if for each neighborhood V of $f(x_0)$, there exists a neighborhood U of x_0 such that $U \subset \overline{f^{-1}(V)}$.

Proof. Obvious.

Proposition 3. Let $f: E \to F$ be a map and let D be a dense subset of E such that the restriction $f \,|\, D$ is continuous on D. Then f is almost continuous at each point of D.

Proof. Let $x_0 \in D$ and let V be an open neighborhood of $f(x_0)$. Since $g = f \,|\, D$ is continuous, there is a neighborhood U of x_0 in D such that $g(U_0) \subset V$. But there exists an open neighborhood U of x_0 in E with $U \cap D = U_0$. Hence $U_0 = U \cap D \subset g^{-1}(V) \subset f^{-1}(V)$. But since D is dense, we have $U \subset \overline{U \cap D} \subset \overline{f^{-1}(V)}$ and so f is almost continuous at x_0.

Note. If it is possible to find a dense subset of a topological space on which the restriction of any given map is continuous, then one has an almost continuous function. In some cases, this is possible as is shown below. Actually what we show is that for any given map on a suitable topological space, the set of points at which the map is almost continuous is everywhere dense. The result is due to Lin [60], which generalizes the result of Husain [47].

Theorem 1. Let E be a Baire space, F a second countable space and f a mapping of E into F. Then the set of points of almost continuity of f is dense in E.

Proof. Let $\{B_n : n \geq 1\}$ be a countable base of the topology on F and set

$$G_n = \overline{f^{-1}(B_n)} \setminus (\overline{f^{-1}(B_n)})^\circ,$$

for each $n \geq 1$. Then $(\overline{G_n})^\circ = \varnothing$ for all $n \geq 1$. Hence $G = \bigcup_{n=1}^\infty G_n$ is of the first category. Now let $x \in E$ at which f is not almost continuous. For some n, $f(x) \in B_n$ and $\overline{f^{-1}(B_n)}$ contains no neighborhood of x. Therefore $x \in G$. This shows that f is almost continuous for all $x \in E \setminus G$. Since G is a subset of the first category in a Baire space E, $E \setminus G$ is dense and at each $x \in E \setminus G$, f is almost continuous.

The following corollaries (due to Husain [47]) are the particular cases of the above theorem.

Corollary 1. Suppose E is a Baire space, F the real line, and $f: E \rightarrow F$. Then the set of points of almost continuity of f is dense in E.

Corollary 2. Suppose E is a metrizable complete space and F the real line. Then each map $f: E \rightarrow F$ has points of almost continuity everywhere dense in E.

Corollary 3. Suppose R is the real line and $f: R \rightarrow R$ any map. Then the set of points of almost continuity of f is dense in R.

Proposition 4. Let f be an almost continuous mapping of a T_1-space E into a topological space F and let x_0 be a limit point of E. Then for any pair U, V of open neighborhoods of x_0 and $f(x_0)$, respectively, there exists an $x \in U \setminus \{x_0\}$ such that $f(x) \in V$.

Proof. Suppose there are open sets U and V, containing x_0 and $f(x_0)$, respectively, such that $U \cap f^{-1}(V) = \{x_0\}$. This means that no point of U

is a limit point of $f^{-1}(V)$. Since the points in a T_1-space are closed, $U \cap \overline{f^{-1}(V)} = \{x_0\}$. But then this shows that f is not almost continuous at x_0, which is a contradiction.

Theorem 2.   Let E, F be first countable spaces, E in addition a T_1-space, and $f: E \rightarrow F$ an almost continuous map at x_0. If x_0 is a limit point of E, then there is a sequence $\{x_n\}$ of distinct points in E converging to x_0 such that $\{f(x_n)\}$ converges to $f(x_0)$.

Proof.   Let $\{U_n\}, \{V_n\}$ be countable bases of open neighborhoods of x_0 and $f(x_0)$, respectively. By Proposition 4, for U_1 and V_1 there exists an $x_1 \in U_1 \backslash \{x_0\}$ such that $f(x_1) \in V_1$. Now consider open sets U_2 and V_2. Clearly $U_2 \backslash \{x_1\}$ is an open neighborhood of x_0 and therefore by Proposition 4, there exists an $x_2 \in U_2 \backslash \{x_0, x_1\}$ such that $f(x_2) \in V_2$. By induction, we define a sequence $x_n \in U_n \backslash \{x_0, x_1, \ldots, x_{n-1}\}$ such that $f(x_n) \in V_n$. Clearly $x_n \rightarrow x_0$ and $f(x_n) \rightarrow f(x_0)$.

We know that the restriction of a continuous function on a subset is always continuous. However, this may fail for almost continuous maps. For example, consider $f: R \rightarrow R$, where $f(x) = 1$ or 0 according to whether x is rational or irrational. Then f is almost continuous, but certainly $f \mid R^+$: $R^+ \rightarrow R$ is not almost continuous at 0, where $R^+ = \{x \in R: x \geq 0\}$. However, the following holds:

Proposition 5.   Let G be an open subset of a topological space E. Then if $f: E \rightarrow F$ is almost continuous, so is $f \mid G: G \rightarrow F$.

Proof.   Let $x_0 \in G$ and let V be a neighborhood of $f(x_0)$ in F. Then there is an open neighborhood U of x_0 in E such that $U \subset \overline{f^{-1}(V)}$ because $f: E \rightarrow F$ is almost continuous. But then it follows that $G \cap U \subset U \subset \overline{f^{-1}(V)}$, where $G \cap U$ is an open neighborhood of x_0 and hence $f \mid G; G \rightarrow F$ is almost continuous.

Example.   A very simple discontinuous function may fail to be almost continuous. For example, $f(x) = 1$ or 0 according to whether $x \geq 0$ or $x < 0$, is not almost continuous at $x = 0$. For if $0 < \varepsilon < 1$, then $\{x: |f(x)| < \varepsilon\}$ is not dense in a neighborhood of 0. On the other hand, some wildly discontinuous function may turn out to be almost continuous. For example. $f(x) = 0$ or 1 according to whether x is rational or irrational is not continuous anywhere but is almost continuous.

It also follows from the last example that the functions of Baire class 1 may fail to be almost continuous (see Husain [49]).

Proposition 6. Let $f: E \to F$ and $g: F \to G$ be almost continuous maps; then:

(1) $g \circ f$ need not be almost continuous,

(2) If A is compact in E, $f(A)$ need not be compact in F. Hence every almost continuous function on a compact space need not be bounded.

(3) If $f_\alpha: E \to F$ ($\alpha \in D$ is a net) and if $f(x) = \lim f_\alpha(x)$, $x \in E$, then f need not be almost continuous.

(4) If F is a linear topological space and $f, g: E \to F$ almost continuous, then $f + g$ need not be almost continuous.

(5) If $f: E \to F$ is almost continuous, where F is a linear topological space, λ real or complex scalar, then λf is almost continuous.

In particular, the set of real-valued almost continuous functions on a topological space remains invariant under scalar multiplication.

Proof. (1)–(4) See the examples at the end of the chapter.

(5) Easy to verify.

§46. Closed Graphs

Another notion which generalizes that of continuous maps is the closed graph. This plays an important role in functional analysis.

Definition 2. Let E, F be topological spaces. A mapping $f: E \to F$ is said to have a *closed graph* if its graph $\{(x, y): y = f(x), x \in E\}$ in the product space $E \times F$ is a closed set.

Proposition 7. Let E, F be topological spaces, where F is Hausdorff. Then a continuous map $f: E \to F$ has a closed graph. But the converse does not hold.

Proof. Let $G = \{(x, y): y = f(x), x \in E\}$ be the graph of f. Let $(x, y) \notin G$; then $y \neq f(x)$. Since F is Hausdorff, there exist disjoint open neighborhoods V_1 and V_2 of $f(x)$ and y respectively. Since f is continuous, there exists an open neighborhood U_1 of x such that $f(U_1) \subset V_1$. Now clearly $U_1 \times V_2$ is an open neighborhood of (x, y) in $E \times F$. Suppose there exists an $x_1 \in U_1$ such that $f(x_1) \in V_2$. Since $f(U_1) \subset V_1$, it follows that $f(x_1) \in V_1$ and hence $V_1 \cap V_2 \neq \emptyset$, a contradiction to the disjointness of V_1 and V_2. Thus $(U_1 \times V_2) \cap G = \emptyset$ and this proves that G is closed.

For the converse, consider $f: R \rightarrow R$, where R is the real line and $f(x) = 1/x$ for $x \neq 0$, and $f(0) = 0$. Clearly f is not continuous at $x = 0$ but its graph is closed in $R^2 = R \times R$ as is easy to verify.

Remark. If F is not Hausdorff in Proposition 7, then even continuity need not imply a closed graph. For example, let E be a a topological space containing more than one point with the indiscrete topology and let $i: E \rightarrow E$ be the identity map. Then i is certainly continuous, but the graph of i is not closed because $E \times E$ has the indiscrete topology and hence the graph of i, being the diagonal set, which is different from the whole space, is not closed.

Proposition 8. Let $f: E \rightarrow F$ be a map. Then the graph of f is closed if and only if for each net $\{x_\alpha : \alpha \in D\}$ in E converging to $x \in E$ and the net $\{f(x_\alpha) : \alpha \in D\}$ converging to $y \in F$, it follows that $y = f(x)$.

Proof. Suppose that the graph G of f is closed. Let $\{x_\alpha : \alpha \in D\}$ be a net converging to x in E such that $\{f(x_\alpha) : \alpha \in D\}$ converges to $y \in F$. Then clearly $(x_\alpha, f(x_\alpha)) \in G$ for all $\alpha \in D$. Hence $\lim_\alpha (x_\alpha, f(x_\alpha)) \in G$, because G is closed. But $\lim_\alpha (x_\alpha, f(x_\alpha)) = (x, y)$. Hence $(x, y) \in G$ or, in other words, $y = f(x)$.

Conversely, assume that the condition holds. Let $(x, y) \in \bar{G}$; then there is a net $\{x_\alpha : \alpha \in D\}$ in E with $f(x_\alpha) \in F$ such that $\{(x_\alpha, f(x_\alpha)) : \alpha \in D\}$ converges to (x, y). This shows that $\{x_\alpha : \alpha \in D\}$ converges to x and $f(x_\alpha)$ converges to y. Hence the assumption implies that $y = f(x)$. In other words, $(x, y) \in G$ and G is closed.

Corollary 4. Let E, F be first countable spaces and $f: E \rightarrow F$ a mapping. Then the graph of f is closed if and only if for each sequence $\{x_n\} \subset E$, $x_n \rightarrow x$ and $f(x_n) \rightarrow y$ imply $y = f(x)$.

Proof. Immediate, since then $E \times F$ is first countable.

Corollary 5. If E, F are metric spaces and $f: E \rightarrow F$ a mapping, then the same characterization as that given in Corollary 4 holds.

Proposition 9. Let a mapping $f: E \rightarrow F$ have a closed graph. Then for each compact subset C of F, $f^{-1}(C)$ is closed.

Proof. Let $x \in \overline{f^{-1}(C)}$. Then there is a net $x_\alpha \in f^{-1}(C)$ such that $f(x_\alpha) = y_\alpha \in C$ and $\{x_\alpha : \alpha \in D\}$ converges to x. Since C is compact, there is a subnet $\{y_\beta : \beta \in B\}$ of $\{y_\alpha : \alpha \in D\}$ which converges to $y \in C$. Clearly $(x_\beta, f(x_\beta)) \in G$, the graph of f. Since G is closed, $\lim_\beta (x_\beta, f(x_\beta)) = (x, y) \in G$. Thus we have $y = f(x)$. Since $y \in C$, it follows that $y = f(x) \in C$. In other words, $x \in f^{-1}(C)$ and so $f^{-1}(C)$ is closed.

Proposition 10. If the map $f: E \to F$ has a closed graph and C is a compact subset of E, then $f(C)$ is closed in F.

Proof. Let $\{y_\alpha : \alpha \in D\}$ be a net in $f(C)$ such that $y_\alpha \to y \in \overline{f(C)}$. There is a net $\{x_\alpha : \alpha \in D\} \subset C$ such that $f(x_\alpha) = y_\alpha$. Since C is compact, there is a subnet $\{y_\beta : \beta \in B\}$ which converges to $x \in C$. Clearly $(x_\beta, f(x_\beta)) \in G$, the graph of f. Since G is closed,

$$\lim_\beta (x_\beta, f(x_\beta)) = \lim_\beta (x_\beta, y_\beta) = (x, y) \in G$$

and hence $y = f(x) \in f(C)$, which proves that $f(C)$ is closed.

Remark. If $f: E \to F$ has a closed graph, then it is not true that the inverse image of each closed subset of F is closed, since then such a function is necessarily continuous, which is not true (see Proposition 7).

Proposition 11. If $f: E \to F$ has a closed graph and if $f^{-1}: F \to E$ exists, then f^{-1} has a closed graph.

Proof. Let G_f, $G_{f^{-1}}$ denote the graphs of f and f^{-1} respectively. Clearly $G_f = \{(x, y): y = f(x), x \in E\}$, whereas $G_{f^{-1}} = \{(x, y): x = f^{-1}(y), y \in F\}$. Since $f^{-1}(y) = x$ if and only if $y = f(x)$, it follows that G_f and $G_{f^{-1}}$ are homeomorphic under the homeomorphism $(x, y) \to (y, x)$ of $E \times F$ onto $F \times E$. Since G_f is closed, so is $G_{f^{-1}}$.

Corollary 6. If $f: E \to F$ is continuous, where F is Hausdorff, and if f^{-1} exists, then the graph of f^{-1} is closed.

Proof. This follows from Propositions 7 and 11.

Proposition 12. Let E, F be topological spaces such that E is compact and let a map $f: E \to F$ have a closed graph. Then f is a closed mapping, i.e., it maps closed sets into closed sets.

Proof.   Since each closed subset C of E is compact (because E is compact), it follows that $f(C)$ is closed by Proposition 10.

Theorem 3 (Little closed-graph theorem).   Let E, F be topological spaces and $f: E \to F$ a mapping. Then:

(a) f is continuous if f has a closed graph and F is compact;

(b) f is open if f is a surjective map with closed graph, E a compact space, and F a Hausdorff space.

Proof.   (a)   It is sufficient to show that for each closed subset $C \subset F$, $f^{-1}(C)$ is closed. But by hypothesis each closed subset of F is compact. Hence by Proposition 9, $f^{-1}(C)$ is closed.

(b)   By hypothesis $f^{-1}: F \to E$ exists. Since f has a closed graph, the graph f^{-1} is closed (Proposition 11). Hence by (a), f^{-1} is continuous or, equivalently, f is open.

Corollary 7.   Let E be a topological space and let u, v be two topologies on E such that $u \supset v$. If E_u is a compact space and if E_v a Hausdorff space, then u is equivalent to v.

Proof.   The identity map $i: E_u \to E_v$ is continuous and surjective and so Theorem 3(b) applies.

Definition 3.   A topological space E_u is said to be *minimal* if there exists no strictly coarser Hausdorff topology other than u.

Proposition 13.   Every compact Hausdorff space is minimal.

Proof.   Immediate from Corollary 7.

Proposition 14.   Let E be a topological space, F a Hausdorff locally compact space and $f: E \to F$ a mapping such that for each compact subset C of F, $f^{-1}(C)$ is closed. Then the graph of f is closed.

Proof.   Let $y \neq f(x)$. Since F is Hausdorff there is a compact (because F is locally compact) neighborhood V of y such that $f(x) \notin V$. But then $f^{-1}(V)$ is closed in E and hence $x \in E \backslash f^{-1}(V)$ which is open and $f(E \backslash f^{-1}(V)) \cap V = \varnothing$. Thus $(E \backslash f^{-1}(V)) \times V$ is a neighborhood of (x, y) which does not meet the graph of f. Hence the graph of f is closed.

Proposition 15. Let $f: E_u \to F_v$ have a closed graph. If $u' \supset u$ and $v' \supset v$, then $f: E_{u'} \to F_{v'}$ also has a closed graph.

Proof. Let G_f be the graph of f; then G_f is closed in $E_u \times F_v$. Since the product topology $u' \times v'$ is finer than the product topology $u \times v$ on $E \times F$, G_f is closed in $E_{u'} \times F_{v'}$.

Proposition 16. Let $f: E_u \to E_v$ be continuous and let v' be a Hausdorff topology on F which is finer than v. Then $f: E_u \to F_{v'}$, has a closed graph.

Proof. The graph G_f of $f: E_u \to F_{v'}$, being the inverse image of the graph of $f: E_u \to F_v$ in $E_u \times F_v$, which is closed under the continuous identity map $i: E_u \times F_{v'} \to E_u \times F_v$, we see that the graph of $f: E_u \to E_{v'}$ is also closed.

Remark. It is not true that the composition of two maps each having a closed graph also has a closed graph. Nor does the sum of two maps having closed graphs have a closed graph. Thus the set of all linear maps from E to F having closed graphs, where E, F are linear topological spaces, need not form a linear space, or even an abelian group.

The case when a linear map (or homomorphism) of a topological linear space (or group) into another topological linear space (or group) with a closed graph is continuous is of great interest and significance. Such theorems are called closed-graph theorems (see §52).

§47. Almost Continuity and Closed Graphs

It is easy to see that an almost continuous map need not have a closed graph. For example, the graph of the almost continuous function $f(x) = 1$ if x is rational and -1 if x is irrational is not closed. Further, a map having the closed graph need not be almost continuous. For example, the map $f(x) = 1/x$ if $x \neq 0$ and $f(0) = 0$ has a closed graph but is not almost continuous at $x = 0$.

On the other hand, each continuous real-valued function has a closed graph, and is almost continuous. In this section we examine how these three notions can be related by implications. Most of the results are from [61].

Proposition 17. Let E, F be two topological spaces and $f: E \to F$ an almost continuous one-to-one map with closed graph. Then E is a Hausdorff space.

Proof. Let $x \neq y$, x, $y \in E$. Then $f(x) \neq f(y)$, because f is one-to-one. Since the graph of f is closed, there are open neighborhoods U and V of x and $f(x)$ respectively such that $f(U) \cap V = \varnothing$. Thus $f^{-1}(V) \subset E \backslash U$ and $\overline{f^{-1}(V)}$ is a neighborhood of y because f is almost continuous. Since $E \backslash U$ is closed, $\overline{f^{-1}(V)} \subset E \backslash U$. But then $x \in U$ and $y \in (\overline{f^{-1}(V)})^{\circ}$ show that E is Hausdorff.

Theorem 4. Let $f: E \to F$ be an almost continuous map, where F is a locally connected Hausdorff space. If f and f^{-1} map connected sets into connected sets, the graph of f is closed.

Proof. Let $x \in E$ and $y \neq f(x)$, $y \in F$. Since F is a Hausdorff locally connected space, there are open connected neighborhoods U and V of y and $f(x)$ respectively such that $U \cap V = \varnothing$. Hence $f^{-1}(U) \cap f^{-1}(V) = \varnothing$. Now we show that $f^{-1}(U) \cap \overline{f^{-1}(V)} = \varnothing$. Let $z \in \overline{f^{-1}(V)} \cap f^{-1}(U)$, then $\{z\} \cup f^{-1}(V)$ is connected since by hypothesis $f^{-1}(V)$ is connected, V being connected. But again by hypothesis $f[\{z\} \cup f^{-1}(V)] = \{f(z)\} \cup V$ is connected, which is impossible because $f(z) \in U$ and $U \cap V = \varnothing$. Now since f is almost continuous, $\overline{f^{-1}(V)}$ is a neighborhood of x and so it follows that $f(\overline{f^{-1}(V)})^{\circ} \cap U = \varnothing$. This proves that the graph of f is closed.

Theorem 5. Let E, F be topological spaces such that F is a Hausdorff locally compact space and $f: E \to F$ an almost continuous map which has a closed graph. Then f is continuous.

Proof. Let $x \in E$ and let V be a compact neighborhood of $f(x)$ (such a neighborhood exists because F is locally compact). Since the graph of f is closed, $f^{-1}(V)$ is closed (Proposition 9, §46) and so $\overline{f^{-1}(V)} = f^{-1}(V)$. But since f is almost continuous, it follows that $\overline{f^{-1}(V)} = f^{-1}(V)$ is a neighborhood of x and thus f is continuous at x.

Remark. In particular, Theorem 5 is valid when F is compact.

Corollary 8. Let f be an almost continuous mapping of the real line into itself with closed graph. Then f is continuous.

Proof. This follows from Theorem 5, because the real line R is locally compact.

The following theorems involve connected spaces and almost continuous maps. We show again the conditions under which an almost continuous map is continuous.

Theorem 6. Let $f: E \to F$ be an almost continuous map, where F is a regular locally connected space and such that for each connected subset $C \subset F, \overline{f^{-1}(C)} \subset f^{-1}(\bar{C})$. Then f is continuous.

Proof. Let $x_0 \in E$ and let V be an open neighborhood of $f(x_0)$. Since F is regular and locally connected, there is a connected open neighborhood W of $f(x_0)$ such that $\overline{W} \subset V$. Thus $f^{-1}(\overline{W}) \subset f^{-1}(V)$. By hypothesis $\overline{f^{-1}(W)} \subset f^{-1}(\overline{W})$ because $\overline{W}$ is connected. Hence $\overline{f^{-1}(W)} \subset f^{-1}(V)$. Since f is almost continuous, $\overline{f^{-1}(W)}$ is a neighborhood of x_0 and so $f^{-1}(V)$ is a neighborhood of x_0, which proves the continuity of f at x_0.

Before we prove the next theorem, we need the following lemma.

Lemma 1. Let E, F be topological spaces and let $f: E \to F$ be a mapping which maps connected sets into connected sets. Then for any connected subset D of $E, f(\overline{D}) \subset \overline{f(D)}$.

Proof. Let D be a connected subset of E. To show that $f(\overline{D}) \subset \overline{f(D)}$, let $y \in f(\overline{D})$. Then there exists $x \in \overline{D}$ such that $f(x) = y$. If $x \in D$, then $f(x) = y \in f(D) \subset \overline{f(D)}$. If x is a limit point of D, then $D \subset D \cup \{x\} \subset \overline{D}$ and $D \cup \{x\}$ is connected and since f is a connected map

$$f(D \cup \{x\}) = f(D) \cup \{f(x)\} = f(D) \cup \{y\}$$

is connected. But any connected set containing more than one point is dense in itself (Exercise 17, Chapter II). Hence y is a limit point of $f(D)$, i.e., $y \in \overline{f(D)}$ and so $f(\overline{D}) \subset \overline{f(D)}$.

Theorem 7. Let E be a Hausdorff space, F a regular locally connected space, and $f: E \to F$ an almost continuous mapping such that f and f^{-1} map connected sets into connected sets. Then f is continuous.

Proof. In view of Theorem 6, it is sufficient to verify that for any connected subset $C \subset F, \overline{f^{-1}(C)} \subset f^{-1}(\bar{C})$. By hypothesis on f^{-1}, if C is connected then $f^{-1}(C)$ is connected. Further, by the hypothesis on f,

$$f(\overrightarrow{\overline{f^{-1}(C)}}) \subset \bar{C}$$

by Lemma 1, whence we have

$$f^{-1}(f(\overline{f^{-1}(C)})) \subset f^{-1}(\bar{C}).$$

Clearly

$$\overline{f^{-1}(C)} \subset f^{-1}(f(\overline{f^{-1}(C)})) \subset f^{-1}(\bar{C})$$

and the result follows.

Under certain stringent conditions it is possible to weaken some conditions on the spaces in the above theorem.

Theorem 8. Let E be a topological space, F a regular locally connected space, and $f: E \to F$ an almost continuous mapping such that for each closed connected subset C in F, $f^{-1}(C)$ is closed. Then f is continuous.

Proof. Let $x_0 \in E$ and let V be an open neighborhood of $f(x_0)$. Since F is regular and locally connected, there is a connected neighborhood W of $f(x_0)$ such that $\bar{W} \subset V$. Thus

$$f^{-1}(W) \subset f^{-1}(\bar{W}) \subset f^{-1}(V).$$

Since $\bar{W}$ is closed and connected, by hypothesis $f^{-1}(\bar{W})$ is closed. Hence

$$\overline{f^{-1}(W)} \subset f^{-1}(\bar{W}) \subset f^{-1}(V).$$

Now since f is almost continuous, $\overline{f^{-1}(W)}$ is a neighborhood of x_0 and so is $f^{-1}(V)$. This proves that f is continuous.

Theorem 9. Let f be an almost continuous mapping of the real line R to a locally connected space F such that for each connected set $C \subset F$, $f^{-1}(C)$ is connected. Then f is continuous.

Proof. Let $x \in R$ and let V be a connected neighborhood of $f(x)$. Then $f^{-1}(V)$ is a connected subset of the real line, hence an interval containing x and so is $\overline{f^{-1}(V)}$. But since f is almost continuous, $\overline{f^{-1}(V)}$ is a neighborhood of x, i.e., x is an interior point of $f^{-1}(V)$. Hence f is continuous.

Corollary 9. An almost continuous real-valued function on the real line satisfying the condition that for any connected set C, $f^{-1}(C)$ is a connected, is continuous.

Proof. This follows immediately from Theorem 9.

Now we give a condition under which a map $f\colon E \to F$ is almost continuous. But first we have:

Definition 4. A mapping $f\colon E \to F$ is called *finitely closed* at $x_0 \in F$ if for each open neighborhood V of $f(x_0)$ there exists an open neighborhood W of $f(x_0)$, $W \subset V$, such that $E \backslash \overline{f^{-1}(W)}$ consists of finitely many components.

Theorem 10. Let E be a topological space and F a regular space. Then a connected and finitely closed map $f\colon E \to F$ at $x_0 \in E$ is almost continuous at x_0.

Proof. Suppose f is not almost continuous at x_0. Then there is an open neighborhood V of $f(x_0)$ such that x_0 is not an interior point of $\overline{f^{-1}(V)}$, i.e., x_0 is a limit point of $E \backslash \overline{f^{-1}(V)}$. Since f is finitely closed, there is an open neighborhood W of $f(x_0)$ such that $W \subset V$ and $E \backslash \overline{f^{-1}(W)} = \bigcup_{i=1}^{n} C_i$, where each C_i is a component.

Since $\overline{f^{-1}(W)} \subset \overline{f^{-1}(V)}$ and since x_0 is not an interior point of $\overline{f^{-1}(V)}$ and so $x_0 \notin (\overline{f^{-1}(W)})^{\circ}$, x_0 is a limit point of $E \backslash \overline{f^{-1}(W)}$ and hence a limit point of some C_i for $1 \le i \le n$. Thus $C_i \cup \{x_0\}$ is connected. But since f maps connected sets into connected sets, it follows that

$$f(C_i \cup \{x_0\}) = f(C_i) \cup \{f(x_0)\}$$

is connected. But since $f(x_0) \in W$, while $f(C_i) \subset F \backslash W$ because F is regular, it follows that $f(C_i) \cup \{f(x_0)\}$ is not connected, thus giving a contradiction. Hence f is almost continuous.

§48. Graphically Continuous Maps

The notion that we are about to introduce was initially called "almost continuous," by Stalling [82]. Since a different notion has already been called "almost continuous," we adopt a more appropriate name for this notion. A number of results presented here are due to Stalling [82].

Definition 5. A map $f\colon E \to F$ is called *graphically continuous* if for each open set $U \subset E \times F$ containing the graph G_f of f, there exists a continuous map $g: E \to F$ such that $G_g \subset U$. In other words, each open neighborhood of the graph of f in $E \times F$ contains the graph of a continuous map.

Proposition 18. Every continuous map is graphically continuous. But the converse does not hold.

Proof. Let $f: E \to F$ be continuous. Then for each open neighborhood U of $G_f \subset E \times F$, indeed, f being continuous and $G_f \subset U$ imply that f is graphically continuous. For the converse, see Example 8.

It is clear from the examples that a graphically continuous function need not be almost continuous, and an almost continuous function need not be graphically continuous. However, the following holds:

Proposition 19. Let $f: E \to F$ be a graphically continuous map and C a closed subset of E. Then the restriction $f \mid C$ of f is also graphically continuous.

Proof. Let G_f denote the graph of f and $G_{f'}$ that of $f' = f \mid C$. Let U' be an open neighborhood of $G_{f'}$ in $C \times F$. Clearly there is an open set U in $E \times F$ such that $U' = U \cap (C \times F)$. The set $U' \cup [(E \setminus C) \times F]$ is open in $E \times F$ and $G_f \subset U' \cup [(E \setminus C) \times F]$. Since f is graphically continuous, there is a continuous map $g: E \to F$ such that $G_g \subset U' \cup [(E \setminus C) \times F]$. But then $G_g \cap (C \times F) \subset U'$. Clearly $g \mid C$ is continuous on C and so $f \mid C$ is graphically continuous.

The composition of two graphically continuous maps may not be graphically continuous, e.g., see Exercise 17. However, we have the following:

Proposition 20. Let E, F, G be three topological spaces and let $f: E \to F$ and $g: F \to G$ be maps such that f is graphically continuous and g is continuous. Then the composition $g \circ f$ is graphically continuous.

Proof. Let U be an open neighborhood of the graph $G_{g \circ f}$ of $g \circ f$ in $E \times G$. The map $h: (x, y) \to (x, g(y))$ of $E \times F$ into $E \times G$ is clearly continuous. Moreover, $h^{-1}(G_{g \circ f})$ is contained in $E \times F$ and $h^{-1}(U)$ is an open neighborhood of $h^{-1}(G_{g \circ f})$. Clearly $G_f \subset h^{-1}(U)$ and hence there is a continuous map $k: E \to F$ such that its graph $G_k \subset h^{-1}(U)$. But then it is clear that $g \circ k: E \to G$ is continuous and $G_{g \circ k} \subset U$. This proves that $g \circ f$ is graphically continuous.

In general, if $f: E \to F$ is continuous and $g: F \to G$ is graphically continuous then $g \circ f$ need not be graphically continuous (see Exercise 17). However, in special cases, it holds. But first we have a lemma.

Lemma 2. Let E be a connected topological space and F a topological space such that $E \times F$ is a completely normal Hausdorff space. If a map $f: E \to F$ is graphically continuous then G_f is connected (cf. [82]).

Proof. Suppose G_f is not connected. Then there are disjoint open sets U, V in $E \times F$ such that $U \cup V \supset G_f$ and $U \cap G_f \neq \varnothing$, $V \cap G_f \neq \varnothing$. Suppose $x, y \in E$ such that $(x, f(x)) \in U$ and $(y, f(y)) \in V$. Let $p_1: E \times F \to E$ be the projection on the first space. Then $p_1^{-1}(\{x\}), p_1^{-1}(\{y\})$ are closed sets in $E \times F$ and hence $U' = U \setminus \{p_1^{-1}(\{y\})\}$ and $V' = V \setminus \{p_1^{-1}(\{x\})\}$ are open subsets in $E \times F$ and so is their union and $G_f \subset U' \cup V'$. Since f is graphically continuous, there is a continuous function $g: E \to F$ such that $G_g \subset U' \cup V'$. Since U', V' are disjoint open sets, G_g cannot be connected. However, g being continuous, G_g is homeomorphic with the connected space E and so it is connected, which is a contradiction. Therefore G_f is connected.

Corollary 10. If C is a closed connected subset of a topological space E and F a topological space such that $E \times F$ is a completely normal Hausdorff space, and if $f: E \to F$ is graphically continuous, then $G_{f|C}$ is also connected.

Proof. This follows by combining Lemma 2 with Proposition 19.

Proposition 21. Let E be a compact Hausdorff space, F a Hausdorff space, and G a topological space. Let $f: E \to F$ be continuous and $g: F \to G$ graphically continuous. Then $g \circ f: E \to G$ is graphically continuous.

Proof. Observe that there is no loss of generality if we assume that f is onto, because otherwise we consider $f(E) \subset F$, which is compact because f is continuous and E is compact; hence $f(E)$ is closed because F is Hausdorff and we apply Proposition 20 to $E \xrightarrow{f} f(E) \xrightarrow{g} G$.

Let U be an open neighborhood of $G_{g \circ f} \subset E \times G$. Let $f': E \times G \to F \times G$ be the map: $f'(x, z) = (f(x), z)$, $x \in E$, $z \in G$. Then $f'(G_{g \circ f}) = G_g$. Clearly for each $y \in F$, $f^{-1}(y)$ is closed and hence compact in a compact space E. Let $x \in f^{-1}(y)$, let U_x be an open neighborhood of x, and let W_x be an open neighborhood of $g \circ f(x) = g(y)$ such that $U_x \times W_x \subset U$. Since $\{U_x\}$ is an open covering of $f^{-1}(y)$ which is compact, we have a finite subcovering U_i ($1 \leq i \leq n$) and we let W_i be the corresponding sets. Let

$$U_y = E \setminus f\left(E \setminus \bigcup_{i=1}^{n} U_i\right) \quad \text{and} \quad W_y = U_y \times \left(\bigcap_{i=1}^{n} W_i\right).$$

Then U_y is an open neighborhood of y. Hence W_y is an open neighborhood of $(y, g(y))$. Clearly $f'^{-1}(W_y) \subset U$. Put $W = \cup \{W_y : y \in F\}$. Then W is an open set in $F \times G$ and $G_g \subset W$. Since g is graphically continuous, there is a continuous map $h : F \to G$ such that $G_h \subset W$. Thus $h \circ f : E \to G$ is continuous and $f'(G_{hf}) = G_h \subset W$. Hence $G_{hf} \subset f^{-1}(W) \subset U$ and $g \circ f$ is graphically continuous.

Recall a point $x \in E$ is said to be a *fixed point* of a mapping $f : E \to E$ if $f(x) = x$. We prove a fixed-point theorem for graphically continuous maps.

Proposition 22.   Let U be an open set in $E \times E$, where E is a Hausdorff space. Suppose any continuous function $g : E \to E$ whose graph G_g lies in U has a fixed point. Then every graphically continuous function $f : E \to E$ whose graph G_f lies in U also has a fixed point.

Proof.   Suppose $f : E \to E$ has no fixed point. Let $\Delta = \{(x, y) : x = y\}$ be the diagonal set in $E \times E$. Then Δ is a closed subset of $E \times E$ because E is Hausdorff. Thus $G_f \subset U \backslash \Delta$ and $U \backslash \Delta$ is open in $E \times E$. Since f is graphically continuous, there is a continuous map $g : E \to E$ with $G_g \subset U \backslash \Delta$. But this contradicts the hypothesis.

§49. Nearly Continuous and w-Continuous Maps

In this section we introduce a notion due to Singal and Singal [77]. This again generalizes continuous maps. We also introduce other kinds of continuities, e.g., w-continuity and θ-continuous Cesaro-type maps indicating the variety of generalizations of continuous maps.

Definition 6.   A map $f : E \to F$ is said to be *nearly continuous* at $x \in E$ if for every neighborhood V of $f(x)$ there is a neighborhood U of x such that $f(U) \subset (\bar{V})^\circ$; f is called nearly continuous if it is so at each $x \in E$.

Remark.   This notion was originally called almost continuous by Singal and Singal [77]. However, the words "nearly continuous" were used by Pták [97] for the notion that we have labeled as "almost continuous" in the earlier sections.

Proposition 23.   Every continuous map $f : E \to F$ is nearly continuous. However, the converse is not true.

Proof.   For each open neighborhood V of $f(x)$, by continuity of f, there exists an open neighborhood U of x such that $f(U) \subset V \subset (\bar{V})^\circ$. Hence f is nearly continuous.

For the converse, let $E = R$, the real line with the topology consisting of sets R, $\varnothing$, and complements of countable sets and let $F = R$ with the usual metric topology. Then the identity map: $i: E \to F$ is nearly continuous but not continuous.

Similar to a characterization of continuous maps, we have the following:

Theorem 11. Let $f: E \to F$ be a mapping. The following statements are equivalent:

(i) f is nearly continuous;

(ii) the inverse image of a regularly open (i.e., $\bar{U}^\circ = U$) subset of F is an open set in E;

(iii) for each regularly open neighborhood V of $f(x) \in F$ there is an open neighborhood U of $x \in E$ such that $f(U) \subset V$;

(iv) for each open subset V of F, $f^{-1}(V) \subset [f^{-1}(\bar{V}^\circ)]^\circ$;

(v) for each closed subset C of F, $[\overline{f^{-1}(\bar{C}^\circ)}] \subset f^{-1}(C)$;

(vi) for any $x \in E$ and for each net $\{x_\alpha\}$ converging to x, the net $\{f(x_\alpha)\}$ is eventually in each regularly open set containing $f(x)$.

Proof. (i) $\Rightarrow$ (ii) For any regularly open subset V of F, if $x \in f^{-1}(V)$ then $f(x) \in V$. Hence there is an open neighborhood U of x such that $f(U) \subset V = \bar{V}^\circ$. Thus $x \in U \subset f^{-1}(V)$ and therefore $f^{-1}(V)$ is a neighborhood of x. Since x is arbitrary, we have shown that $f^{-1}(V)$ is open.

(ii) $\Rightarrow$ (iii) Let V be a regularly open neighborhood V of $f(x) \in F$. Then, V being a regularly open set, by (ii) $f^{-1}(V)$ is open and $x \in f^{-1}(V)$. Thus there is an open neighborhood U of x such that $U \subset f^{-1}(V)$ or $f(U) \subset V$, proving (iii).

(iii) $\Rightarrow$ (iv) For each open subset V of F and $x \in f^{-1}(V)$, $\bar{V}^\circ$ is a regularly open neighborhood of $f(x)$. Hence by (iii) there is an open neighborhood U of x such that $f(U) \subset \bar{V}^\circ$. Hence $x \in U \subset f^{-1}(\bar{V}^\circ)$. Thus we have $f^{-1}(V) \subset [f^{-1}(\bar{V}^\circ)]^\circ$.

(iv) $\Rightarrow$ (v) For each closed subset C of F, $F\backslash C$ is open and hence by (iv),

$$f^{-1}(\overline{F\backslash C}) \subset [f^{-1}(\overline{F\backslash C})^\circ]^\circ,$$

or

$$[f^{-1}(F)\backslash f^{-1}(C)] \subset [f^{-1}(\overline{F\backslash C})^\circ]^\circ \subset f^{-1}(F)\backslash[\overline{f^{-1}(\bar{C}^\circ)}]$$

or

$$f^{-1}(C) \supset \overline{f^{-1}(\bar{C}^\circ)}.$$

(v) $\Rightarrow$ (vi)   Let $x \in E$ and let $\{x_\alpha\}$ be a net in E such that $x_\alpha \to x$. Let V be a regularly open set of F containing $f(x)$. Then $F \backslash V$ is closed and by (v), $f^{-1}(\overline{F \backslash V})^\circ \subset f^{-1}(F \backslash V)$. But V being regularly open, $\overline{f^{-1}(F \backslash V)} \subset f^{-1}(F) \backslash f^{-1}(V)$. This means that $f^{-1}(V) \subset [f^{-1}(V)]^\circ$ and so $f^{-1}(V)$ is an open neighborhood of x because $[f^{-1}(V)]^\circ \subset f^{-1}(V)$. Since $x_\alpha \to x$, there exists α_0 such that for all $\alpha \geq \alpha_0$, $x_\alpha \in f^{-1}(V)$. This proves that $f(x_\alpha)$ is eventually in V.

(vi) $\Rightarrow$ (i)   Let f be a map from E to F satisfying (vi). Suppose f is not nearly continuous at $x \in E$. Then there is an open set V containing $f(x)$ such that for each open neighborhood U of x, $f(U) \cap (F \backslash (\overline{V})^\circ) \neq \varnothing$. This implies that $U \cap f^{-1}(F \backslash \overline{V}^\circ) \neq \varnothing$ for each open neighborhood U of x. The family of all open sets containing x is directed by inclusion. Thus for each open neighborhood U of x there is $x_U \in U \cap f^{-1}(F \backslash \overline{V}^\circ)$. Thus the net $\{x_U\}$ converges to x and is such that $f(x_U) \notin \overline{V}^\circ$. This means that $f(x_U)$ is not eventually in a regularly open set $\overline{V}^\circ$, a contradiction, and this proves (i).

Proposition 24.   A nearly continuous map of a topological space E into a regular space F is continuous.

Proof.   Let V be an open neighborhood of $f(x)$, $x \in E$. Since F is regular, there is an open neighborhood W of $f(x)$ such that $\overline{W}^\circ \subset V$. Since f is nearly continuous, there is an open neighborhood U of $f(x)$ such that $f(x) \in f(U) \subset \overline{W}^\circ \subset V$. Hence f is continuous at an arbitrary x and therefore at each x.

Corollary 11.   Each nearly continuous real-valued function on a topological space is continuous.

Proof.   This follows from Proposition 24, since the real line is regular.

Remark.   If $N(X)$ denotes the set of all nearly continuous real-valued functions on a topological space and $C(X)$ as usual the set of all continuous real-valued functions on X, then $C(X) = N(X)$.

Proposition 25.   (a)   The restriction of a nearly continuous map $f : E \to F$ on any subset A of E is nearly continuous.

(b)   Let $\{P_\alpha\}$ be an open covering of a topological space E, F a topological space, and f a mapping of E into F. If for each α, the restriction $f | P_\alpha : P_\alpha \to F$ is nearly continuous then f is nearly continuous.

(c) If E is a disconnected (i.e., there are closed or open disjoint non-empty subsets E_1, E_2 of E such that $E = E_1 \cup E_2$) space, F a topological space, and f a mapping of E into F and if $f \mid E_i, 1 \leq i \leq 2$ is nearly continuous, then f is nearly continuous.

(d) If $f\colon E \to F$ is open and continuous and if $g\colon F \to G$ is a mapping, then $g \circ f$ is nearly continuous iff g is nearly continuous.

Proof. (a) Let A be a subset of E and $f_A = f \mid A$. For any regularly open subset V of F, $f_A^{-1}(V) = A \cap f^{-1}(V)$. Since $f^{-1}(V)$ is open in E because f is nearly continuous, it follows that $f_A^{-1}(V)$ is relatively open in A and so f_A is nearly continuous.

(b) Let $x \in E$. Then there exists α such that $x \in P_\alpha$. Let V be a regularly open subset of F containing $f(x)$. Since the restriction $f \mid P_\alpha = f_\alpha$ is nearly continuous, there is an open (in P_α) neighborhood U of x such that $f(U) \subset V$. But then there is an open subset W of E such that $x \in U = W \cap P_\alpha$. But since U is an open set in E (because W and P_α are open in E), it follows that $f\colon E \to F$ is nearly continuous.

(c) Let $E = E_1 \cup E_2$. Since E_i $(i = 1, 2)$ are closed, and $E_1 \cap E_2 = \varnothing$, E_i $(i = 1, 2)$ are open. Hence by (b), f is nearly continuous.

(d) Suppose $g \circ f$ is nearly continuous. Let V be a regularly open subset of G. Then $(g \circ f)^{-1}(V) = f^{-1}(g^{-1}(V))$ is open. But since f is open, $f(f^{-1}(g^{-1}(V))) = g^{-1}(V)$ is open and so g is nearly continuous.

Conversely, suppose g is nearly continuous. Let V be a regularly open subset of G. Then $g^{-1}(V)$ is an open subset of F and so $f^{-1}(g^{-1}(V))$ is open in E because f is continuous. Hence $(g \circ f)^{-1}(V) = f^{-1}(g^{-1}(V))$ is open and so $g \circ f$ is nearly continuous.

Definition 7. A map $f\colon E \to F$ is said to be *weakly* or *w-continuous* if for each $x \in E$ and each open neighborhood V of $f(x)$ there is an open neighborhood U of x such that $f(U) \subset \bar{V}$ (Levin [58(a)]).

Proposition 26. Every nearly continuous map $f\colon E \to F$ is weakly continuous. But the converse does not hold.

Proof. Obviously, if V is an open neighborhood of $f(x)$, $x \in E$, then because of near continuity of f, $f^{-1}(V)$ contains a neighborhood U of $x \in E$ such that $f(U) \subset \bar{V}^\circ \subset \bar{V}$ and so f is weakly continuous.

For the converse let $E = \mathbb{R}$ be the space considered following Proposition 23 and $F = \{0, 1, 2\}$ with the topology $\mathscr{C}^* = \{\varnothing, F, \{0\}, \{2\}, \{0, 2\}\}$.

Let $f: E \to F$ be the function defined as follows:

$$f(x) = \begin{cases} 0 & \text{if } x \text{ is rational;} \\ 1 & \text{if } x \text{ is irrational.} \end{cases}$$

Then f is weakly continuous but not nearly continuous.

Proposition 27. An open (weakly or) w-continuous map $f: E \to F$ is nearly continuous.

Proof. If V is an open neighborhood of $f(x) \in F$, then there is an open neighborhood U of x such that $f(U) \subset \bar{V}$. Since f is open, $f(U)$ is an open set and so $f(U) \subset \bar{V}^{\circ}$. This proves that f is nearly continuous.

Proposition 28. Every continuous map $f: E \to F$ is weakly continuous. But the converse is not true.

Proof. For each open neighborhood V of $f(x)$ there exists an open neighborhood U of x such that $f(U) \subset V \subset \bar{V}^{\circ}$. Hence f is weakly continuous. For the converse see Levine's example [58(b)].

Proposition 29. A map $f: E \to F$ is weakly continuous if and only if for each open set V in F, $f^{-1}(V) \subset [f^{-1}(\bar{V})]^{\circ}$.

Proof. If f is weakly continuous, then for each open set V of F, $f(x) \in V$, there is an open set U, $x \in U$ such that $f(U) \subset \bar{V}$, which implies $U \subset f^{-1}(\bar{V})$ and hence $x \in U \subset [f^{-1}(\bar{V})]^{\circ}$. Since $x \in f^{-1}(V)$, it follows that $f^{-1}(V) \subset [f^{-1}(\bar{V})]^{\circ}$. Conversely, if for each open set $V \subset F$, $f^{-1}(V) \subset [f^{-1}(\bar{V})]^{\circ}$, then by putting $U = [f^{-1}(\bar{V})]^{\circ}$, we see that $f(x) \in f(U) \subset f(f^{-1}(\bar{V})) = \bar{V}$, which shows that f is w-continuous.

Proposition 30. Let F be a regular space. A mapping $f: E \to F$ is weakly continuous if and only if f is continuous.

Proof. Suppose $f: E \to F$ is continuous; then it is weakly continuous by Proposition 28. Now assume f is weakly continuous. Let V be an open neighborhood of $f(x)$, where $x \in E$. Since F is regular, there is an open neighborhood W of $f(x)$ such that $\bar{W} \subset V$. Since F is weakly continuous there is an open neighborhood U of $x \in E$ such that $f(U) \subset \bar{W} \subset V$. Thus f is continuous.

Remark. Since the real line is a regular topological space, each real-valued function is weakly continuous iff it is continuous.

Corollary 11'. An open mapping $f: E \to F$ is nearly continuous iff it is w-continuous.

Proof. The "if" part follows from Proposition 27. For the "only if" part, let f be an open nearly continuous map. Let V be an open neighborhood of $f(x)$. Since f is nearly continuous, there is an open neighborhood U of x such that $f(U) \subset (\bar{V})^{\circ}$ or $x \in U \subset f^{-1}(\bar{V})^{\circ}$. Since f is open, $f(U)$ is open and so $f(U) \subset \bar{V}^{\circ} \subset \bar{V}$ proves that f is w-continuous.

Definition 8. A map $f: E \to F$ is said to be w^{-1}-*continuous*, if for each open set V in F, $f(x) \in V$, we have that $f^{-1}(\partial V)$ is closed in E, where ∂V is the boundary of V.

Proposition 31. A continuous mapping $f: E \to F$ is w^{-1}-continuous.

Proof. If f is continuous and if V is an open subset of F, then $f^{-1}(\partial V)$ is clearly closed.

Remark. The converse of the above proposition is not true, as is easy to see.

Proposition 32. A map $f: E \to F$ is w-continuous and w^{-1}-continuous iff f is continuous.

Proof. The "if" part is clear (Propositions 28, 31). For the "only if" part, let $x \in E$ be an arbitrary point. Since f is w-continuous, there is an open neighborhood U of x such that $f(x) \in f(U) \subset \bar{V}$. But since f is w^{-1}-continuous, $f^{-1}(\partial V) = f^{-1}(\bar{V} \backslash V)$ is closed and so $U \backslash f^{-1}(\partial V)$ is open. Further, $f(x) \notin \partial V$ implies $x \notin f^{-1}(\partial V)$. The proof will be complete if we show that $f(x) \in f(U \backslash f^{-1}(\partial V)) \subset V$. Let $y \in U \backslash f^{-1}(\partial V)$; then $f(y) \in \bar{V}$. But $y \notin f^{-1}(\partial V)$ and so $f(y) \notin \partial V = \bar{V} \backslash V$, which implies that $f(y) \in V$.

Definition 9. A map $f: E \to F$ is said to be of *Césaro-type* if there exist nonempty open sets $U \subset E$ and $V \subset F$ such that for $y \in V$, $U \subset \overline{f^{-1}(y)}$. See [80].

Example. The function $f(x) = 0$ if $x \notin [0, 1]$ and $f(x) = \overline{\lim} \, (a_1 + \cdots + a_n)n^{-1}$ for $x \in [0, 1]$, where $x = 0. a_1 a_2 a_3 \ldots$ (nonterminating binary expansion of x) is of Césaro-type. It assumes every value between 0 and 1 but is not continuous. See [80].

Proposition 33. A Césaro-type map $f\colon E \to F$ is discontinuous if F is a connected nontrivial T_1-space.

Proof. Let U, V be nonempty open sets such that $y \in V$ implies $U \subset \overline{f^{-1}(y)}$. Since a singleton in a connected nontrivial T_1-space cannot be open, V must contain at least two points y_1, y_2 with $V \setminus \{y_2\}$ being an open neighborhood of y_1. Suppose f is continuous; then $f^{-1}(V \setminus \{y_2\})$ is open and $U \cap f^{-1}(V \setminus \{y_2\}) \neq \varnothing$ and is open, which is impossible because $f^{-1}(y_2)$ is dense in U. Hence f is discontinuous.

Theorem 12. Let f be a real-valued function on the real line. Then f is continuous iff

  (i) f is graphically continuous;
 (ii) f is almost continuous;
(iii) f is not of Césaro-type.

Proof. If f is continuous, then (i) and (ii) are clear and (iii) follows from Proposition 33. For the "if" part, suppose (i)–(iii) hold but f is not continuous at x_0. Then there is an open interval J containing $f(x_0)$ but for all open sets V containing x_0, $f(x) \notin J$, for some $x \in V$. Choose s, $t \in J$ such that $s < f(x_0) < t$ and $I = (s, t)$. By (ii) there is an open set N such that $x_0 \in N$ and $f^{-1}(I)$ is dense in N. But there exists some $x \in N$ such that $f(x) \notin J$. Let $K = (q, r)$ be an open interval such that $f(x) \in K$ and $\bar{K} \cap I = \varnothing$. We may assume that $s < t < q < r$. By (ii) there is an open set M such that $x \in M$ and $f^{-1}(K)$ is dense in M. Consider the open sets $U = N \cap M$ and $V = \{y\colon t < y < q\}$. Since $x \in N$, $M \cap U \neq \varnothing$, and clearly $V \neq \varnothing$. Hence by (iii) there exists a $y_0 \in V$ and some open interval $L \subset U$ such that $y_0 \notin f(L)$. Since $f^{-1}(I)$ is dense in N, we may choose a, $b \in L$ such that $f(a)$, $f(b) \in I$ and $a < b$. Let $P = \{(u, v);\ a < u < b,\ y < v\}$. Since $f^{-1}(K)$ is dense in M, $P \cap G_f \neq \varnothing$, where G_f is the graph of f. By construction $\overline{P^c} \cap G_f \neq \varnothing$. Hence P and $\overline{P^c}$ separate G_f. By (i) and Lemma 2, §48, this is a contradiction. Hence f is continuous (see [80]).

§50. Semicontinuous Maps

Let (E, d) be a metric space. Let $\mathscr{C}$ denote the set of all nonempty compact subsets of E. For each $C \in \mathscr{C}$ and each $\varepsilon > 0$, we define $U_\varepsilon(C) = \{x \in E\colon d(x, c) < \varepsilon$ for some $c \in C\}$. We see that $U_\varepsilon(C)$ behaves like an open neighborhood of C. Let f be a set-valued function from E to $\mathscr{C}$. Clearly a function from E to $\mathscr{C}$ can be viewed as a set-valued function from E to E, since for each $x \in E$, $\{f(x)\}$, being a singleton in $\mathscr{C}$ is a compact subset of E.

Definition 10. A set-valued function $f: E \to \mathscr{C}$ is *upper-semicontinuous* (or *lower-semicontinuous*) at $x \in E$ if to each $\varepsilon > 0$ there exists a neighborhood V of x such that

$$\bigcup_{y \in V} f(y) = f(V) \subset U_\varepsilon(f(x)) \text{ [or } f(x) \subset U_\varepsilon(f(y)), \ y \in V].$$

A set-valued function $f: E \to \mathscr{C}$ is called continuous if and only if f is upper-semicontinuous and lower-semicontinuous.

We shall show that any upper-semicontinuous set-valued function is continuous on a very large subset of E. For this, we first define two numbers:

(a) $M_\varepsilon(C) = \{$largest positive integer m for which there exist m distinct points $\{x_1, \ldots, x_m\} \subset C$ with $d(x_i, x_j) > \varepsilon$ for $i \neq j\}$.

(b) $N_\varepsilon(C) = \{$smallest positive integer n for which there exist n ε-neighborhoods of points of E which cover $C\}$. Since C is compact, $M_\varepsilon(C)$ and $N_\varepsilon(C)$ are finite numbers.

Proposition 34. Let E be a metric space, $\mathscr{C}$ the set of all nonempty compact subsets, and let $f: E \to \mathscr{C}$ be a lower-semicontinuous (or upper-semicontinuous) set-valued function. Then f is continuous at each point of a residual set (Definition 11, §41). (Cf. [36b].)

Proof. We prove the proposition for lower-semicontinuous f, since other case is similar. We show that there is a residual set in E on which f is continuous. First, to obtain the residual set, let n be a positive integer and $\varepsilon > 0$. We define $\bar{B}_n(\varepsilon) = \{x \in E: M_\varepsilon(f(x)) \leq n$ and for each ε', $0 < \varepsilon'$ $< 3\varepsilon$ and a neighborhood V of x there exists $y \in V$ such that $f(y) \not\subset U_{\varepsilon'}(f(x))\}$.

We show that $\bar{B}_n(\varepsilon)$, $n \geq 1$ is closed. Since f is lower-semicontinuous, it is easy to verify that $M_\varepsilon(f(y)) \geq M_\varepsilon(f(x))$ for all y sufficiently close to x. Let $z \in \text{Cl } \bar{B}_n(\varepsilon)$. Then $M_\varepsilon(f(z)) \leq n$. Let V be a neighborhood of z and $0 < \varepsilon'$ $< 3\varepsilon$. Choose ε'' such that $\varepsilon' < \varepsilon'' < 3\varepsilon$. Then again by lower-semicontinuity of f, $U_{\varepsilon'}(f(z)) \subset U_{\varepsilon''}(f(x))$ for all x sufficiently close to z. Let $x \in V \cap \bar{B}_n(\varepsilon)$ such that $U_{\varepsilon'}(f(z)) \subset U_{\varepsilon''}(f(x))$. Since $x \in \bar{B}_n(\varepsilon)$ and V is a neighborhood of x, there exists $y \in V$ for which $f(y) \not\subset U_{\varepsilon''}(f(x))$. Thus $f(y) \not\subset U_{\varepsilon'}(f(z))$ and we see that $z \in \bar{B}_n(\varepsilon)$. Hence each $\bar{B}_n(\varepsilon)$ is closed for each $n \geq 1$.

Now to show that $\bar{B}_n(\varepsilon)$ is nondense, we prove that $\bar{B}_n(\varepsilon)$ cannot contain a nonempty open subset. For each $x \in \bar{B}_n(\varepsilon)$, there exists y arbitrarily close to

x for which $M_\varepsilon(f(y)) \geq M_\varepsilon(f(x)) + 1$. Let $m = M_\varepsilon(f(x))$ for $x \in \bar{B}_n(\varepsilon)$. Then there are m points $x_1, \ldots, x_m$ in $f(x)$ such that $d(x_i, x_j) > \varepsilon$ for $i \neq j$. Put $2\delta = \min\{d(x_i, x_j) - \varepsilon,\ i \neq j\}$ and $\eta = \min(\delta, \varepsilon)$. Let V be a neighborhood of x. There exists $y \in V$ such that $f(x) \subset U_\eta(f(y))$ and $f(y) \not\subset U_{2\varepsilon}(f(x))$. Choose $\{y_1, \ldots, y_m\}$ in $f(y)$ such that $d(y_i, x_i) < \eta$ for $1 \leq i \leq n$ and let $y_{m+1} \in f(y) \setminus U_{2\varepsilon}(f(x))$. Then for $j \neq k, 1 \leq j, k \leq m+1$, $d(y_j, y_k) > \varepsilon$. Thus $M_\varepsilon(f(y)) \geq m + 1$.

Put $B = \bigcup_{n \geq 1} \{\bar{B}_n(\varepsilon): \varepsilon \text{ rational} > 0\}$. Then B is of the first category and hence $G = E \setminus B$ is a residual set. We show that $f: G \to \mathscr{C}$ is upper-semicontinuous at the points of G. Let $x \in E \setminus B$ and let $\varepsilon > 0$. Choose ε', rational, $0 < \varepsilon' < \varepsilon/3$ and let $m = M_{\varepsilon'}(f(x))$. Since $x \notin \bar{B}_n(\varepsilon')$, there is a positive number $\varepsilon'' < 3\varepsilon'$ and a neighborhood V of x such that $f(y) \subset U_{\varepsilon''}(f(x))$ for all $y \in V$. Thus $f(y) \subset U_\varepsilon(f(x))$ for all $y \in V$ and so f is upper-semicontinuous. Since f is lower-semicontinuous, f is continuous.

Now we wish to specialize this situation to real-valued functions instead of set-valued functions.

Definition 11.   Let E be a topological space and R the real line with its usual metric topology. A map $f: E \to R$ is said to be *upper-semicontinuous* (respectively *lower-semicontinuous*) if for each $a \in R$, $\{x \in E: f(x) < a\}$ (respectively $\{x \in E: f(x) > a\}$) is open in E.

Proposition 35.   (a)   A continuous map $f: E \to R$ is upper-semicontinuous and lower-semicontinuous.

(b) $f: E \to R$ is continuous if and only if f is upper-semicontinuous and lower-semicontinuous.

(c)   But an upper-semicontinuous or lower-semicontinuous map $f: E \to R$ need not be continuous.

Proof.   (a) and (b) follow from the definitions.

(c)   Consider

$$f(x) = \begin{cases} 1 & \text{if } x \geq 0 \\ 0 & \text{if } x < 0. \end{cases}$$

Then f is not continuous at $x = 0$. However, it is upper-semicontinuous.

Proposition 36.   Let $f: E \to R$ and define $(-f)(x) = -f(x)$. Then f is lower-semicontinuous if and only if $-f$ is upper-semicontinuous. Thus f is continuous if and only if both f and $-f$ are upper- or lower-semicontinuous.

Proof.   Observe that for $a \in R$,

$$\{x \in E: f(x) > a\} = \{x: -f(x) < -a\}.$$

Hence this proposition follows from Proposition 35.

Proposition 37.   Let $f_\alpha (\alpha \in A)$ be a family of continuous maps from a topological space E into the extended real line $R \cup \{+\infty, -\infty\}$.

Then $\bar{f}(x) = \sup_\alpha f_\alpha(x)$ is lower-semicontinuous and $\underline{f}(x) = \inf_\alpha f_\alpha(x)$ is upper-semicontinuous. If A is a finite set of indices, then $\bar{f}$ and $\underline{f}$ are continuous.

Proof.   It is clear that for $a \in R$,

$$\{x \in E: \bar{f}(x) > a\} = \bigcup_\alpha \{x: f_\alpha(x) > a\}.$$

Hence $\bar{f}$ is lower-semicontinuous. Further, we see that $\underline{f}(x) = -\sup_{\alpha \in A} (-f_\alpha(x))$. Hence Proposition 36 applies and $\underline{f}$ is upper-semicontinuous. The remainder of the proposition follows from the observation that $\{x: \bar{f}(x) < a\} = \bigcap_\alpha \{x \in E: f_\alpha(x) < a\}$ and the fact that the finite intersection of open sets is open.

Proposition 38.   Let $f_n(x)$ be a sequence of real-valued continuous functions on a topological space E such that $\{f_n(x)\}$ converges to a function $f: E \to R$ uniformly. Then f is continuous.

Proof.   Given $\varepsilon > 0$, there exists n_0 such that for all $n \geq n_0$,

$$|f_n(x) - f(x)| < \varepsilon/3$$

for all $x \in E$, because $\{f_n(x)\}$ is uniformly convergent. Now let $x_0 \in E$. Then

$$|f(x) - f(x_0)| \leq |f(x) - f_n(x)| + |f_n(x) - f_n(x_0)| + |f_n(x_0) - f(x_0)|$$

$$\leq \frac{2\varepsilon}{3} + |f_n(x) - f_n(x_0)|$$

for $n \geq n_0$. Since each f_n is continuous, there is a neighborhood U of x_0 such that for all $x \in U$,

$$|f_n(x) - f_n(x_0)| < \frac{\varepsilon}{3},$$

whenever $n \geq n_0$ is fixed. Thus we obtain that for $x \in U$, $|f(x) - f(x_0)| < \varepsilon$, which proves that f is continuous.

Corollary 12. Let $f_n\colon E \to R$ be a sequence of continuous functions such that $|f_n(x)| \leq M_n$ for all $x \in E$, where M_n is a sequence of positive numbers such that $\sum_{n=1}^{\infty} M_n < \infty$. Then $f(x) = \sum_{n=1}^{\infty} f_n(x)$ exists and is continuous.

Proof. Put $g_n(x) = \sum_{k=1}^{n} f_k(x)$ and apply Proposition 38.

Remark. The corollary is usually called the Weierstrass M-test for uniform convergence of a series of functions.

Proposition 39. Let $f, g\colon E \to R$ be two upper-semicontinuous (respectively lower-semicontinuous) maps and $\lambda > 0$ (respectively $\lambda < 0$). Then $f + g$, fg and λf are upper-semicontinuous (respectively lower-semicontinuous).

Proof. Easy to verify.

§51. Approximately Continuous Functions

Definition 12. Let S be a subset of the real line R, endowed with the Lebesgue measure [42], denoted by m. Let $x \in S$ and let $I_\delta = I_\delta(x)$ denote the interval $[x - \delta, x + \delta]$. The number

$$\varrho_x(S) = \lim_{\delta \downarrow 0} \frac{m(S \cap I_\delta)}{m(I_\delta)}$$

is called the *metric density* of S at x.

Remark. Obviously the metric density of a set on each of its isolated points is zero. Also a countable set has metric density zero at each of its points.

Sometimes one defines upper and lower metric densities of a set S at a point x and then the metric density is the number at which the upper and lower densities coincide.

It is clear that $0 \leq \varrho_x(S) \leq 1$, since $m(S \cap I_\delta) \leq m(I_\delta)$. The sets of metric density equal to 1 play an important role.

Definition 13. A function $f\colon I \to R$, where I is a subset of R, is called *approximately continuous at* $x_0 \in I$ if for a given $\varepsilon > 0$, the set $\{x \in I : |f(x) - f(x_0)| < \varepsilon\}$ has metric density equal to 1. A function f is said to be *approximately continuous* if it is so at each point of I.

Proposition 40. Every continuous function $f: I \to R$, where I is an open interval, is approximately continuous. But the converse is not true.

Proof. Let $f: I \to R$ be continuous at $x_0 \in I$. Given $\varepsilon > 0$, there is a $\delta > 0$ such that $|f(x) - f(x_0)| < \varepsilon$ whenever $|x - x_0| < \delta$. Since

$$I_\delta = (x_0 - \delta, x_0 + \delta) \subset \{x \in I: |f(x) - f(x_0)| < \varepsilon\},$$

it follows that $\varrho_{x_0}(I_\delta) = 1$ and hence f is approximately continuous. For the converse, see Exercise 20.

An interesting characterization of approximately continuous functions is given by:

Theorem 13. A function $f: I \to R$ is approximately continuous at $x_0 \in I$ (an open interval) if and only if there is a set $G \subset I$ having metric density 1 at x_0 and such that $f \mid G$ is continuous at x_0.

Proof. "*If*" *part*: Suppose f satisfies the condition. Given $\varepsilon > 0$ there is a $\delta > 0$ such that

$$S = G \cap (x_0 - \delta, x_0 + \delta) \subset \{x: |f(x) - f(x_0)| < \varepsilon\}.$$

Since the metric density of S is 1, $\varrho_{x_0}(S) = 1$. Hence f is approximately continuous.

"*Only if*" *part*: Assume that for any given $\varepsilon > 0$, the metric density of the set

$$G_\varepsilon(x_0) = \{x \in I: |f(x) - f(x_0)| < \varepsilon\}$$

is equal to 1. Let $\{\varepsilon_n\}$ be a monotonically decreasing sequence of positive numbers tending to zero. For each ε_n there is a $\delta_n > 0$ such that

$$m(G_{\varepsilon_n}(x_0)) \cap (x_0 - \delta_n, x_0 + \delta_n) > 2\delta_n(1 - \varepsilon_n).$$

We may choose δ_n's such that $\delta_{n+1} < \delta_n$ for all $n \geq 1$ and $\delta_{n+1}/\delta_n < 1/\varepsilon_n$. Put

$$F_n = G_{\varepsilon_n}(x_0) \cap [(x_0 + \varepsilon_n\delta_{n+1}, x_0 + \delta_n) \cup (x_0 - \delta_n, x_0 - \varepsilon_n\delta_{n+1})]$$

for $n \geq 1$. Then

$$F_{n+1} \subset G_{\varepsilon_{n+1}}(x_0) \subset G_{\varepsilon_n}(x_0).$$

Let $G = \bigcup_{n=1}^{\infty} F_n \cup \{x_0\}$. Then $\varrho_{x_0}(G) = 1$, because $m(F_n) > 2\delta_n(1 - 2\varepsilon_n)$. Now to show that $f \mid G$ is continuous at x_0, we take an arbitrary sequence $\{x_n\}$ in G converging to x_0. Clearly $x_n \in F_{m(n)}$ for some m depending upon

n and $m(n) \to \infty$ as $n \to \infty$ with $|f(x_n) - f(x_0)| < \varepsilon_{m(n)}$. Hence $\{f(x_n)\}$ converges to $f(x_0)$ and so f is continuous at x_0 with respect to the points of G, which has metric density 1 at x_0. Hence f is approximately continuous.

Proposition 41. An almost continuous real function need not be approximately continuous. Nor need an approximately continuous real function be almost continuous.

Proof. Let $f(x) = 1$ if x is rational and $= -1$ if x is irrational. Clearly $f: R \to R$ is almost continuous at each rational point. But it is not approximately continuous since the metric density of rational numbers at each of its points is zero. For the second part, see Exercise 5(d).

§52. Applications of Almost Continuity

We have seen earlier that among many generalizations of continuity, almost continuity plays a significant role. First of all one sees that on certain spaces (for example, Baire and second countable spaces), each self map is almost continuous on a residual subset. In particular, each self map of the real numbers has points of almost continuity everywhere dense. Now it is clear that by strengthening conditions on the map, one may conclude that each self map on suitable spaces is almost continuous.

For this purpose, we recall that a map $f: E \to F$, where E, F are groups, is called a homomorphism if $f(x + y) = f(x) + f(y)$ $\big(\text{or } f(xy) = f(x) f(y)\big)$ if the groups are additive (or multiplicative), where $x, y \in E$. We show:

Proposition 42. Let E be a Baire topological group and let F be (a) a separable or (b) a Lindelöf Hausdorff topological group. Then each homomorphism f of E into F is almost continuous.

Proof. (a) Assume F is separable and let $\{y_n\}$ be a countable dense subset of F. Let W be a neighborhood of the identity in F. Choose a symmetric neighborhood V of the identity e in F with $V^2 \subset W$. Then $F = \bigcup_{n=1}^{\infty} y_n V$. Hence

$$E = f^{-1}(F) = \bigcup_{n=1}^{\infty} f^{-1}(y_v V)$$

and so

$$E = \bigcup_{n=1}^{\infty} \overline{f^{-1}(y_n V)}.$$

Since E is a Baire space, for at least one n_0, $\overline{f^{-1}(y_{n_0}V)}$ has an interior point. Since $y_{n_0}V$ and V are homeomorphic sets in F, it is clear that $\overline{f^{-1}(y_{n_0}V)}$ and $\overline{f^{-1}(V)}$ are homeomorphic and so $\overline{f^{-1}(V)}$ has an interior point x_0, say. But then $e = x_0 x_0^{-1}$ is an interior point of $\overline{f^{-1}(V)}\,\overline{f^{-1}(V)} \subset \overline{f^{-1}(V^2)} \subset \overline{f^{-1}(W)}$. Thus f is almost continuous at the identity of E. Since the translations in a topological group are homeomorphisms, it follows that f is almost continuous at each point of E.

(b) Assume F is Lindelöf. Let V and W be open neighborhoods of the identity e in F such that $V^2 \subset W$. Clearly $f(E) \subset \overline{f(E)} \subset f(E)V$ (Exercise 14, Chap. V). That means $\{f(x)V: x \in E\}$ is an open covering of a Lindelöf space $f(E)$, since every subspace of a Lindelöf Hausdorff space is a Lindelöf space. Thus only a countable subcovering, say $\{f(x_n)V: x_n \in E\}$, covers $f(E)$, i.e., $f(E) = \bigcup_{n=1}^{\infty} f(x_n)V$. Hence

$$E = f^{-1}(f(E)) = \bigcup_{n=1}^{\infty} \overline{f^{-1}(f(x_n)V)}.$$

Since E is a Baire space, at least for one n_0, $\overline{f^{-1}(f(x_{n_0})V)}$ has an interior point. Now by the same argument as in (a), it follows that f is almost continuous.

Corollary 13.   Let E be a metrizable complete topological group and F a separable or Lindelöf topological group. Then each homomorphism $f: E \to F$ is almost continuous.

Proof.   Since each complete metric space is a Baire space, the corollary follows from Proposition 42.

Definition 14.   A map $f: E \to F$ where E, F are topological spaces, is said to be *almost open* if for each open neighborhood U of x, $\overline{f(U)}$ is a neighborhood of $f(x)$.

By the same argument as in Proposition 42 and Corollary 13, we obtain:

Proposition 43.   Let E be a Baire or, in particular, a complete metrizable topological group and F a separable or Lindelöf Hausdorff topological group. Then each homomorphism of F onto E is almost open.

Let E, F be two real or complex linear spaces. A mapping $f: E \to F$ is called *linear* if $f(\alpha x + \beta y) = \alpha f(x) + \beta f(y)$, where $x, y \in E$ and α, β are real or complex scalars. Recall that a subset $M \subset E$ is said to be *circled* if for all $x \in M$ and all scalars λ, $|\lambda| \leq 1$, $\lambda x \in M$ (see Definition 16, §31).

Theorem 14. Let E be a Baire topological vector space and F a topological vector space. Then each linear mapping of E into F is almost continuous.

Proof. Let W and V be circled neighborhoods of the origin $0 \in F$ such that $V + V \subset W$. Clearly $\bigcup_{n \geq 1}^{\infty} nV = F$, because if $y \in F$, then $(1/n)y \to 0$ as $n \to \infty$ and hence for large n, $y/n \subset V$ or $y \in nV$. Hence

$$E = f^{-1}(F) = \bigcup_{n=1}^{\infty} nf^{-1}(V)$$

(because f is linear). Since E is a Baire space, for at least one n_0, $\overline{n_0 f^{-1}(V)} = n_0 \overline{f^{-1}(V)}$ has an interior point. Thus $\overline{f^{-1}(V)}$ has an interior point, since $\overline{f^{-1}(V)}$ and $n_0 \overline{f^{-1}(V)}$ are homeomorphic. Let x_0 be the interior point of $\overline{f^{-1}(V)}$. Then $-x_0$ is the interior point of $-\overline{f^{-1}(V)} = \overline{f^{-1}(-V)} = \overline{f^{-1}(V)}$ because V is circled and hence symmetric. Thus $0 = x_0 + (-x_0)$ is an interior point of

$$\overline{f^{-1}(V)} + \overline{f^{-1}(V)} \subset \overline{f^{-1}(V) + f^{-1}(V)} \subset \overline{f^{-1}(W)}.$$

Hence f is almost continuous at 0. Since translations are homeomorphisms, the result follows.

Corollary 14. Let F be a topological vector space and E a Frechét space a Banach space, or a Hilbert space and $f: E \to F$ is a linear mapping. Then f is almost continuous.

Proof. This is a particular case of Theorem 14.

Proposition 44. Let E be a topological vector space, F a Baire topological vector space or, in particular, a Fréchet space, Banach space, or Hilbert space and f a linear mapping of E onto F. Then f is almost open.

Proof. Similar to Theorem 14.

The above theorem and propositions hold even for a larger class of locally convex spaces than the Baire class of topological locally convex spaces. For example see Husain [45].

We have seen earlier that for certain classes of topological spaces and certain maps, almost continuity implies continuity. Similarly, for homomorphisms on topological groups, we have the following:

Theorem 15. (a) (Closed graph theorem). Let E be a topological group and F a complete metrizable topological group. Let $f: E \to F$ be an almost continuous homomorphism, the graph of which is closed. Then f is continuous.

(b) (Open mapping theorem). Let E, F be the same as in (a) and let $g: F \to E$ be a continuous and almost open homomorphism. Then g is open.

Proof. (a) Let $\{V_n\}$ be a countable base of closed neighborhoods of the identity in F such that each V_n is symmetric (i.e., $-V_n = V_n$) and $V_{n+1} + V_{n+1} \subset V_n$ for $n \geq 1$, and for each $a \in F$ and V_n, there exists V_m with $-a + V_n + a \subset V_m$. First assume f is $1 : 1$. Since f is almost continuous, for each n, $U_n = f^{-1}(V_n)$ is a neighborhood of the identity 0 in E. To show that f is continuous, it is sufficient to show that $f^{-1}(V_n)$ is a neighborhood of the identity in E for each n. Consider $U_n = \overline{f^{-1}(V_n)}$. We claim that $f(U_{n+1}) \subset V_n$, i.e., $x \in U_{n+1}$ implies $f(x) \in V_n$. Since $U_{n+1} = \overline{f^{-1}(V_{n+1})}$, there exists $x_1 \in f^{-1}(V_{n+1})$ or $f(x_1) = y_1 \in V_{n+1}$ such that $x - x_1 \in U_{n+2}$. By induction there exists $x_m \in f^{-1}(V_{n+m})$ or $f(x_m) = y_m \in V_{n+m}$ such that $x - \sum_{i=1}^{m} x_i \in U_{n+m+1}$. Clearly

$$\sum_{i=k}^{m} y_i \in \sum_{i=k}^{m} V_{n+i} \subset V_{n+k-1} \qquad (m \geq k \geq 1).$$

Since $\{V_n\}$ is a decreasing sequence, we see that $\{\sum_{i=k}^{m} y_i\}$ forms a Cauchy sequence in F. Since F is complete, there exists $z \in F$ such that $z = \sum_{i=1}^{\infty} y_i$. Clearly $\sum_{i=1}^{m} y_i \in V_n$ for each m and since V_n is closed, $z \in V_n$. Also

$$z = \sum_{i=1}^{\infty} y_i = \sum_{i=1}^{\infty} f(x_i) \in V_n.$$

Observe that

$$x - \sum_{i=1}^{m} x_i \in U_{n+m+1} \subset U_{n+k+1}$$

for all $m \geq k$. Since U_{n+k+1} is strictly decreasing as $k \to \infty$, the sequence $\{x - \sum_{i=1}^{m} x_i\}_{m \geq 1}$ converges to 0 as $m \to \infty$. Hence

$$x - \sum_{i=1}^{\infty} x_i \in U_{n+k+1} \qquad \text{for all } k \geq 1,$$

because U_{n+k+1} is closed. Therefore

$$x - \sum_{i=1}^{\infty} x_i \in \bigcap_{k=1}^{\infty} U_{n+k+1}.$$

Observe that

$$\left(x + \sum_{i=1}^{m} x_i, f(x) - \sum_{i=1}^{m} f(x_i)\right) \in G_f,$$

the graph of f. Hence

$$\left(x - \sum_{i=1}^{\infty} x_i, f(x) - z\right) \in \bar{G}_f = G_f.$$

We show that $x - \sum_{i=1}^{\infty} x_i = 0$. It is sufficient to show that $\bigcap_{m=1}^{\infty} U_m$ $= \bigcap_{k=1}^{\infty} U_{n+k+1} = \{0\}$. Let $x' \in U_m = \overline{f^{-1}(V_m)}$ for all $m \geq 1$. Then there exists $x_m' = f^{-1}(y_m')$, $y_m' \in V_m$ such that $x_m' \in x' + U$ for any arbitrary neighborhood U of 0. Observe that $\{x_m'\}$ converges to x' and $y_m' \in V_m$ implies that $\{y_m'\}$ converges to $0 \in F$ because $\{V_m\}$ is a decreasing sequence of a fundamental system of neighborhoods of $0 \in F$. Since $(x_n', f(x_n'))$ $\in (x' + U, V_n) \cap G_f$, it follows that $(x', 0) \in G_f$ because G_f is closed. Hence $f(x') = 0$. Since f is 1 : 1, $x' = 0$. This proves that $\bigcap_{k=1}^{\infty} U_{n+k+1} = \{0\}$, which shows that $x = \sum_{i=1}^{\infty} x_i$ and so $f(x) - z = 0$ or $z = f(x) \in V_n$ or $x \in f^{-1}(V_n)$. Since x is an arbitrary point of U_{n+1}, we have shown that $f(U_{n+1}) \subset V_n$. This completes the proof of (a), provided f is 1 : 1.

Now to remove the additional condition that f is 1 : 1, we observe that $f^{-1}(0)$, when $0 \in F$ is the identity, is a closed (because the graph of f is closed) normal subgroup of E and hence $E/f^{-1}(0)$ is again a topological group. Moreover, the induced homomorphism $\varphi: E/f^{-1}(0) \to F$ is one-to-one, almost continuous, and has closed graph; therefore it is continuous by the special case proved above. Also the canonical homomorphism $\varphi: E \to E/f^{-1}(0)$ is always continuous and open. Hence $f: E \to F$, being the composition of continuous homomorphisms $\varphi: E \to E/f^{-1}(0) \to F$, is also continuous.

(b) Since $g: F \to E$ is continuous, $g^{-1}(0)$ is a closed normal subgroup of F and hence $F/g^{-1}(0)$ is a complete metrizable topological group. Since the canonical map $\varphi: F \to F/g^{-1}(0)$ is continuous and open, and since the map $g': F/g^{-1}(0) \to E$ is one-to-one and therefore g'^{-1} is a one-to-one map from E to $F/g^{-1}(0)$, which is easily seen to be almost continuous and has closed graph, by (a), g'^{-1} is continuous and so g' is open. Hence g is open because φ is open. This completes the proof of part (b).

Now we wish to consider the case for topological vector spaces and indicate that on more general locally convex spaces than Banach spaces, a linear map is automatically almost continuous.

Definition 15. (a) Let E be a locally convex space. A subset B of E is called a *barrel* if B is convex, circled, absorbing, and closed (see §31, Definition 16).

(b) A subset A is said to *absorb* a subset B if there exists $\alpha > 0$ such that $\lambda B \subset A$ for all. $| \lambda | \leq \alpha$, $\lambda \neq 0$.

(c) A subset M of a topological vector space E is called *bounded* if for each neighborhood U of 0 in E there exists $\alpha > 0$ such that $\lambda M \subset U$ for all $| \lambda | \leq \alpha$, $\lambda \neq 0$, i.e., M is absorbed by every neighborhood of 0.

It was pointed out earlier (§28) that in each locally convex space there exists a base of neighborhoods of 0, which are convex, circled, absorbing, and closed, i.e., each one of them is a barrel. But it is not always true that a barrel is a neighborhood of 0. See, e.g., Husain [45] or Bourbaki [14]. All locally convex spaces in the sequel are Hausdorff.

Definition 16. (a) A locally convex space E is said to be *barreled* if each barrel in E is a neighborhood of 0.

(b) A locally convex space is said to be *quasibarreled* if each barrel which absorbs each bounded subset is a neighborhood of 0.

(c) A locally convex space E is said to be *bornological* if each convex set which absorbs each bounded subset of E is a neighborhood of 0.

Each barreled and each bornological space is quasibarreled. But the converses do not hold. In general, a barreled space is not necessarily bornological nor is a bornological space necessarily barreled, Bourbaki [14] and Husain [45].

Proposition 45. Every Baire locally convex space E is barreled and hence quasibarreled.

Proof. Let B be a barrel in E. Then $E = \bigcup_{n=1}^{\infty} nB$, since for each $x \in E$, there exists $\lambda > 0$ such that $x \in \lambda B$ because B is absorbing and by choosing a positive integer n with $\lambda/n \leq 1$ we obtain $x \in n(\lambda/n)B \subset nB$, because B is circled. Since E is a Baire space, for at least one n_0, $\overline{n_0 B} = n_0 \bar{B} = n_0 B$ (because B is closed) has an interior point. Since $n_0 B$ is homeomorphic with B, B has an interior point x_0, say. But then $-x_0$ is an interior point of $-B = B$. Since B is convex, $0 = \frac{1}{2}(x_0) + \frac{1}{2}(-x_0)$ is an interior point of $\frac{1}{2}B + \frac{1}{2}B = B$ (because B is convex). This proves that B is a neighborhood of 0 and so E is barreled.

In particular, every Fréchet locally convex space, each Banach space, and in particular each Hilbert space is barreled. Also each metrizable locally convex space is bornological and so quasibarreled (cf. Husain [45] or Köthe [55] or Kelley and Namioka [98]).

Theorem 16. Let E be a barreled space and F a locally convex space. Then:

    (a)   Each linear mapping $f: E \to F$ is almost continuous;

    (b)   each linear surjective mapping $g: F \to E$ is almost open.

Proof. (a) Let U be a convex circled neighborhood of 0 in F. Then $f^{-1}(U)$ is convex and circled. Since each neighborhood of 0, in particular U, is absorbing (observe: For $x \in E$, $\lambda x \to 0$ as $\lambda \to 0$. It follows that $\lambda x \in U$ for small $\lambda > 0$ and hence $x \in \mu U$, for sufficiently large $\mu > 0$), $f^{-1}(U)$ is absorbing. Since the closure operation preserves convexity and the circling and absorbing properties, we see that $f^{-1}(U)$ is a barrel, hence a neighborhood of 0 in E because E is barreled. This proves that f is almost continuous at 0 and hence everywhere, because f is linear.

    (b)   Similar proof to (a).

In view of Proposition 45, it follows from Theorem 16 that each linear map from a Baire locally convex space to any other locally convex space is almost continuous and a linear map from a locally convex space onto a Baire space is almost open. Using these facts and the duality theory of locally convex spaces, it is possible to extend Theorem 15 to more general linear topological spaces than the metrizable complete ones. For example, see Husain [45], Schaefer [99], etc.

Thus we observe that the category argument used as in the form of Baire spaces or as in the Fréchet spaces is extended by the notion of almost continuity. Another application of the category argument is found in:

Theorem 17 (Banach–Steinhaus). Let $\{f_n\}$ be a sequence of linear continuous maps from a Banach space E into a normed space F such that for each $x \in E$ there exists a real number $M_x > 0$ (depending upon x) such that $\| f_n(x) \| \le M_x$ for all $n \ge 1$. Then there is a real number $M > 0$ such that

$$\sup\{\| f_n(x) \|: \| x \| \le 1\} = \| f_n \| \le M$$

for all $n \ge 1$. In particular, if $F = \mathbb{R}$, the real line, and $f(x) = \lim_{n \to \infty} f_n(x)$, then f is a continuous linear functional.

Proof. For each $m > 0$ (integer), consider $E_m = \{x \in E: \| f_n(x) \| \le m$ for all $n \ge 1\}$. Clearly E_m is a closed subset of E. Also if $x \in E$ then there exists

M_x such that $\|f_n(x)\| \le M_x$ for all $n \ge 1$. Choose a large enough positive integer m; then $x \in E_m$. Hence $E = \bigcup_{m=1}^{\infty} E_m$. Since E is a Baire space (because each Banach space is a Baire space), at least one E_{m_0} contains a ball. Let $\delta > 0$ and $x_0 \in E_{m_0}$ such that

$$B_\delta(x_0) = \{x \in E \colon \|x - x_0\| < \delta\} \subset E_{m_0}.$$

Thus $\|f_n(x)\| \le m_0$ for all $x \in B_\delta(x_0)$ and $n \ge 1$. Put $y = x - x_0$. Then for $\|y\| < \delta$ we have

$$\|f_n(y)\| = \|f_n(x - x_0)\| \le m_0 + \|f_n(x_0)\| \le M,$$

for all $n \ge 1$. But then for $\|z\| \le 1$, where $z = y/\delta$, we have

$$\|f_n(z)\| = \|f_n(y/\delta)\| \le M\delta^{-1}$$

for all $n \ge 1$. This shows that

$$\sup_{\|z\| \le 1} \|f_n(z)\| \le \delta^{-1} M$$

or $\|f_n\| \le \delta^{-1} M$ for all $n \ge 1$. This proves the first part of the theorem. For the second part, it is easy to check that f is linear and

$$|f(x)| \le \|f_n\| \, \|x\| \le \delta^{-1} M \|x\|,$$

which proves that f is continuous.

Now we give a sample of applications of the category argument to analysis.

Theorem 18. There exists a nondifferentiable, everywhere continuous function on $I = [0, 1]$.

Proof. Let $C(I)$ denote the set of all real-valued continuous functions on I. For $f, g \in C(I)$, put

$$d(f, g) = \sup_{x \in I} |f(x) - g(x)| = \|f - g\|.$$

Then d is trivially a metric on $C(I)$. To complete the proof we will need the following three results.

Proposition 46. $C(I)$ is complete in this metric.

Proof. Let $\{f_n\}$ be a Cauchy sequence in $C(I)$. Then $d(f_n, f_m) \to 0$ as $n, m \to \infty$. Hence for each $x \in I$, $\{f_n(x)\}$ is a Cauchy sequence of real numbers and hence $f(x) = \lim_{n\to\infty} f_n(x)$ exists. Thus f is a real-valued function on I. We show that $f \in C(I)$. Consider the obvious inequality

$$|f(x) - f(x_0)| \leq |f(x) - f_n(x)| + |f_n(x) - f_n(x_0)| + |f_n(x_0) - f(x_0)|.$$

Given $\varepsilon > 0$, there exists n_0 such that for $n \geq n_0$, $|f(x) - f_n(x)| < \varepsilon/3$ and $|f_n(x_0) - f(x_0)| < \varepsilon/3$. But then for any fixed $n \geq n_0$, since f_n is continuous there exists $\delta > 0$ such that for all x, $|x - x_0| < \delta$, we have $|f_n(x) - f_n(x_0)| < \varepsilon/3$. Thus for $|x - x_0| < \delta$, we have $|f(x) - f(x_0)| < \varepsilon$ and so f is continuous. It its easy to see that $d(f_n, f) \to 0$ as $n \to \infty$.

Corollary 15. $C(I)$ is a Baire space.

Proof.   This follows from §41, Theorem 18.

Theorem 19.   The set of all $f \in C(I)$, where f is differentiable, forms a set of the first category in $C(I)$.

Proof.   Let n be a positive integer. Define

$$E_n = \{f \in C(I) \colon \exists x, 0 \leq x \leq 1 - 1/n \text{ such that } \left|\frac{f(x+h) - f(x)}{h}\right| \leq n$$
$$\text{for all } h, 0 < h \leq 1/n\}.$$

It is clear that any f which is differentiable at even one point of $[0, 1]$ is in E_n. Put $E = \bigcup_{n=1}^{\infty} E_n$. If we show that each E_n is nondense, then E is of the first category. Since $C(I)$ is a Baire space, $C(I) \setminus E \neq \varnothing$.

First we show that each E_n is closed. Clearly

$$E_n = \bigcap_{0 < h < 1/n} \left\{f \in C(I) \colon \exists x, 0 \leq x \leq 1 - 1/n \text{ such that } \left|\frac{f(x+h) - f(x)}{h}\right| \leq n\right\}.$$

Now for any fixed h and for any sequence $f_m \in C(I)$ such that $|[f_m(x + h) - f_m(x)]/h| \leq n$, where $f_m \to f$ in the norm of $C(I)$, we conclude that

$$\left|\frac{f(x+h) - f(x)}{h}\right| \leq n \qquad \text{for all } x, 0 \leq x \leq 1 - \frac{1}{n}.$$

Thus for any fixed h, the set

$$\left\{f \in C(I) \colon \exists x, 0 \leq x \leq 1 - \frac{1}{n} \ni \left|\frac{f(x+h) - f(x)}{h}\right| \leq n\right\}$$

is closed and so is its intersection E_n for each n.

Now to show that E_n has empty interior, we take $f \in E_n$. It is enough to show that each ball $B_\varepsilon(f)$ in $C(I)$ contains a function $g \in C(I)$, $\| g - f \| < \varepsilon$ such that for all x, $0 \le x \le 1 - 1/n$, there exists h, $0 \le h \le 1/n$ such that $| [g(x + h) - g(x)]/h | > n$.

Since polynomials are dense in $C(I)$ (Corollary 7, §64), we can find a polynomial p on I such that $\| p - f \| < \varepsilon/2$. Clearly the derivative p' of p is in $C(I)$. Let $M = \sup_{x \in I} | p'(x) |$. Now take a continuous function q which is piecewise linear as follows: The slope of the line segment is $\pm M + n + 1$ and its ordinate never goes more than $\varepsilon/2$ beyond the x axis. Then $\| q(x) \| \le \varepsilon/2$ and $p + q$ is a continuous function and

$$\| f - (p + q) \| \le \| f - p \| + \| q \| < \varepsilon$$

so that $p + q \in B_\varepsilon(f)$ and

$$\left| \frac{p(x + h) + q(x + h) - p(x) - q(x)}{h} \right|$$
$$\ge \left\| \left| \frac{q(x + h) - q(x)}{h} \right| - \left| \frac{p(x + h) - p(x)}{h} \right| \right\|$$

for each $x \in [0, 1 - 1/n]$. But then we can choose $h \in [0, 1/n]$ such that the right-hand side is $\ge M + m + 1 - M = n + 1 > n$. Thus E_n is nondense. This completes the proof of Theorem 19.

Since the set of differentiable functions in the Baire space $C(I)$ is of the first category, it shows the existence of a continuous but nondifferentiable function on I, and the proof of Theorem 18 is now complete.

Similar arguments show the following:

Proposition 47. The set of all analytic functions on the complex plane is of the first category in the Fréchet space C^∞ of all infinitely differentiable functions, with the countable seminorms $\{p_i\}$ defined by: For positive integers n, j, let $\{p_{nj}\}$ be rearranged in a sequence $\{p_i\}$, where

$$p_{nj}(f) = \sup\{| f^{(i)}(x) | : | x | \le j, 0 \le i \le n\} \qquad (f^{(0)} = f),$$

and then the metric is given by

$$d(f, g) = \sum_{i=0}^{\infty} \frac{1}{2^i} \frac{p_i(f - g)}{1 + p_i(f - g)},$$

which is sometimes called the Fréchet metric.

Examples and Exercises

1. Let $I = [0, 1]$ and let $f: I \to R$ be a real-valued function. Show that f is almost continuous iff for each open subset P of the reals

$$f^{-1}(P) \subset [(\overline{(f^{-1}(P))^c})]^\circ.$$

2. Show that for any ordinal $\alpha > 1$ of first or second class there exists f: $I \to R$ which is almost continuous and of Baire class α on I (see [19]).

3. Show that there exists an almost continuous $f: I \to R$ which is not Lebesgue measurable (see [19]).

4. Let u be the topology on the real line R such that each almost continuous $f: (R, u) \to (R, u)$ is continuous. Show that u is the discrete topology.

5. A mapping $f: I \to R$ is called *quasicontinuous* at $x_0 \in I$ if for each neighborhood U of x_0 and each neighborhood V of $f(x_0)$ there exists a nonempty open set $P \subset U$ such that $f(P) \subset V$. Show the following:
    (a) $f: I \to R$ is quasicontinuous iff for every open set $P \subset R$,

$$f^{-1}(P) \subset [\overline{f^{-1}(P)}]^\circ.$$

    (b) If $f: I \to R$ is almost continuous and quasicontinuous then f is continuous (see [19]).
    (c) There exists an almost continuous function $f: I \to R$ which is not quasicontinuous, and a quasicontinuous function $g: I \to R$ which is not almost continuous.
    (d) There exists a Darboux continuous function $f: I \to R$ (i.e., if $x_1 \neq x_2$ and if $f(x_1) \leq y \leq f(x_2)$ then there exists $x_3 \in I$ such that $f(x_3) = y$) which is not almost continuous and an almost continuous function which is not Darboux [19].

6. There exists an approximately continuous function which is not almost continuous.

7. Let

$$f(x) = \begin{cases} x, & x \text{ rational,} \\ -x, & x \text{ irrational.} \end{cases}$$

    Show the following:
    (a) f is almost continuous;
    (b) f is not graphically continuous;
    (c) f is not a connected map, i.e., it does not map connected sets into connected sets.
    (d) f is not a connectivity map, i.e., the graph function: $g(x) = (x, f(x))$ from R to R^2, is not connected.

8. Let

$$f(x) = \begin{cases} \sin\dfrac{1}{x}, & x \neq 0, \\ 0, & x = 0. \end{cases}$$

Show the following:
   (a) f is graphically continuous;
   (b) f is not almost continuous;
   (c) f is connected;
   (d) f is a connectivity function.

9. Let H be a Hamel basis of the real numbers R and e any function from H to R. Show that the function f defined by

$$f(rh) = e(h) \text{ for } h \in H$$

with $r \neq 0$ rational and $f(0) = 0$ can be extended to a function g on R by additivity such that g is Darboux continuous but discontinuous.

10. Show that a Darboux continuous function which is in Baire class one has the connected graph [17].

11. If f is a Darboux continuous function and if $\{y:f^{-1}(y) \text{ is infinite}\}^{\circ} = \emptyset$, then f is continuous.

12. Let $f: I \to R$ be a lower-semicontinuous function. Then f is Darboux continuous iff for each $x \in I$, $f(x) = \underline{\lim}_{z \to x^+} f(z) = \underline{\lim}_{z \to x^-} f(z)$ (see [18]).

13. If f is approximately continuous, g a derivative, and f or g is bounded on I, then $f + g$ is Darboux continuous.

14. A map $f: E \to F$ is said to be σ-continuous if for each neighborhood V of $f(x)$, where $x \in E$ is arbitrary, there is a neighborhood U of x such that $f(\bar{U}) \subset \bar{V}$. Show that:
   (a) Each σ-continuous map is weakly continuous;
   (b) every nearly continuous map is σ-continuous;
   (c) almost continuity does not imply σ-continuity (see Example 7).

15. Let $\{E_n\}$ be a countable collection of pairwise disjoint sets in $[0, 1]$ such that each E_n is dense everywhere in $[0, 1]$ and $[0, 1] = \bigcup_{n=1}^{\infty} E_n$. (Such a decomposition exists, as is easy to see.) Put $f(x) = n$ if $x \in E_n$. Then f is almost continuous but not bounded on $[0, 1]$ (see [49]).

16. Let f_n be defined on $[-1, -1]$ as follows:

$$f_n(x) = \begin{cases} 1, & \text{if } \tfrac{1}{2} \leq x \leq 1; \\ n(x - \tfrac{1}{2}) + 1, & \text{if } \tfrac{1}{2} - 1/n \leq x \leq \tfrac{1}{2}; \\ 0, & \text{if } -1 \leq x \leq \tfrac{1}{2} - 1/n. \end{cases}$$

Then $f(x) = \lim_{n \to \infty} f_n(x) = 1$ if $\tfrac{1}{2} \leq x \leq 1$ and $= 0$ if $-1 \leq x < \tfrac{1}{2}$. Show that each f_n is almost continuous but f is not (see [49]).

17. Show that if f, g are almost (or graphically) continuous, then fg, f/g and $f \circ g$ need not be almost (or graphically) continuous.

18. Show that the function

$$f(x, y) = \frac{2xy}{x^2 + y^2} \sin \frac{\pi}{(x^2 + y^2)^{1/2}}, \qquad (x, y) \neq (0, 0),$$

and $f(0, 0) = 0$ is continuous in each variable separately, connected but not continuous in both variables together.

19. Let f be the function defined by

$$f(x) = \overline{\lim_{n \to \infty}} \frac{a_1 + \cdots + a_n}{n}$$

for $x \in [0, 1]$ and $f(x) = 0$ for all $x \notin [0, 1]$, where $x = 0 \cdot a_1 a_2 \cdots a_n \cdots$ (nonterminating binary expansion of x). Show that: (i) f is not continuous; (ii) f is graphically continuous; (iii) the graph of f is a connected subset of $[0, 1] \times R$ (see [80]).

20. Let $\{U\}$ denote the collection of all open sets of the real line and $\{N\}$ the collection of nowhere dense subsets. Then $\mathscr{P} = \{U \backslash N : U \in \{U\}, N \in \{N\}\}$ defines a topology $\mathscr{C}'$ strictly finer than the metric topology $\mathscr{C}$ on the real line R. Show that the identity mapping i of $(R, \mathscr{C})$ onto $(R, \mathscr{C}')$ is not continuous. Is i approximately continuous? Is it almost continuous?

21. Is a connectivity map graphically continuous?

22. Show that a homomorphism $f: E \to F$ of a Hausdorff topological group E into a compact topological group F with a closed graph is continuous.

23. Let E be a real Hausdorff locally convex space and E' the set of all continuous linear real-valued maps on E. [E' is called the (*topological*) *dual* of E.] Let $\sigma(E', E)$ denote the weak* topology on E', i.e., the coarsest topology on E' which makes E' a locally convex space such that for each $x \in E$ each map $f \to f(x)$ of E' into the reals is continuous.
Show that $(E', \sigma(E', E))$ is a Hausdorff locally convex space.

24. Let E be a normed space with the norm $\| \cdots \|$ and E' its dual. For each $f \in E'$, put $\| f \| = \sup\{| f(x) |\} : \| x \| \leq 1\}$. Show that E' with $\| f \|$ is a Banach space. Show that the unit ball $\{f \in E' : \| f \| \leq 1\}$ is $\sigma(E', E)$-compact. (Alaoglu's theorem: see [49] or Bourbaki [14], etc.).

25. Let $(E, \| \cdots \|)$ be an infinite-dimensional real (or complex) Banach space and E^* the set of all linear real- (or complex-) valued maps

(called the *algebraic dual*). Let ω denote the *finest* locally convex topology on E, i.e., the topology with respect to which each map $x \to f(x)$ (for $f \in E^*$) is continuous from E to $\mathbb{R}$ (or C). Show that the identity map $i: (E, \omega) \to (E, \| \cdots \|)$ is continuous (hence almost continuous and has closed graph), almost open but not open, and $i^{-1}: (E, \| \cdots \|) \to (E, \omega)$ has closed graph and is almost continuous but not continuous.

26. (a) Let E, F be topological vector spaces and f a continuous linear mapping of E into F. Show that each bounded subset of E is mapped onto a bounded subset of F under f. Give an example showing that a noncontinuous linear map may map bounded sets into bounded sets.

    (b) Let u, v be two topologies on a linear space E such that (E, u) and (E, v) are topological vector spaces and $u \supset v$. Show that each u-bounded subset is also v-bounded.

    (c) A linear map of E into F which maps bounded subsets into bounded subsets is continuous iff E is bornological.

27. Let E be a Hausdorff topological vector space. Then a subset $A \subset E$ is bounded iff for each sequence $\{\lambda_n\}$ of scalars such that $\lambda_n \to 0$ and each sequence $\{x_n\} \subset A$, $\lambda_n x_n \to 0$ in the topology of E.

28. Show that each compact subset of a topological vector space is bounded.

29. (A barreled space in which each closed and bounded set is compact is called a *Montel space*.) Show that a normed linear space is a Montel space iff it is finite-dimensional.

30. Give an example of an infinite-dimensional Montel Fréchet space. Is the space of all holomorphic function on the open disk $|z| < 1$ of the complex numbers a Montel space? (Answer: yes, use Ascoli's theorem, Chapter VIII.)

31. A map $f: E \to F$ is said to be *inversely compact* if for each singleton $\{y\} \subset F$, $f^{-1}\{y\}$ is a compact subset of E. Show that a continuous map need not be inversely compact, nor need an inversely compact map be continuous, nor closed, nor open.

32. A continuous, closed, inversely compact surjective map $f: E \to F$ is called a *perfect* (or *proper*) map. Let $f: E \to F$ be a perfect map. Show that:

    (a) If E is Hausdorff, so is F.
    (b) If E is regular, so is F.
    (c) If E is completely regular, so is F provided f, in addition, is open.
    (d) If E is metrizable, so is F.
    (e) If E is second countable, so is F.

    (f)   If F is compact, so is E.

    (g)   If F is countably compact, so is E.

    (h)   If F is Lindelöf, so is E.

    (i)   If F is paracompact, so is E.

    (j) E is locally compact iff F is locally compact.

    (k)   Show that the composition of perfect maps is perfect.

33.   Show that:

    (a)   For each function $f\colon [0, 1] \to R$ there is an everywhere dense subset $E \subset [0, 1]$ such that f is continuous on E relative to E (see [96]).

    (b)   There exist maps of $I = [0, 1]$ onto itself such that for each everywhere dense subset A of E, f and f^{-1} are not a homeomorphisms between A and $f(A)$ (see [40]).

VIII

Function Spaces

In this chapter we study the set of all maps from one topological space to another with various topologies. These various topologies, when induced on certain subsets of the set of all maps from one topological space to another, give information about the convergence of sequences or nets that are used on these subsets. The main subset of the set of all maps from one topological space to another is the set of all continuous maps. The study of this set is very important for applications.

§53. The Set of All Maps

Let E, F be any two sets; then F^E denotes the set of all maps from E into F. Clearly if the set E is finite or, equivalently, the cardinality of E is equal to a positive integer n, then F^E is the set of all finite sequences $\{a_i\}_{i=1}^{n}$, where $a_i \in F$. If Car $E = \aleph_0$ (aleph naught), then $F^{\aleph_0}$ consists of all infinite sequences $\{a_i\}_{i \geq 1}$ such that $a_i \in F$ for $i \geq 1$. These two special cases are of great interest in practically every field of mathematics.

Definition 1. Let E, F be two topological spaces and let F^E denote the set of all maps from E to F. For each open set U in F and for a subset K of E, we define

$$T(K, U) = \{f \in F^E : f(x) \in U \text{ for all } x \in K\}.$$

Proposition 1. The collection $\{T(x, U): x \in E, U \text{ open in } F\}$ of subsets in F^E forms a subbase of a topology in F^E.

Proof. Let $T(x_1, U_1)$ and $T(x_2, U_2)$ be two members of the collection, where x_1, $x_2 \in E$ and U_1, U_2 are open subsets of F. Then it is easy to check that

$$T(\{x_1, x_2\}, U_1 \cap U_2) \subset T(x_1, U_1) \cap T(x_2, U_2).$$

Since $U_1 \cap U_2$ is open, it follows that the family $\{T(K, U): K, \text{ a finite subset}$ of E and U open in $F\}$ forms a base of a topology.

Definition 2. The topology defined by the collection $\{T(x, U): x \in E,$ U an open set in $F\}$ in F^E is called the *point–open topology* and is denoted by $\mathscr{C}_p$ on F^E.

Proposition 2. Let E be a set, F a topological space, and let F^E be endowed with the point–open topology $\mathscr{C}_p$. Then for each $x \in E$, the *projection map* or *evaluation map*: $f \to f(x)$ of F^E into F is continuous. Furthermore, $\mathscr{C}_p$ is the coarsest topology on F^E which makes the mapping $f \to f(x)$ continuous.

Proof. Let U be an open neighborhood of $f(x) \in F$. The set $T(x, U)$ $= \{g \in F^E : g(x) \in U\}$ is an open subset of F^E in the topology $\mathscr{C}_p$ by definition. Also clearly $f \in T(x, U)$. Hence the projection mapping: $f \to f(x)$ is continuous.

Now suppose $\mathscr{C}_{p_1}$ is another topology on F^E which makes each projection map: $f \to f(x)$ of F^E into F continuous. Then for each open neighborhood U of $f(x)$, there is a $\mathscr{C}_{p_1}$-open neighborhood V of f in F^E such that for all $g \in V$, $g(x) \in U$. But then it follows that $V \subset T(x, U)$ and so $\mathscr{C}_p$ is coarser than $\mathscr{C}_{p_1}$.

Definition 3. A net $(f_\alpha)_{\alpha \in D}$ in the topological space F^E is said to *converge simply* to $f \in F^E$ if for each $x \in E$, the net $\{f_\alpha(x)\}_{\alpha \in D}$ in F converges to $f(x)$.

Proposition 3. Let E be a set and F a topological space. Let F^E be endowed with a topology $\mathscr{C}$; then the following statements are equivalent:

(a) $\mathscr{C} = \mathscr{C}_p$ is the point–open topology on F^E;

(b) $\mathscr{C}$ is the product topology on F^E;

(c) $\mathscr{C}$ is the coarsest topology on F^E which makes each projection map: $f \to f(x)$ of F^E into F, for each $x \in E$, continuous;

(d) For each net $(f_\alpha)_{\alpha \in D}$ in F^E, $f_\alpha \to f$ in the $\mathscr{C}$ topology if and only if $\{f_\alpha(x)\}_{\alpha \in D}$ converges to $f(x)$ for each $x \in E$

Proof. This follows by combining the results on products (cf. §22, Proposition 3).

Theorem 1. Let E be any set and F a topological space. Let F^E be endowed with the point–open topology. Then:

(a) F^E is a T_0-space, T_1-space, T_2-space, regular, or completely regular if F respectively is;

(b) F^E is a compact space if F is compact;

(c) F^E is a locally compact space if F is locally compact and E is a finite set;

(d) F^E is a metrizable space if F is metrizable and if E is a countable set.

Proof. Since F^E is the product space with the product topology being equal to the point–open topology, (a)–(d) follow from Proposition 3 and Theorem 3, §22.

Theorem 2 (Criterion for compactness in the $\mathscr{C}_p$-topology). Let F be a Hausdorff space, E any set, and let F^E be endowed with the topology $\mathscr{C}_p$. Let A be a subset of F^E. Then A is $\mathscr{C}_p$-compact if and only if

(i) A is $\mathscr{C}_p$-closed;

(iii) for each $x \in E$, $\overline{A(x)} = \mathrm{Cl}\{f(x): f \in A\}$ is a compact subset of F.

Proof. First we observe that by hypothesis, F^E is a Hausdorff space [Theorem 1(a)] in the $\mathscr{C}_p$ topology. Now assume that A is compact in F^E; then by Proposition 4, §22, A is $\mathscr{C}_p$-closed in F^E. Hence (i) holds. Further, since the map: $f \to f(x)$ is continuous and A is compact, it follows that $A(x) = \{f(x): f \in A\} = \overline{A(x)}$ is compact in F. This establishes (ii).

For the converse, assume (i) and (ii) hold. Observe that A, being a subset of F^E, is a subset of $G = \prod\{\overline{A(x)}: x \in E\}$. By (ii), G is compact via Tychonoff's theorem (§35, Theorem 3) and so A is relatively $\mathscr{C}_p$-compact. By (i), A is $\mathscr{C}_p$-closed and so A is $\mathscr{C}_p$-compact. This completes the proof.

Corollary 1. Let F be the real line R in Theorem 2. Then a subset A of R^E is $\mathscr{C}_p$-compact iff A is $\mathscr{C}_p$-closed and for each $x \in E$, $\{f(x): f \in A\}$ is closed and bounded in R.

Remark. If F is not Hausdorff, then (i) and (ii) are only sufficient conditions for compactness of A, because then F^E is not Hausdorff and hence a $\mathscr{C}_p$-compact set need not be $\mathscr{C}_p$-closed.

The following examples deal with some of the most well-known and useful function spaces in analysis and topology.

Example 1. Let $F = E = \{0, 1\}$, the two-point set with the discrete topology. Then F^E is a compact Hausdorff space under the $\mathscr{C}_p$-topology.

Example 2. Let $F = \{0, 1\}$, $E = [0, 1]$. Then F^E is a compact Hausdorff space under the $\mathscr{C}_p$-topology. (This is called the *Cantor's space*.)

Example 3. Let $E = F = [0, 1]$. Then F^E is a compact Hausdorff space under the $\mathscr{C}_p$-topology.

Example 4. Let E be any set and $F = [0, 1]$. Then F^E is a compact Hausdorff space under the $\mathscr{C}_p$-topology.

Example 5. Let E be a completely regular space and $\mathbb{R}$, the real line. Then $\mathbb{R}^E$ is a completely regular space under the $\mathscr{C}_p$-topology and $C(E, \mathbb{R})$ $= C_R(E)$, the set of all continuous real-valued functions, is also a Hausdorff space in the induced $\mathscr{C}_p$-topology.

§54. Compact–Open Topology and the Topology of Joint Continuity

In this section we shall assume that E, F both are topological spaces. Let $\{T(C, U)\}$ denote the collection of subsets in F^E, where C runs over compact subsets of E and U over open subsets of F, where $T(C, U) = \{f: \in F^E: f(x) \in U, \text{ for all } x \in C\}$.

Proposition 4. The collection $\{T(C, U)\}$ defines a topology (called the *compact–open topology*) denoted by $\mathscr{C}_c$ on F^E, which is finer than the point–topology $\mathscr{C}_p$.

Proof. The fact that $\{T(C, U)\}$ defines a subbase for a topology can be verified by observing that

$$T(C_1 \cup C_2, U_1 \cap U_2) \subset T(C_1, U_1) \cap T(C_2, U_2),$$

where C_i ($i = 1, 2$) is compact in E and U_i ($i = 1, 2$) is open in F.

Since each singleton is a compact set, it follows that $T(\{x\}, U)$ is a member of the subbase of the $\mathscr{C}_c$-topology. Hence $\mathscr{C}_c$ is finer than $\mathscr{C}_p$.

Proposition 5. Let E and F be topological spaces and let F^E be endowed with the compact–open topology $\mathscr{C}_c$. Then F^E is Hausdorff iff E is Hausdorff.

Proof. Suppose F is Hausdorff. Let $f \neq g, f, g \in F^E$. Thus there exists $x \in E$ such that $f(x) \neq g(x)$. Since F is Hausdorff, there are open disjoint neighborhoods U and V of $f(x)$ and $g(x)$ in F. Now let C be a compact subset of E. Then $T(C, U)$ and $T(C, V)$ are $\mathscr{C}_c$-neighborhoods of f and g respectively in the $\mathscr{C}_c$-topology. If $h \in T(C, U) \cap T(C, V)$, then $h(x) \in U \cap V$ for $x \in C$, which is a contradiction because U, V are disjoint. Thus f and g have disjoint open neighborhoods $T(C, U)$ and $T(C, V)$, respectively, and so F^E is Hausdorff. The other part follows similarly.

Let E and F be two topological spaces and let φ be the map of $F^E \times E$ into F defined by: $\varphi(f, x) = f(x)$, where $f \in F^E$, $x \in E$. The question whether or not there exixts a coarsest topology on F^E which makes φ continuous is of some interest. Indeed, there is a topology on F^E, viz., the discrete topology, which makes φ continuous. For if V is an open neighborhood of $f(x), f \in F^E$, then for any neighborhood U of x such that $f(U) \subset V$, the subset $T = \{g \in F^E : g(U) \subset V\}$, being a subset of a discrete space, is open. Hence $\varphi(T, U) \subset V$ shows that φ is continuous.

On the other hand, the topology $\mathscr{C}_p$ on F^E does not make φ continuous in general. Actually there may not exist a smallest topology which makes φ continuous.

Definition 4. A topology $\mathscr{C}$ on F^E, where E, F are topological spaces. is said to be *a topology of joint continuity* if the map φ of $F^E \times E$ into F, where $\varphi(f, x) = f(x)$, is continuous. A topology on F^E is said to be a *topology of joint continuity on compacta* if for each compact subset C of E, the restriction map φ of $F^E \times C$ into E, where $\varphi(f, x) = f(x)$, $f \in F^E$, $x \in C$, is continuous.

Proposition 6. Any topology $\mathscr{C}$ of joint continuity on compacta in F^E is finer that the compact–open topology $\mathscr{C}_c$.

Proof. Let C be a compact subset of E and U an open subset of F. Then we show that $T(C, U)$ is a $\mathscr{C}$-open set in F^E. Let $\varphi(f, x) = f(x)$, where $f \in F^E$, $x \in E$. Since φ is a continuous map of $F^E \times C \to F$, for each compact set $C \subset E$ and for each open neighborhood V of $f(x)$, there is a $\mathscr{C}$-neighborhood G of f in F^E and compact neighborhood K of x in C such that $\varphi(G, K) \subset V$. But then it implies that for each $g \in G$, $g(K) \subset V$, which shows that $G \subset T(K, V)$ and hence $T(G, K)$ is a $\mathscr{C}$-neighborhood of f. Thus $\mathscr{C}$ is finer than the compact–open topology $\mathscr{C}_c$.

Corollary 2. Let $\mathcal{C}_k$ be a topology of joint continuity on compacta, $\mathcal{C}_c$ the compact–open topology, and $\mathcal{C}_p$ the point–open topology on F^E. Then $\mathcal{C}_p \subset \mathcal{C}_c \subset \mathcal{C}_k$.

Proof. Clear from Propositions 4 and 6.

Corollary 3. Let F be a Hausdorff space and E a topological space such that F^E with $\mathcal{C}_k$ is compact. Then the three topologies $\mathcal{C}_k$, $\mathcal{C}_c$, $\mathcal{C}_p$ on F^E coincide.

Proof. By hypothesis F^E is a $\mathcal{C}_k$-compact Hausdorff space because $\mathcal{C}_k \supset \mathcal{C}_p$ and the $\mathcal{C}_p$-topology is Hausdorff. Now $\mathcal{C}_p \subset \mathcal{C}_k$ on F^E implies that the identity map $(F^E, \mathcal{C}_k) \to (F^E, \mathcal{C}_p)$ is continuous. Hence it is open by Proposition 12, §35, because $(F^E, \mathcal{C}_k)$ is compact. Thus $\mathcal{C}_k = \mathcal{C}_p$ on F^E and hence all the three topologies coincide by Corollary 2.

§55. Subsets of F^E with Induced Topologies

Let E, F be two topological spaces. Then we have seen that the set F^E of all maps from E to F can be endowed with at least three topologies, viz., $\mathcal{C}_k$, $\mathcal{C}_c$, $\mathcal{C}_p$ as described in the previous sections. Now we wish to consider subsets of F^E with the induced topologies $\mathcal{C}_k$, $\mathcal{C}_c$, and $\mathcal{C}_p$.

Two of the important subsets of F^E are the following:

(1) $K(E, F)$, the set of all $f \in F^E$ such that the restriction $f \mid C$ on each compact subset C of E to F is continuous.

(2) $C(E, F)$, the set of all continuous mappings of E into F.

It is clear that $C(E, F) \subset K(E, F) \subset F^E$.

Proposition 7. Let E and F be two topological spaces such that F is Hausdorff. Then $K(E, F)$ and $C(E, F)$ are Hausdorff spaces under the point–open topology induced from F^E.

Proof. All the statements follow from the fact that F^E is Hausdorff under the topology $\mathcal{C}_p$ when F is Hausdorff and every subspace of a Hausdorff space is Hausdorff (Theorem 1, §20).

Similarly, $K(E, F)$, $C(E, F)$ are Hausdorff spaces under the compact–open topology $\mathcal{C}_c$ induced from F^E as well as under a topology $\mathcal{C}_k$ of joint continuity on compacta induced from F^E.

We have observed before that a topology of joint continuity on compacta on F^E is strictly finer than the compact–open topology in general and $\mathcal{C}_c$

is not a topology of joint continuity. However, on the subset $K(E, F)$ of F^E, the situation is different. We have:

Proposition 8. Let E and F be topological spaces such that E is Hausdorff or regular. Then the compact–open topology $\mathscr{C}_c$ on $K(E, F)$ is a topology of joint continuity on compacta.

Proof. Let C be a compact subset of E, $x \in E$, U an open subset in F containing $f(x)$, and let $\varphi : (f, x) \to f(x)$ be the mapping of $K(E, F) \times E$ into F. Assume $(f, x) \in \varphi^{-1}(U)$. Since f is continuous on C, there is a closed hence compact (because E is Hausdorff or regular) neighborhood P of x in C such that $f(P) \subset U$. Now let $T(P, U) = \{g \in K(E, F): g(P) \subset U\}$. Then $T(P, U) \times P$ is a neighborhood of (f, x) in $K(E, F) \times C$ and is contained in $\varphi^{-1}(U)$. Since $T(P, U)$ is a member of the $\mathscr{C}_c$-topology in $K(E, F)$, the proposition is proved.

Corollary 4. Let E be a locally compact Hausdorff space and F any topological space. Then a topology on $K(E, F) = C(E, F)$ is a topology of joint continuity on compacta if and only if it is a topology of joint continuity.

Proof. Certainly if $\varphi : K(E, F) \times E \to F$ is continuous, where $\varphi(f, x) = f(x)$, $f \in K(E, F)$, $x \in E$, then so is $\varphi : K(E, F) \times C \to F$ for any compact subset C of E. Now assume that $K(E, F)$ is endowed with a topology of joint continuity on compacta. Let U be an open set in F and let $(f, x) \in \varphi^{-1}(U)$. Since f is continuous there is a compact (because E is locally compact) neighborhood P of x such that $f(P) \subset U$. Now by the arguments used in Proposition 8, the corollary follows.

Now we have a criterion for compactness of subsets of $K(E, F)$ in the $\mathscr{C}_c$-topology.

Theorem 3. Let E and F be two Hausdorff topological spaces. (E could be taken to be regular or Hausdorff.) Let $\mathscr{C}_c$ and $\mathscr{C}_p$ be the topologies on $K(E, F)$. A subset $A \subset K(E, F)$ is $\mathscr{C}_c$-compact if and only if

(i) A is $\mathscr{C}_c$-closed;

(ii) $\overline{A(x)} = \mathrm{Cl}\{f(x): f \in A\}$ is compact in F for each $x \in E$;

(iii)   the topology $\mathscr{C}_p$ on $\mathrm{Cl}_p A \subset F^E$ is a topology of joint continuity on compacta.

Proof. *"Only if" part*: Assume A is $\mathscr{C}_c$-compact. Observe that by hypothesis F is Hausdorff, so $\mathscr{C}_c$ is a Hausdorff topology on $K(E, F)$ and hence

A is $\mathscr{C}_c$-closed. This proves (i). Since for each $x \in E$ the map: $f \to f(x)$ of $K(E, F)$ into F is $\mathscr{C}_p$-continuous and so $\mathscr{C}_c$-continuous (because $\mathscr{C}_c \supset \mathscr{C}_p$), it follows that $\overline{A(x)}$ is compact in F. This proves (ii). Since A is $\mathscr{C}_c$-compact, $\mathscr{C}_c \supset \mathscr{C}_p$ and $\mathscr{C}_p$ is Hausdorff, $\mathscr{C}_p$ and $\mathscr{C}_c$ coincide on A. Hence A is $\mathscr{C}_p$-compact in F^E. Thus by Proposition 8, $\mathscr{C}_c$ (and hence $\mathscr{C}_p$) is a topology of joint continuity on compacta. This proves (iii).

"*If*" *part*: Assume that (i), (ii), and (iii) hold for a subset $A \subset K(E, F)$. For each $x \in E$, $\overline{A(x)}$ is $\mathscr{C}_p$-compact in F. Clearly $\mathrm{Cl}_p A \subset \prod [\overline{A(x)}: x \in E]$. Since the right-hand side is compact, so is $\mathrm{Cl}_p A$. Since, by (iii), the topology $\mathscr{C}_p$ on $\mathrm{Cl}_p A$ is a topology of joint continuity on compacta, $\mathscr{C}_p$ is finer than $\mathscr{C}_c$ on $K(E, F)$ by Proposition 6 and hence $\mathscr{C}_c$ and $\mathscr{C}_p$ coincide on $\mathrm{Cl}_p A$, because $\mathrm{Cl}_p A$ is compact. By (i), A is $\mathscr{C}_c$-closed in $K(E, F)$ and hence $\mathscr{C}_p$-closed in $\mathrm{Cl}_p A$. Thus $\mathrm{Cl}_p A = A$ and A is $\mathscr{C}_c$-compact.

Recall that a space E is said to be a k-space (§40, Definition 8) if a set U is open whenever $U \cap C$ is open in C for each compact subset C of E. We have seen that on a k-space E every map f on E into a topological space F is continuous if and only if its restriction $f \mid C$ on each compact set $C \subset E$ is continuous (Theorem 16, §40).

We have also shown that every locally compact space or a metric space (or more generally a first countable space) is a k-space (§40, Proposition 35). Thus we have shown that for a k-space E and any topological space F, $C(E, F) = K(E, F)$. If $F = \mathbb{R}$, the real line, then E is a k_r-space iff $C(E, \mathbb{R}) = K(E, \mathbb{R})$.

Thus in this particular case Theorem 3 reads as follows:

Theorem 4.   Let E be a Hausdorff k_r-space and F any Hausdorff topological space. Let $\mathscr{C}_c$ and $\mathscr{C}_p$ be the usual topologies on $C(E, F)$. A subset $A \subset C(E, F)$ is $\mathscr{C}_c$-compact if and only if

(i) A is $\mathscr{C}_c$-closed;

(ii) $\overline{A(x)} = \mathrm{Cl}\{f(x): f \in A\}$ is compact in F for each $x \in E$;

(iii) the topology $\mathscr{C}_p$ on $\mathrm{Cl}_p A \subset F^E$ is a topology of joint continuity on compacta.

§56. The Uniformities on F^E

We have considered the cases of the space F^E, when E is a set and F a topological space, or when E and F are both topological spaces. Now we wish to consider the case of F^E when E is a set or a topological space and F

a uniform space. We recall that if F is a uniform space and E is any set, then F^E is the product space or, in other words, the set of all maps from E to F. We have seen that if F is a uniform space the product space has a uniformity on $F^E \times F^E$ by defining a subbase: for $f, g \in F^E$, the sets

$$[U] = \{(f, g): f, g \in F^E, (f(x), g(x)) \in U\}$$

for each $x \in E$, where $\{U\}$ is the uniformity for F. Also recall that F^E is a complete uniform space iff F is a complete uniform space (§29, Proposition 15).

Now we wish to investigate the uniformities on F^E and $C(E, F)$, where E is a topological space and F a uniform space which will give us the $\mathscr{C}_c$ topology. Indeed we know the uniformity which gives us the $\mathscr{C}_p$-topology on $C(E, F)$, which is the induced uniformity from F^E. There exists a largest uniformity on F^E, which is usually called the uniformity of uniform convergence. All other uniformities lie between the uniformities of uniform convergence and pointwise convergence.

Proposition 9. The collection $\{[V]\}$, where V runs over a subbase for the uniformity $\{V\}$ of F, is a subbase for a uniformity on F^E.

Proof. Recall that for $V_1, V_2 \in \{V\}$, $V_1 \circ V_2 \in \{V\}$, $V^{-1} \in \{V\}$ and there exists $V_3 \in \{V\}$, such that $V_1 \cap V_2 \supset V_3$. Now then, by definition, we see that $[V_1 \circ V_2] = [V_1] \circ [V_2]$, $[V^{-1}] = [V]^{-1}$, and $[V_1 \cap V_2] = [V_1] \cap [V_2] \supset [V_3]$. Thus $\{[V]: V \in \{V\}\}$ is a subbase for a uniformity.

Remark. Observe that the topology on F^E induced by the uniformity of uniform convergence has for a base the following sets (let $f \in F^E$).

$$[V](f) = \{g \in F^E: (g, f) \in [V]\}.$$

The following theorem gives the relevant properties:

Theorem 5. (a) The uniformity of uniform convergence is finer than the uniformity of pointwise convergence.

(b) The topology defined by the uniformity of uniform convergence is finer than the topology $\mathscr{C}_p$ defined by the uniformity of pointwise convergence.

(c) Let $\{f_\alpha: \alpha \in D\}$ be a net in F^E. If (f_α) converges to f in the uniform topology then (f_α) converges to f pointwise. But the converse does not hold.

(d) If the uniformity of F is generated by bounded pseudo-metrics or gauges $\{d\}$, then the uniformity of the uniform convergence on F^E is generated by gauges $\{d'\}$, where d' is defined by

$$d'(f, g) = \sup\{d(f(x), g(x)), x \in E\}.$$

(e)   A net (f_α) converges to f uniformly in F^E if and only if (f_α) is a Cauchy net in the uniformity of the uniform convergence and $(f_\alpha(x))$ converges to $f(x)$ in F for each $x \in E$.

(f)   If F is a complete uniform space, so is F^E in the uniformity of uniform convergence.

Proof.   (a)   Observe that $[V] = \{(f, g): (f(x), g(x)) \in V$ for *each* $x \in E\}$ and each $V \in \{V\}$ is a member of the uniformity of pointwise convergence (where $\{V\}$ is the uniformity of F) and $[V]$ is a member of the uniformity of uniform convergence if for *all* $x \in E, (f(x), g(x)) \in V$ for each V in $\{V\}$. Thus (a) is clear.

(b)   From (a) it follows that the respective topologies bear the same relation as that for their uniformities.

(c) (f_α) converges to f uniformly implies that for all $x \in E$, $(f_\alpha(x))$ converges to $f(x)$. Thus (f_α) converges to f in the topology of pointwise convergence. For the failure of the converse, see Exercise 16, Chapter VII.

(d)   First of all we observe that there is no loss of generality in assuming that the uniformity of F is generated by bounded pseudometrics because, if d is any gauge on F, then $d* = \min(1, d)$ is a bounded pseudometric which gives an equivalent uniformity. Now let $\varepsilon > 0$. The sets $V = \{(y, z): y, z \in F, d(y, z) < \varepsilon\}$ form a base for the uniformity of F. But

$$\{(f, g): d'(f, g) < \varepsilon\} = \{(f, g): d(f(x), g(x)) < \varepsilon \text{ for all } x \in E\}$$
$$= [V]$$

as defined above. Thus d' is a member of the gauge of the uniformity for the uniform convergence. The converse follows easily.

(e)   Indeed, if a net $\{f_\alpha\}$ converges uniformly to f then $\{f_\alpha\}$ is a Cauchy net in the uniformity of uniform convergence and $(f_\alpha(x))$ converges to $f(x)$ for each $x \in E$. Conversely, assume (f_α) is a Cauchy net in the uniformity of uniform convergence and $(f_\alpha(x))$ converges to $f(x)$ for each $x \in E$. Let V be a closed symmetric member of the uniformity of F. Clearly for each $x \in E$, $(f_\alpha(x))$ is a Cauchy net in F. Given $V \in \{V\}$, there exists α_0 depending upon V only such that for $\alpha, \beta \geq \alpha_0$, $(f_\alpha(x), f_\beta(x)) \in V$ or $f_\alpha(x) \in V[f_\beta(x)]$ for all $x \in E$.

Since $V[f_\beta(x)]$ is closed, it follows that $f(x) \in V[f_\beta(x)]$ for all $x \in E$

and $\beta \geq \alpha_0$. But by symmetry of V, we have $f_\beta(x) \in V[f(x)]$ for $\beta \geq \alpha_0$ and all $x \in E$. This proves that (f_α) converges to f uniformly.

(f) Since F is a complete uniform space, we have seen (Proposition 15, §29) that F^E is a complete space in the uniformity of pointwise convergence. But then from (e) we have (f).

Theorem 6. Let E and F be two topological spaces such that F is also a uniform topological space. Let $C(E, F)$ denote the set of all continuous maps from E to F, endowed with the uniformity of uniform convergence induced from F^E. Then:

(i) $C(E, F)$ is a closed subspace of F^E;

(ii) if F is complete, then $C(E, F)$ is a complete space under the induced uniformity of uniform convergence;

(iii) the topology of uniform convergence on $C(E, F)$ is a topology of joint continuity.

Proof. (i) Let $f \in F^E \setminus C(E, F)$. That means f is not continuous at some $x \in E$. Thus there is a member V of the uniformity $\{V\}$ of F such that $f^{-1}(V[f(x)])$ is not a neighborhood of x. Now choose a symmetric member W of $\{V\}$ such that $W \circ W \circ W \subset V$. Let $g \in F^E$ such that $(g(y), f(y)) \in W$ for each $y \in E$. Then $\{g\} \in W[f]$, i.e., $\{g(y)\} \subset W[f(y)]$ for all $y \in E$ and

$$g^{-1}(y) \subset f^{-1}(W^{-1}[y]) = f^{-1}(W[y]) \qquad \text{for } y \in W[g(x)].$$

Thus

$$g^{-1}(W[g(x)]) \subset f^{-1}(W \circ W \circ W[f(x)]) \subset f^{-1}(V[f(x)])$$

shows that $g^{-1}(W[g(x)])$ is not a neighborhood of x, i.e., $g \in F^E \setminus C(E, F)$ for all $g \in F^E$ such that $(g(y), f(y)) \in W$ for each $y \in E$. This proves that $F^E \setminus C(E, F)$ is open and hence $C(E, F)$ is closed.

(ii) Since a closed subset of a complete uniform space is complete (Proposition 15, §29), (ii) follows from (i).

(iii) Let $\varphi: (f, x) \to f(x)$ be the mapping of $C(E, F) \times E$ to F. Let V be a member of the uniformity on F and let $V[f(x)]$ be a neighborhood of $f(x) \in F$, for $x \in E$. Let $y \in f^{-1}(V[f(x)])$. If $g(z) \in V[f(z)]$ for all $z \in E$, then

$$\varphi(g, y) = g(y) \in V[f(y)] \subset V \circ V[f(x)],$$

which is easy to check. Hence φ is continuous and (iii) is proved.

§57. $\mathfrak{S}$-Uniformities and $\mathfrak{S}$-Topologies

In the last sections we have seen that the uniformity of uniform convergence is the largest and the uniformity of pointwise convergence is the coarsest of the uniformities on F^E. In this section we wish to describe an intermediate situation.

Definition 5. Let E be a set and F a uniform space with uniformity $\mathcal{V} = \{V\}$ on F. Let $\mathfrak{S} = \{S\}$ be a family of subsets of E. The family of subsets

$$[V] = \{(f, g): f, g \in F^E, (f(x), g(x)) \in V \text{ for all } x \in S\}$$

where $S \in \mathfrak{S}$, forms a subbase for a uniformity when V runs over $\mathcal{V}$ and S over $\mathfrak{S}$. This uniformity on F^E is called an $\mathfrak{S}$-*uniformity* and the topology induced by an $\mathfrak{S}$-uniformity is called an $\mathfrak{S}$-*topology*.

If $\mathfrak{S}_1 \subset \mathfrak{S}_2$, then the $\mathfrak{S}_2$-topology is finer than the $\mathfrak{S}_1$-topology.

Proposition 10. For each $S \in \mathfrak{S}$, let R_s denote the restriction map which carries $f \in F^E$ into the restriction $f \mid S$ of f, for each S, i.e., $R_s(f) = f \mid S$: $S \to F$ for each $S \in \mathfrak{S}$. Then the $\mathfrak{S}$-uniformity on F^S is the coarsest uniformity which makes each R_s uniformly continuous for each $S \in \mathfrak{S}$, where F^S is given the pointwise uniformity.

Proof. Simple verification.

Remark. There are three important particular cases of the $\mathfrak{S}$-uniformity: (a) $\mathfrak{S}$ consists of all singletons of E. This is nothing but the uniformity of pointwise convergence; (b) $\mathfrak{S}$ consists of the whole set E. Then we get the uniformity of uniform convergence; (c) A third important case is obtained when $\mathfrak{S}$ consists of compact subsets of a topological space E, etc.

Definition 6. When $\mathfrak{S}$ consists of compact subsets of a topological space E, the $\mathfrak{S}$-uniformity is called the *uniformity of uniform convergence on compacta* and the topology obtained from this uniformity is called the *uniform topology of compact convergence*.

A few basic facts about the $\mathfrak{S}$-uniformity on F^E are described in the following:

Theorem 7. Let E be a topological space, $(F, \mathcal{V})$ a uniform space, and $\mathfrak{S}$, the family of all compact subsets of E. Let F^E and $C(E, F)$ be endowed with the $\mathfrak{S}$-uniformity. Then:

(i)   The ⑤-uniformity is finer than the uniformity of pointwise convergence and coarser than the uniformity of uniform convergence.

(ii)   A net (f_α) in F^E converges in the ⑤-topology if and only if it is a Cauchy net in the ⑤-uniformity and $(f_\alpha(x))$ uniformly converges to $f(x)$ on each $S \in ⑤$, where $x \in S$.

(iii)   If $(F, \mathscr{V})$ is a complete space so is F^E in the ⑤-uniformity.

(iv) $C(E, F)$ is a closed subspace of F^E in the ⑤-topology. Thus if F is complete, then the ⑤-topology is a topology of joint continuity on each member $S \in ⑤$ and $C(E, F)$ is complete in the ⑤-topology.

Proof.   The proofs are similar to those for Theorems 5 and 6.

We have seen that the compact–open topology is useful for the space of all continuous maps from a topological space E into another topological space F. We wish to relate this to the ⑤-topology, where ⑤ consists of all compact subsets of E.

Theorem 8.   Let E be a topological space, $(F, \mathscr{V})$ a uniform space, and ⑤ the collection of all compact subsets of E. Then the ⑤-topology $\mathscr{C}$ on $C(E, F)$ is the same as the compact–open topology.

Proof.   Let $\mathscr{C}_c$ denote the compact–open topology. First we show that $\mathscr{C}_c \subset \mathscr{C}$. Let K be a compact subset of E, U an open subset of F, and $f \in C(E, F)$ such that $f(K) \subset U$. Since $f(K)$ is compact, there exists a member V of the uniformity $\mathscr{V}$ on F such that $V[f(K)] \subset U$ by Proposition 10. Now if $g \in C(E, F)$ is such that $g(x) \in V[f(x)]$ for each $x \in K$, then $g(K) \subset U$. Thus each set $\{f \in C(E, F): f(K) \subset U\}$ is open in the $\mathscr{C}$-topology.

Now to show $\mathscr{C} \subset \mathscr{C}_c$, let K be any compact subset of E and V in $\mathscr{V}$. Now choosing a symmetric closed member W of $\mathscr{V}$ such that $W \circ W \circ W \subset V$ and x_i $(1 \le i \le n)$ in K such that $\{W[f(x_i)]\}$ cover $f[K]$, put

$$K_i = K \cap f^{-1}(W[f(x_i)]),$$

and let

$$U_i = [W \circ W[f(x_i)]]^\circ.$$

Now if $g(K_i) \subset U_i$ for each $i = 1, \ldots, n$, then for each $x \in K$ there is an m such that $x \in K_m$ and hence

$$g(x) \in W \circ W[f(x_m)].$$

But since $f(x) \in W[f(x_m)]$, it follows that

$$(g(x), f(x)) \in W \circ W \circ W \subset V \text{ for } x \in K.$$

This proves the theorem.

Corollary 5. Let E be a k-space (§40), $(F, \mathscr{V})$ a complete uniform space, and $C(E, F)$ the space of all continuous maps from E to F. Let $\mathscr{C}$ denote the $\mathfrak{S}$-uniformity, where $\mathfrak{S}$ consists of all compact subsets of E. Then $C(E, F)$ is complete under $\mathscr{C}$.

Proof. It is an easy consequence of Theorem 8.

A particular case of Corollary 5 is:

Corollary 6. Let E be a locally compact or a first countable space in the above corollary. Then the same conclusion holds.

§58. Equicontinuous Maps

Let E be a topological space and $(F, \mathscr{V})$ a uniform space. Then we have seen that F^E can be endowed with the uniformity of pointwise convergence.

Definition 7. A subset $A \subset F^E$ is said to be *equicontinuous at a point* $x_0 \in E$ if for each member V of the uniformity $\mathscr{V}$ on F, there exists a neighborhood U of x_0 such that $(f(x), f(x_0)) \in V$ for all $x \in U$ and $f \in A$ or equivalently $f(x) \in V[f(x_0)]$ for all $x \in U$ and $f \in A$.

A subset $A \subset F^E$ or a family A of functions from E to F is said to be *equicontinuous* if it is so at each point of E.

Proposition 11. (i) A subset $A \subset F^E$ is equicontinuous at $x \in E$ if and only if for each member V of the uniformity $\mathscr{V}$ on F, the set $\cap \{f^{-1}(V[f(x)]): f \in A\}$ is a neighborhood of x.

(ii) If $A \subset F^E$ is equicontinuous and $B \subset A$, then B is also equicontinuous.

(iii) If A is a family of equicontinuous maps then each $f \in A$ is continuous.

(iv) If A is a family of equicontinuous maps of a uniform space $(E, \mathscr{U})$ into another uniform space $(F, \mathscr{V})$, then each f is uniformly continuous.

Proof. (i) This follows from the definition.

(ii) Since $\cap \{f^{-1}(V[f(x)]): f \in A\} \subset \cap \{f^{-1}(V[f(x)]), f \in B\}$, (ii) follows.

(iii) Since the neighborhood $\cap \{f^{-1}(V[f(x)]): f \in A\}$ is clearly contained in $f^{-1}(V[f(x)])$ for each f, it follows that f is continuous.

(iv) Following (iii), there is a member U of the uniformity $\{U\}$ on E such that $U[x] \subset f^{-1}(V[f(x)])$. In other words, for all $(x, y) \in U$ we have $(f(x), f(y)) \in V$, which proves the uniform continuity of f.

Proposition 12. If $A \subset F^E$ is equicontinuous, then $\mathrm{Cl}_p A$ is also equicontinuous, where $\mathrm{Cl}_p A$ is the closure of A in the topology $\mathscr{C}_p$ of pointwise convergence.

Proof. Let V be a closed member of the uniformity $\mathscr{V}$ on F and $x \in E$. Then there exists a neighborhood U of x such that for all $y \in U$ and all $f \in A$, $(f(y), f(x)) \in V$. Let $g \in \mathrm{Cl}_p A$. Then $(g(y), g(x)) \in V$ for $y \in U$, because V is closed. This shows that $\mathrm{Cl}_p A$ is equicontinuous.

We have shown that on F^E the topology of pointwise convergence is coarser than the compact–open topology, which is coarser than the $\mathfrak{S}$-topology, where $\mathfrak{S}$ is the family of all compact sets. However, on $C(E, F)$, the latter two are the same. In the following theorem we show that on equicontinuous subsets even the topology of pointwise convergence coincides with the other. Precisely, we have:

Theorem 9. Let A be an equicontinuous set of mappings of a topological space E into a uniform space $(F, \mathscr{V})$. Then on A the point–open topology coincides with the compact–open topology.

Proof. We have seen that the point–open topology is not a topology of joint continuity, i.e., a topology which makes the mapping $\varphi: (f, x) \to f(x)$ of $F^E \times E$ into F continuous. Now first we show that the point–open topology on A is a topology of joint continuity. For this let V, W be members of the uniformity $\mathscr{V}$ of F such that $V \circ V \subset W$, $x \in E$ and let U be a neighborhood of x such that $f(U) \subset V[f(x)]$ for all $f \in A$. Clearly $B = \{g \in F^E : g(x) \in V[f(x)]\}$ is a $\mathscr{C}_p$-neighborhood of f in F^E. Thus for $g \in B \cap A$ and $y \in U$, we have $g(y) \in V[g(x)]$ and $g(x) \in V[f(x)]$. This shows that $(g(y), g(x)) \in V \circ V \subset W$ for all $g \in B$ and $y \in U$ and so the joint continuity of φ follows. Since each topology of joint continuity is finer than the compact–open topology and *a fortiori* the point–open topology, we have shown that the point–open topology on A, being a topology of joint continuity, is finer than the compact–open topology and hence they coincide on A, because in general the inclusion $\mathscr{C}_c \supset \mathscr{C}_p$ holds.

Corollary 7. Let A be an equicontinuous set of maps from a topological space E into a uniform space $(F, \mathscr{V})$. Then A is $\mathscr{C}_p$-compact if and only if it is $\mathscr{C}_c$-compact.

Proof. Clearly any set which is $\mathscr{C}_c$-compact is also $\mathscr{C}_p$-compact (because $\mathscr{C}_c \supset \mathscr{C}_p$). The converse follows from Theorem 9.

Corollary 8. Let A be an equicontinuous $\mathscr{C}_p$-closed set of maps from a topological space E into a uniform space $(F, \mathscr{V})$. If for each $x \in E$, $\overline{A(x)}$ is compact in F, then A is $\mathscr{C}_c$-compact.

Proof. Observe that each f can be identified with the vector $\{f(x): x \in E\}$. Thus $A \subset \prod\{\overline{A(x)}: x \in E\}$, which is compact by Tychonoff's theorem (because $\overline{A(x)}$ is compact by hypothesis). Thus A, being $\mathscr{C}_p$-closed by hypothesis, is $\mathscr{C}_p$-compact and Corollary 7 applies.

Now we wish to present a condition under which a given family of maps is equicontinuous.

Theorem 10. Let A be a $\mathscr{C}$-compact set of maps from a topological space $(F, \mathscr{V})$, where $\mathscr{C}$ is a topology of joint continuity. Then A is equicontinuous.

Proof. Let $x \in E$ be fixed but arbitrary. Let W' be a member of the uniformity $\mathscr{V}$ on F. Choose $V \in \mathscr{V}$ so that $V \circ V \subset W'$. Since the map φ: $(f, x) \to f(x)$ of $A \times E$ into F is continuous, there is a neighborhood U of x and a $\mathscr{C}|_A$-open neighborhood W of f in A such that for all $g \in W$ and $y \in U$, $\varphi(g, y) = g(y) \in V[f(x)]$ or, in other words, $(g(y), f(x)) \in V$ for all $g \in W$, $y \in U$. But then it follows that

$$(g(y), g(x)) \in V \circ V \text{ or } g(y) \in V \circ V[g(x)]$$

for all $y \in U$ and $g \in W$. Obviously $\{W\}$ is a $\mathscr{C}|_A$-open covering of the $\mathscr{C}$-compact set A. Thus there is only a finite subcovering W_i ($1 \leq i \leq n$) of A. Let U_i ($1 \leq i \leq n$) be the corresponding neighborhoods of x. Then $A = \bigcup_{i=1}^{n} W_i$ and $U = \bigcap_{i=1}^{n} U_i$ is a neighborhood of x and for all $y \in U$ and $g \in A$, we have $(g(y), g(x)) \in V \circ V \subset W'$. This proves that A is equicontinuous.

Remark. Observe that whenever we consider an equicontinuous subset of maps in F^E, we are actually talking about equicontinuous subsets of $C(E, F)$ because of the Proposition 11 (iii). Now we have a general theorem about compactness.

Theorem 11. (Ascoli). Let E be a locally compact Hausdorff topological space, $(F, \mathscr{V})$ a Hausdorff uniform space, and let $C(E, F)$ be endowed with the $\mathfrak{S}$-topology $\mathscr{T}$, where $\mathfrak{S}$ consists of all compact subsets of E. Then a subset A of $C(E, F)$ is $\mathscr{T}$-compact if and only if

(i′) A is $\mathscr{T}$-closed;

(ii′)   for each $x \in E$, $\overline{A(x)}$ is compact in F;

(iii′) A is equicontinuous.

Proof. Assume A is $\mathscr{T}$-compact. By Theorem 8, the $\mathfrak{S}$-topology $\mathscr{T}$ on $C(E, F)$ is the same as the compact–open topology $\mathscr{T}_c$ and on a $\mathscr{T}$-compact set A, the topology which is the same as $\mathscr{T}_c$ is a topology of joint continuity. Hence by a criterion of compactness (Theorem 4, §55), we have (i′), (ii′), and (iii′) from Theorem 10. Conversely, if A satisfies (i′)–(iii′) then A is $\mathscr{T}_c$-compact by Corollary 8. Hence from (i′) A is $\mathscr{T}$-compact.

A relatively more general Ascoli theorem is as follows. But first we need the following:

Definition 8. Let E be a topological space and F a uniform space and let F^E be endowed with an $\mathfrak{S}$-uniformity, where $\mathfrak{S}$ is a collection of subsets of E which cover E. A subset $A \subset F^E$ is said to be $\mathfrak{S}$-*equicontinuous* if for each subset $S \in \mathfrak{S}$, the set $\{f \mid S : f \in A\}$ of restriction maps is an equicontinuous set of maps in F^S, where F^S is given the relative topology induced from F^E.

Theorem 12. Let E be a Hausdorff k-space, $(F, \mathscr{V})$ a Hausdorff uniform space, and let $C(E, F)$ be endowed with the $\mathfrak{S}$-topology $\mathscr{T}$, where $\mathfrak{S}$ consists of all compact subsets of E. Then a subset A of $C(E, F)$ is $\mathscr{T}$-compact if and only if

(a) A is $\mathscr{T}$-closed;

(b)   for each $x \in E$, $\overline{A(x)}$ is compact in F;

(c) A is $\mathfrak{S}$-equicontinuous.

Proof. Since E is a k-space, we have $C(E, F) = K(E, F) = \{f \in F^E : f \mid K$ is continuous for each compact set $K\}$. Since the theorem holds for subsets in $C(K, F)$ for each $K \in \mathfrak{S}$ by Theorem 11, it holds also for subsets of $C(E, F)$ as well.

§59. Equicontinuity and Metric Spaces

So far we have discussed maps from a set or a topological space E to a uniform space F and studied notions of $\mathfrak{S}$-compactness and equicontinuity. In this section, we study a particular situation in which F is a metric space. Thus the equicontinuity of a set of maps will read as follows:

Proposition 13. Let E be a topological space, (F, d') a metric space, and A a set of maps from E into F. Then A is equicontinuous at $x_0 \in E$ iff given $\varepsilon > 0$ there exists a neighborhood U of x_0 such that $d'(f(x), f(x_0)) < \varepsilon$ for all $x \in U$ and all $f \in A$.

Proof. This is an easy consequence of the definition.

Remark. If in addition E is also a metric space with metric d, then a set A of maps from E to a metric space F is equicontinuous at x_0 if and only if for each $\varepsilon > 0$ there is a $\delta > 0$ depending upon ε only such that for all $f \in A$, $d'(f(x), f(x_0)) < \varepsilon$ whenever $d(x, x_0) < \delta$. Similarly one interprets the equicontinuity on the whole space.

Observe that if a subset $A \subset F^E$ is equicontinuous then $A \subset C(E, F)$.

Hence all the theorems proved earlier, when F is a uniform space, have corresponding interpretations for metric spaces F.

Proposition 14. Let E be a topological space and (F, d) a metric space. Let $CB(E, F) = \{f \in C(E, F): \delta(f(E)) < \infty\}$, where $\delta(G)$ denotes the diameter of the set $G \subset F$. Then $CB(E, F)$ can be given a metric topology. If (F, d) is complete, then $CB(E, F)$ is a complete metric space.

Proof. For each pair $f, g \in CB(E, F)$, denote

$$d^+(f, g) = \sup\{d(f(x), g(x)): x \in E\}.$$

Since the right-hand side is finite because f, g have bounded ranges, it follows that $d^+(f, g)$ is a finite number for $f, g \in CB(E, F)$. Furthermore, $d^+(f, g) \geq 0$ obviously. Also $f = g$ iff $f(x) = g(x)$ for all $x \in E$ iff $d(f(x), g(x)) = 0$ for all $x \in E$ iff $d^+(f, g) = 0$. The fact that $d^+(f, g) = d^+(g, f)$ is trivial and the triangular inequality for d^+ is an immediate consequence of the triangular inequality for d. This proves the proposition in view of Theorem 5, § 56.

Now we study the situation where $CB(E, F) = C(E, F)$. There are certain conditions on E or F which make this equality possible.

Proposition 15. Let E be a topological space and F a compact metric space. Then $C(E, F) = CB(E, F)$.

Proof. $CB(E, F) \subset C(E, F)$ always. However since F is a compact metric space, the diameter of F and hence of any of its subsets is finite. Thus $C(E, F) = CB(E, F)$.

Proposition 16. Let E be a compact space and F a metric space. Then $C(E, F) = CB(E, F)$.

Proof. Again we need to show only that $C(E, F) \subset CB(E, F)$. We observe that for each $f \in C(E, F)$, $f(E)$ is a compact subset of a metric space F and hence it is bounded or, equivalently, the diameter of $f(E)$ is finite (§37).

Proposition 17. Let E be a noncompact space and F a metric space. If the metric d on F is bounded, i.e., $\delta(F) < \infty$, then $C(E, F) = CB(E, F)$.

Proof. Again by assumption the diameter $\delta(F)$ of the range space is finite. Hence the image of E under every continuous map has finite diameter and the proposition follows.

Remark. An example of a bounded metric is

$$d_1(x, y) = \frac{|x - y|}{1 + |x - y|} \text{ on the real line } R.$$

Thus for this metric d_1, $C(R, R) = CB(R, R)$.
However the metric d^+ defined by

$$d^+(f, g) = \sup\left\{\frac{|f(x) - g(x)|}{1 + |f(x) - g(x)|} : x \in R\right\} \text{ on } C(R, R)$$

gives a topology different from the metric topology that one can define on $C(R, R)$ by taking the usual Euclidean metric $|\cdots|$, the absolute value on R, and then putting

$$d^*(f, g) = \sum_{i=1}^{\infty} \frac{1}{2^i} \frac{\max\limits_{|x| \le i} |f(x) - g(x)|}{1 + \max\limits_{|x| \le i} |f(x) - g(x)|}$$

for $f, g \in C(R, R)$.

Thus it is to be noted that the metric on $CB(E, F)$ depends on the metric given on F if F is not a compact space. If F is a compact metric space then it does not matter what metric topology is chosen on F, the metric on $CB(E, F)$ is then invariant.

Since there always exists the $\mathscr{C}_c$-topology (compact-open topology) on $C(E, F) \supset CB(E, F)$, it is of interest to know the conditions under which the $\mathscr{C}_c$-topology coincides with the metric topology defined above for $CB(E, F)$.

Proposition 18. Let E be a compact Hausdorff space and (F, d) a metric space. Then on $C(E, F) = CB(E, F)$, the compact–open topology $\mathscr{C}_c$ coincides with the metric topology defined by

$$d^+(f, g) = \sup\{d(f(x), g(x)): x \in E\}.$$

Proof. Let $\varepsilon > 0$ be given and let $f \in C(E, F)$. The open ball $B_\varepsilon^+(f) = \{g \in C(E, F): d^+(g, f) < \varepsilon\}$ is a neighborhood of f in the metric topology. Since $f(E)$ is a compact subset of the metric space F, we can cover it by a finite number of balls of diameter $\varepsilon/4$. Let U_i $(1 \leq i \leq n)$ be a finite family of nonempty open sets in E such that $\delta(f(\bar{U}_i)) < \frac{1}{4}\varepsilon$, because f is continuous. Now choose an open set $V_i \neq \varnothing$ (for each $i = 1, \ldots, n$) such that $\delta(V_i) \leq \frac{3}{4}\varepsilon$. Let $g \in \bigcap_{i=1}^{n} T(\bar{U}_i, V_i)$ which is a $\mathscr{C}_c$ neighborhood of f. But then $g(\bar{U}_i) \subset V_i$ for $i = 1, \ldots, n$ implies that

$$d^+(g, f) = \sup\{d(g(x), f(x)): x \in E\} \leq \delta(V_i) + \delta(g(\bar{U}_i)) \leq \frac{3}{4}\varepsilon + \frac{1}{4}\varepsilon = \varepsilon.$$

Thus $g \in B_\varepsilon^+(f)$. Hence $B_\varepsilon^+(f)$ is a $\mathscr{C}_c$-neighborhood of f.

Conversely, assume $f \in T(K, V)$, a subbasic $\mathscr{C}_c$-neighborhood of f, where K is compact in E and V open in F. Since $f(K)$ is compact and $f(K) \subset V$, $d(f(K), F \setminus V) = \varepsilon > 0$. Now for any $g \in B_{\varepsilon/2}^+(f)$, we have $d^+(f, g) < \frac{1}{2}\varepsilon$ and hence $g(K) \subset V$, i.e., $B_\varepsilon^+(f) \subset T(K, V)$. Hence $T(K, V)$ is a neighborhood in the metric topology and the proof is completed.

Proposition 19. Let E be a σ-compact Hausdorff space and (F, d) a metric space. Then $C(E, F)$ with $\mathscr{C}_c$ is a metric space and the compact–open topology $\mathscr{C}_c$ coincides with the metric topology.

Proof. Let K_i $(i \geq 1)$ be a sequence of compact sets such that $K_i \subset K_{i+1}$ for each i. For $f, g \in C(E, F)$, define:

$$d_i(f, g) = \min\left(\frac{1}{i}, d_i^+(f, g)\right), \text{ where } d_i^+(f, g) = \sup\{d(f(x), g(x)) x \in K_i\}.$$

Then d_i is a metric on $C(K_i, F)$ and $d^+(f, g) = \sup_{i \geq 1} d_i(f, g)$ gives a metric topology which coincides with the compact–open topology because each compact subset of E is contained in some K_n.

Corollary 9. Let E be a Hausdorff locally compact space which is the union of an increasing sequence of compact sets and F, a metric space. Then the compact–open topology $\mathscr{C}_c$ on $C(E, F)$ coincides with the metric topology.

Proof. Let K_i $(i \geq 1)$ be an increasing sequence of compact sets such that $E = \bigcup_{i=1}^{\infty} K_i$. For each i, on $C(K_i, F) = CB(K_i, F)$, the compact–open topology coincides with the metric topology. Let d_i^+ be a metric on $C(K_i, F)$. We define d^+ on $C(E, F)$ as follows: For $f, g \in C(E, F)$

$$d^+(f, g) = \sum_{i=1}^{\infty} \frac{1}{2^i} \frac{d_i^+(f, g)}{1 + d_i^+(f, g)},$$

where $f, g \in C(E, F)$ and

$$d_i^+(f, g) = d_i^+(f \mid K_i, g \mid K_i).$$

It is easy to check that d^+ is a metric. Since E, being locally compact, is a k-space, we have $C(E, F) = K(E, F)$ (Theorem 12, §58). Observe that the compact–open topology on each $C(K_i, F)$ is the induced compact–open topology from $CB(E, F)$ if we embed $C(K_i, F)$ as a subspace of $CB(E, F)$ by using the extension theorem (Theorem 8, §61) since E, being a Hausdorff locally compact space, is completely regular.

Corollary 10. Let E be a compact Hausdorff space and R the real line. The space $C(E, R)$ is a complete metric space with the uniform metric

$$d^+(f, g) = \sup\{|f(x) - g(x)| : x \in E\},$$

where R is given the usual absolute-value metric.

Proof. By Theorem 7, §57, $C(E, R)$ is a complete space under the compact–open topology which coincides with the metric topology on $C(E, R)$ by Proposition 18.

Corollary 11. Let E be a metric locally compact space which is the countable union of an increasing sequence of compact sets and let R be the real line. Then $C(E, R)$ is a complete metric space and the metric topology coincides with the compact–open topology.

Proof. This is a particular case of Corollary 9.

We have seen that in general when E is a topological space and F a uniform space, then the compact–open topology on $C(E, F)$ is not necessarily a topology of joint continuity, i.e., a topology for which the map $\varphi: C(E, F) \times E \to F$, where $\varphi(f, x) = f(x)$, $f \in C(E, F)$, $x \in E$ is continuous. But in the particular case when F is a metric space with a bounded metric, the situation is different.

Proposition 20. Let E be a topological space, (F, d) a metric space with a bounded metric d. Let $C(E, F)$ be the set of all continuous maps f from E to F, endowed with the metric topology. Then the evaluation map $\varphi: (f, x) \to f(x)$ of $C(E, F) \times E$ into F is continuous.

Proof. Let $\varepsilon > 0$ and let $B_\varepsilon(f_0(x_0))$ be an open ball containing $f_0(x_0)$, where $x_0 \in E$, and $f_0 \in C(E, F)$. Since $f_0 \colon E \to F$ is continuous,

$$U = f_0^{-1}(B_{\varepsilon/2}(f_0(x_0)))$$

is a neighborhood of x_0. Let $B_{\varepsilon/2}(f_0)$ be the ball in $C(E, F)$ containing f_0. Then for all $f \in B_{\varepsilon/2}(f_0)$ and $x \in U$, we have $d^+(f, f_0) < \tfrac{1}{2}\varepsilon$ and $d(f(x), f_0(x)) < \tfrac{1}{2}\varepsilon$. But then for $f \in B_{\varepsilon/2}^+(f_0)$ and $x \in U$,

$$\begin{aligned} d(\varphi(f, x), \varphi(f_0, x_0)) = d(f(x), f_0(x_0)) &\leq d(f(x), f_0(x)) \\ &+ d(f_0(x), f_0(x_0)) < \tfrac{1}{2}\varepsilon + \tfrac{1}{2}\varepsilon = \varepsilon. \end{aligned}$$

This shows the continuity of φ.

Let E be a Hausdorff real or complex locally convex space (Definition 17, §31). Then E is a completely regular space (Proposition 21, §31). Let E^* denote the set of all linear mappings of E into its scalar field (real or complex field K). Let E' denote the subset of E^* which consists of all continuous maps. Then clearly we have:

$$E' \subset [C(E, K) \cap E^*] \subset K^E.$$

We may endow E' with one of the three important $\mathfrak{S}$-topologies:

(i)   the $\mathfrak{S}$-topology (Definition 6, §57), where $\mathfrak{S}$ consists of finite subsets of E (called the w*-topology);

(ii)   the $\mathfrak{S}$-topology $\mathscr{T}_c$, where $\mathfrak{S}$ consists of all compact subsets of E (called the *compact–open* topology as usual);

(iii)   the $\mathfrak{S}$-topology $\mathscr{T}_\beta$, where $\mathfrak{S}$ consists of all bounded (cf. Definition 16, §52) subsets of E (called the *strong topology*).

Since each finite subset is compact and each compact set is bounded, we have

$$\mathscr{T}_\beta \supset \mathscr{T}_c \supset \mathscr{T}_p.$$

(See the remark following Definition 5, §57.) Hence each $\mathscr{T}_\beta$-compact subset in E' is $\mathscr{T}_c$-compact and hence $\mathscr{T}_p$-compact. Furthermore, on each equicontinuous subset of E' [more generally of $C(E, \mathbb{R})$], $\mathscr{T}_p$ and $\mathscr{T}_c$ coincide (Theorem 9, §58).

Further, since bounded sets are preserved under continuous linear maps [Exercise 26(a), Chapter VII], each $\mathscr{C}_\beta$-bounded subset is $\mathscr{C}_c$-bounded and hence $\mathscr{C}_p$-bounded.

Theorem 13. Let E be a Hausdorff locally convex space and E' its dual. The following statements are equivalent:

(a) E is barreled (Definition 17, §52);
(b)   each $\mathscr{C}_p$-bounded subset of E' is equicontinuous.

Proof. (a) $\Rightarrow$ (b)   Let E be barreled and M a $\mathscr{C}_p$-bounded subset of E'. Let $M^\circ = \{x \in E: |f(x)| \leq 1$ for all $f \in M\}$, the *polar* of M. Then M° is convex because if $x, y \in M^\circ$ and $0 \leq \lambda \leq 1$, then by linearity of f,

$$|f(\lambda x + (1 - \lambda)y)| \leq \lambda |f(x)| + (1 - \lambda)|f(y)| \leq \lambda + 1 - \lambda = 1.$$

Also M° is circled, because for $|\lambda| \leq 1$ and $x \in M^\circ$,

$$|f(\lambda x)| = |\lambda|\ |f(x)| \leq 1$$

and so $\lambda x \in M^\circ$. M° is closed because f is continuous and $[-1, 1]$ is closed; hence $f^{-1}[-1, 1] = \{x \in E: |f(x)| \leq 1\}$ is a closed subset of E and so is $M^\circ = \cap\{|f|^{-1}[-1, 1], f \in M\}$. Now to show that M° is absorbing, let $x \in E$. Since M is $\mathscr{C}_p$-bounded, i.e., $\{f(x): f \in M\}$ is a bounded subset of the real or complex field, i.e., there exists $\alpha > 0$ such that $|f(x)| \leq \alpha$ for all $f \in M$ or $|\alpha^{-1}f(x)| = |f(\alpha^{-1}x)| \leq 1$ (because f is linear) shows that $\alpha^{-1}x \in M^\circ$ or $x \in \alpha M^\circ$ and M° is absorbing. Thus M° is a barrel of E. Since E is barreled, M° is a neighborhood of 0, i.e., there exists a neighborhood U of 0 such that $U \subset M^\circ$, i.e., for all $x \in U$ and $f \in M$, $|f(x)| \leq 1$. Hence M is equicontinuous at 0 and thus everywhere because translations in E are homeomorphisms.

(b) $\Rightarrow$ (a)   Let B be a barrel in E. Consider $B^\circ = \{f \in E': |f(x)| \leq 1$ for $x \in B\}$ (called the polar of B). We show that B° is $\mathscr{C}_p$-bounded. Let $x \in E$. Then since B is absorbing there exists $\alpha > 0$ such that $x \in \alpha B$ and so for all $f \in B^\circ$,

$$|f(x)| = |f(\alpha y)| = |\alpha f(y)| = \alpha |f(y)| \leq \alpha,$$

because $y \in B$ and so $|f(y)| \leq 1$ (because $f \in B^\circ$). Thus B° is $\mathscr{C}_p$-bounded and so equicontinuous by (b). Hence in E, $B^{\circ\circ}$ is a neighborhood of 0 in E, as shown in (a) $\Rightarrow$ (b). But it is easy to see that $B \subset B^{\circ\circ}$. Since B is convex and $\mathfrak{S}$-closed in E, where $\mathfrak{S}$ runs over all finite subsets of E' as shown in (a) $\Rightarrow$ (b), it can be shown that $B = B^{\circ\circ}$ (cf. [45], Chapter II, and Exercise 13, Chapter IX) and so B is a neighborhood of 0 in E. This proves that E is barreled.

Similarly, one proves the following with the observation that a subset $M \subset E'$ is $\mathscr{C}_\beta$-bounded iff its polar B° in E absorbs bounded subsets of E ([45], Chapter II).

Theorem 14. Let E be a Hausdorff locally convex space and E' its dual. Then E is quasibarreled (Definition 17, §52), iff each $\mathscr{C}_\beta$-bounded subset of E' is equicontinuous.

Corollary 12. Let E be a Fréchet (in particular Banach) space and E' its dual. The following statements are equivalent for any subset $A \subset E'$:

  (i) A is $\mathscr{C}_p$-bounded;
  (ii) A is $\mathscr{C}_\beta$-bounded;
  (iii) A is equicontinuous.

Proof. Since each Fréchet space, being a Baire space (Corollary 9, §41) is barreled and quasibarreled (Proposition 45, §52), Theorems 13 and 14 prove the corollary, because every equicontinuous set is always strongly bounded (cf. [45], Chapter II).

Corollary 13. Let E be a Hausdorff locally convex space and E' its dual. If a subset $M \subset E'$ is equicontinuous, then $\mathrm{Cl}_p M$ in K^E is also equicontinuous. Hence $\mathrm{Cl}_p M$ is $\mathscr{C}_p$-compact.

Proof. By Proposition 12, §58, $\mathrm{Cl}_p M$ is equicontinuous and hence $\mathrm{Cl}_p M \subset E'$. But then it is enough to prove that a $\mathscr{C}_p$-closed equicontinuous subset $A \subset E'$ is $\mathscr{C}_p$-compact. Since for each x, $\overline{A(x)} = \mathrm{Cl}\{f(x): f \in A\}$ is a closed and bounded (because A is equicontinuous) subset of the reals, $\overline{A(x)}$ is compact (Theorem 7, §37), it follows that A is $\mathscr{C}_p$-compact by Corollary 8, §58.

Corollary 14. Let $(E, \|\cdots\|)$ be a Banach space and E' its dual. Then the unit ball $B = \{f \in E': \|f\| \le 1\}$ is equicontinuous and $\mathscr{C}_p$-closed.

Proof. For a fixed $x \in E$,

$$|f(x)| \le \|f\| \, \|x\| \le \|x\|$$

for all $f \in B$ shows that B is $\mathscr{C}_p$-bounded. Since each Banach space is a Fréchet space, by Corollary 12, B is equicontinuous. We show that B is $\mathscr{C}_p$-closed.

Let $\{f_\alpha\}$ be a net in B such that $f_\alpha \to f$ is the $\mathscr{C}_p$-topology, i.e., for each $x \in E$, $f_\alpha(x) \to f(x)$. Since for each x,

$$\| x \| \leq 1, \qquad |f_\alpha(x)| \leq \|f_\alpha\| \, \|x\| \leq 1,$$

it follows that $|f(x)| \leq 1$ for all x, $\|x\| \leq 1$. Hence $\sup_{\|x\| \leq 1} |f(x)| = \|f\| \leq 1$ shows that $f \in B$ and the proof is completed.

§60. Sequential Convergence in Function Spaces

It has been observed before that sequential closure does not necessarily give a closed set in a general topological space which is not first countable. Similarly, sequential compactness does not imply compactness. But, even so, sequential convergence is important in analysis. In this section, we wish to relate sequential convergence of a sequence of continuous functions and pointwise convergence.

Definition 9. Let $f_n \in C(E, F)$ be a sequence of continuous maps from a topological space E into a topological space F. Then $\{f_n\}$ is said to *converge sequentially pointwise* to $f \in C(E, F)$ if for each $x \in E$, $\{f_n(x)\}$ converges to $f(x)$ in F.

Remark. Observe that the limit map f in the above definition is assumed to be continuous. In general the pointwise limit of a sequence or a filter in $C(E, F)$ need not be in $C(E, F)$, as classical examples in real analysis show (e.g., Exercise 16, Chapter VII). In other words, sequential convergence does not give a closure in $C(E, F)$ in general. Indeed, when $C(E, F)$ is metrizable in the uniform topology as described in the previous sections, the situation becomes different. More precisely, the following holds:

Proposition 21. Let E and F be topological spaces and let $C(E, F)$ be endowed with the pointwise convergence topology $\mathscr{C}_p$ [written $C_p(E, F)$] induced from the product space F^E. Then:

(i)   For each fixed $x_0 \in E$, the evaluation map: $f \to f(x_0)$ from $C_p(E, F)$ to F is continuous.

(ii)   If f is a continuous map from E to G, then the induced map from $C_p(G, F)$ to $C_p(E, F)$ is continuous, where G is any topological space.

Proof. (i)   Observe that the evaluation map: $f \to f(x_0)$ is the projection from the product space F^E to F and hence continuous.

(ii)   If $f: E \to G$ is continuous, then for any $g \in C(G, F)$, we see that the composition $g \circ f: E \to F$ is continuous and hence $g \circ f \in C(E, F)$. Thus f induces a map: $g \to g \circ f$ of $C(G, F)$ into $C(E, F)$ and is continuous in the topologies of pointwise convergence.

Remark.   We have seen that if each $(E_\alpha)_{\alpha \in A}$ is a metrizable space and A is a countable set, then $\prod_{\alpha \in A} E_\alpha$ is metrizable and thus the pointwise convergence on $C(A, F)$ is metrizable if A is a countable set with the discrete topology and $F = E_\alpha$ (for all $\alpha \in A$) is a metric space. In general this may fail, as is shown in the following:

Proposition 22.   Let E be a completely regular space, F a metric space such that the cardinality of $E > \aleph_0$, and F is not totally pathwise disconnected (i.e., there exists a nonconstant continuous function l of $[0, 1]$ into F). Then there exists no real function d on $C(E, F) \times C(E, F)$ with the following properties:

(a) $d(f, g) \geq 0$ for all $f, g \in C(E, F)$;

(b) $d(f_n, g) \to 0$ if and only if $f_n \to g$ in the induced $\mathscr{C}_p$-topology on $C(E, F)$. [Consequently, $C_p(E, F)$ is not metrizable.]

Proof.   Let d' denote the metric on F and let $l: [0, 1] \to F$ be a nonconstant continuous map. Put

$$d^\times = \sup\{d'(l(0), l(t)): t \in [0, 1]\},$$

which is ≥ 0. Since $d'(l(0), l(t))$ is a continuous map on the connected space $[0, 1]$, for each $b \leq d^\times, b \geq 0$ there is a $t_0 \in [0, 1]$ with $d'(l(0), l(t_0)) = b$. Since E is completely regular, there is a continuous function $g: E \to [0, 1]$ with $g(x) = t_0$ and $g(F \setminus U) = 0$, where U is an open neighborhood of $x \in E$. But then $f = l \circ g$ is a continuous map of E into F such that $f(F \setminus U) = l(0)$ and $d'(l(0), f(x)) = b$ for any $x \in E$ and any open neighborhood U of x. (Observe that $l \circ g$ depends upon x and U.)

Now suppose there exists a map $d: C(E, F) \times C(E, F)$ with (a) and (b). We show a contradiction. Take $g \in C(E, F)$ to be the constant map $x \to l(0)$. For each fixed $x_0 \in E$, let

$$\alpha_n(x_0) = \sup\{d'(f(x_0), g(x_0)): f \in C(E, F), d(f, g) < 1/n\}.$$

Then $\alpha_n(x_0) \to 0$ as $n \to \infty$ for a fixed x_0; for otherwise, there is a subsequence $\{n_k\}$ such that $\alpha_{n_k}(x_0) \geq \varepsilon > 0$ for a given fixed $\varepsilon > 0$. Thus the subsequence $\{f_{n_k}\}$ is such that $d(f_{n_k}, g) < 1/n_k$ while $d'(f_{n_k}(x_0), g(x_0)) \geq \frac{1}{2}\varepsilon$, which is a contradiction in view of (b).

Now put $E_n = \{x \in E: \alpha_n(x) < d^\times/2\}$. Then clearly $E = \bigcup_{n=1}^{\infty} E_n$ and at least one of E_n, say E_{n_0}, is infinite since the cardinality of $E > \aleph_0$.

Since E is completely regular and thus regular, it is possible to choose countable family of open sets $\{U_n\}$ of E by induction such that $\bar{U}_n \cap \bar{U}_m = \varnothing$ for $n \neq m$ and $U_n \cap E_{n_0} \neq \varnothing$ for all $n \geq 1$. Now let $x_n \in U_n \cap E_{n_0}$, $n \geq 1$ and let $f_n \in C(E, F)$ with $f_n(E \setminus U_n) = l(0)$ and

$$d^+\big(l(0), f_n(x_n)\big) = \alpha_{n_0}(x_n) + \tfrac{1}{2}d^\times.$$

Clearly $f_n \to g$ pointwise, but since

$$d'\big(f_n(x_n), g(x_n)\big) > \alpha_{n_0}(x_n)$$

it follows that $d(f_n, g) \geq 1/n_0$, which is a contradiction.

Examples and Exercises

1. Let $\{n\}$ be a finite set of integers and R, the set of real numbers. Show that $R^{\{n\}}$ is the Euclidean space which is a linear topological space.

2. Let A be any set and R, the real line. The product space R^A is a complete topological vector space.

3. If E is a set and F a normal space, then F^E is not necessarily normal.

4. Let E, F be two topological spaces and $f: E \to F$ a continuous map. Let G be a topological space and consider the map $f^*: G^F \to G^E$ defined by $f^*(h) = h \circ f$, where $h \in G^F$. Then f^* is continuous in h and is also continuous in each variable separately.

   (a) If F is a Hausdorff locally compact space and E, G Hausdorff spaces, then the map: $F^E \times G^F \to G^E$, defined by $q(f, g) = g \circ f$, is continuous, where $f \in F^E$ and $g \in G^E$.

   (b) Suppose $F = Q$, rational numbers with the induced topology from the reals, $G = I = [0, 1]$ and $q: I^Q \times Q \to I$ the evalution map. If $f_0: Q \to I$ is the map which maps Q onto 0 and $r_0 \in Q$, an arbitrary point, then q is not continuous at (f_0, r_0). (The continuity of q fails because Q is not locally compact.)

5. Let (F, d) be a metric space and E any topological space. Then each finite subset of $C(E, F)$ is equicontinuous.

6. The set of all real-valued functions on the closed interval $[a, b]$, $-\infty < a \leq b < \infty$, whose derivatives are uniformly bounded on (a, b), is equicontinuous.

7. Let (F, d) be a metric space and E a metric space or, more generally, first countable. Then each sequence $\{f_n\}$ in F^E converges continuously to $f \in F^E$ [i.e., $f_n(x_n) \to f(x)$ for each $x \in E$ and each sequence $x_n \to x$] iff $\{f_n\}$ converges to f uniformly on each compact subset.

8. Let E be any space and F a Hausdorff space. Then the embedding mapping $x \to (f(x))_f$ of E into F^E is closed.

9. If $f: E \to F$ is an onto map then $f^*: G^F \to G^E$, defined in Exercise 4, is always one-to-one.

10. If E, F, G, are any topological spaces, then $(F \times G)^E$ is isomorphic with $F^E \times G^E$.

11. Let $\mathscr{E}$ be an equicontinuous family of maps from a compact space E into a compact uniform space F. Then $\mathscr{E}$ is relatively compact in F^E with the $\mathscr{C}_p$-topology.

12. Let E be a k-space and (F, d) a metric space. Let $\{f_n\}$ be a sequence of equicontinuous functions such that $\lim_{n\to\infty} f_n(x) = f(x)$, for each $x \in E$. Then f is continuous and $\{f_n\}$ converges to f in the $\mathscr{C}_c$-topology.

13. Show that each uniformly bounded and equicontinuous subset of continuous real-valued functions on the Euclidean space R^n is sequentially compact.

14. A topology $\mathscr{E}$ on $C(E, F)$ is said to be *splitting* if for every topological space G, the continuity of $f: G \times E \to F$ implies the continuity of $f: G \to (C(E, F), \mathscr{E})$. And a topology $\mathscr{E}$ on $C(E, F)$ is called *conjoining* if for every topological space G, the continuity of $g: G \to (C(E, F), \mathscr{E})$ implies the continuity of $g: G \times E \to F$, where $g(x) = g(x, y)$, $x \in G$, $y \in E$. Show that:
    (a) $\mathscr{E}$ is conjoining iff the evaluation map: $(f, x) \to f(x)$ from $C(E, F) \times E \to G$ is continuous;
    (b) the discrete topology on $C(E, F)$ is conjoining;
    (c) any metric topology on $C(E, F)$ is conjoining;
    (d) the compact–open topology $\mathscr{C}_c$ on $C(E, F)$ is conjoining when E is locally compact;
    (e) the $\mathscr{C}_c$-topology on $C(Q, I)$ is not conjoining, when $Q =$ set of rational numbers and $I = [0, 1]$;
    (f) any topology finer than a conjoining topology on $C(E, F)$ is conjoining;
    (g) any conjoining topology is finer than any splitting topology;
    (h) the indiscrete topology on $C(E, F)$ is splitting;
    (i) the $\mathscr{C}_c$-topology on $C(E, F)$ is always splitting;
    (j) any topology coarser than the splitting topology is splitting;
    (k) there always exists a unique finest splitting topology.

15. Let R be the real line and let $C(R, R)$ denote the set of all continuous real-valued functions on R. Show that $\mathscr{C}_c$ is strictly finer than $\mathscr{C}_p$ on $C(R, R)$. Show also that $\mathscr{C}_c$ is a metric topology whereas $\mathscr{C}_p$ is not.

16. Let $A(C, C)$ denote the set of all analytic complex functions on the complex plane C. Show that it is a complete metric space with the metric

$$d(f, g) = \sum_{i=1}^{\infty} \frac{1}{2^i} \frac{\max\limits_{|z| \leq i} |f(z) - g(z)|}{1 + \max\limits_{|z| \leq i} |f(z) - g(z)|}.$$

Show that $A(C, C)$ is a closed subspace of $C(C, C)$.

17. Let E be a locally convex Hausdorff space. It is possible there exists another locally convex topology on E having the same dual E'. Such topologies are called *compatible with duality*. The finest locally convex compatible (with duality) topology on E is called the *Mackey topology*. A locally convex space with the Mackey topology is called a *Mackey space*. Show that:

   (a) E is a Mackey space iff every convex circled and $\sigma(E', E)$-compact subset of E' is equicontinuous.

   (b) Every metrizable locally convex space is a Mackey space.

   (c) Each barreled (more generally quasibarreled) space is a Mackey space. In particular, every Baire or Fréchet or Banach space is a Mackey space (see [45], [13], or [55b]).

IX

Extensions of Mappings

In this chapter we wish to study various extension theorems. An example of an extension theorem has already been met with, viz., Tietze's extension theorem (Theorem 8, §19). We wish here to extend this theorem to a general situation and also to consider the extension of some particular maps, for instance, linear ones on linear topological spaces. We prove the Hahn–Banach extension theorem and the Bishop Theorem.

§61. Extensions of Maps on Completely Regular and Metric Spaces

In this section we wish to give a few generalizations of the Tietze's extensions which hold for normal spaces. Also a specialization to metric spaces is given. Recall Tietze's theorem: Let X be a normal topological space and Y a closed subspace of X. If $f: Y \rightarrow [0, 1]$ is a continuous map then there exists a continuous map $\tilde{f}: X \rightarrow [0, 1]$ such that $\tilde{f} \mid Y = f$ (Theorem 8, §19).

Now by means of the Stone–Čech compactification (Definition 20, §44), which exists for each completely regular space, we may extend the above theorem as follows:

Theorem 1 (Extension theorem). Let X be a completely regular space and f a continuous function on a compact subset K of X into $[0, 1]$. Then there exists a continuous extension $\tilde{f}$ of f, i.e., a continuous function $\tilde{f}$ on X into $[0, 1]$ such that $\tilde{f}(X) = f(x)$ for all $x \in K$.

Proof. Let βX denote the Stone–Čech compactification of X. Then βX, being a compact Hausdorff space, is normal. Further, K, being a compact subset of X, is a closed subset of βX. Hence by Tietze's extension theorem (Theorem 8, §19), there is a continuous extension g of f from βX into $[0, 1]$. But then the restriction $\tilde{f} = g \mid X$ of g to X is the required continuous extension of f.

In categorical language Tietze's extension theorem says the following:

Corollary 1. Let *Com* denote the category of compact Hausdorff spaces and continuous maps as morphisms. Then the closed unit interval $[0, 1]$ is an injective (see §6) in *Com*. Actually each $[a, b]$, $-\infty < a < b < +\infty$, is an injective.

Remark. (1) Since $[0, 1]$ is homeomorphic with $[a, b]$, $-\infty < a < b < \infty$ under the map: $t \to (1 - t)a + tb$, one can replace $[0, 1]$ by $[a, b]$ in the above theorem.

(2) Since $[a, b]$ is continuously embedded in $R = \{x: -\infty < x < \infty\}$, it follows that one can replace $[0, 1]$ by R in the above theorem.

(3) Further, since R is continuously embedded in the n-dimensional Euclidean space R^n, we may replace $[0, 1]$ by R^n in the above theorem. Observe that the map: $x \to f(x)$ from X into R^n can be written as $x \to (f_i(x)) \in R^n$, where $f(x) = (f_1(x), \ldots, f_n(x)) = (f_i(x))$, in which each f_i is a map: $x \to f_i(x)$ from X into R. Since $x \to f(x)$ is continuous if and only if each $x \to f_i(x)$ $(1 \le i \le n)$ is continuous, the remark follows.

These remarks show that the range space in Tietze's extension theorem can be taken to be an n-dimensional Euclidean space. Obviously one is motivated to replace the range space by a more general linear space. For this one asks the following:

Let X be any topological space and Y a closed subspace of X. If f is a continuous map of Y into a topological space Z or in particular a locally convex space (or more generally a topological vector space), does there exist a continuous map $\tilde{f}$ from X into Z such that $\tilde{f} \mid Y = f$, i.e., such that $\tilde{f}$ extends f?

Indeed, in general, the answer is in the negative. For example, if i is the identity map of the set Q of rational numbers with the metric topology induced from the reals R onto Q, then there is no continuous map f of R into Q such that $f \mid Q = i$, as is easy to see.

Even though for the general case the extension theorem may not be valid, the problem of finding the particular cases in which it is valid remains and is of great interest. We shall show that the extension of continuous maps from a subspace of a metric space into a locally convex space is always possible. We shall also show that if X is a topological (real or complex) vector space and Y a subspace of X then every continuous linear mapping of Y into the real or complex number field extends to a continuous linear mapping on X to the real or complex field (Hahn–Banach theorem). Before we prove the extension theorems we recall some facts dealt with before:

Fact 1. Every metric space is paracompact (see Theorem 22, §43).

Fact 2. Let X be a metric space, A a closed subspace of X. Then there exists an open covering $\{P\}$ of $X \backslash A$ such that

(i) $\{P\}$ is a locally finite open covering;

(ii) any neighborhood of $a \in A \backslash A^\circ$ contains infinitely many members of $\{P\}$;

(iii) for any neighborhood W of $a \in A$, there exists a neighborhood $W' \subset W$ of a such that $P \cap W' \neq \varnothing$ implies $P \subset W$.

Proof of this fact is contained in the proof of Theorem 22, §43. It suffices to remark that $X \backslash A$ is a metric space and hence paracompact. Now for each $x \in A$, let U_x be open neighborhood of x such that $\delta(U_x) < \frac{1}{2}d(x, A)$. Clearly $\{U_x\}$ is an open covering of $X \backslash A$. Thus there is a locally finite refinement by open sets, say, $\{P\}$ and all the conditions (i)–(iii) are satisfied.

Definition 1. A covering for $X \backslash A$ satisfying (i)–(iii) in Fact 2 is called a *canonical* covering.

Definition 2. (i) A collection of finite number of points in an Euclidean space is said to be a set of *vertices*. (ii) The set of straight line-segments joining vertices is called the set of *edges*. (iii) The convex compact subset enclosed by a finite set of vertices and the edges joining them is called a *cell* or *simplex*. (iv) The point-set composed of an arbitrary collection of closed cells is called a *polytope*, provided (a) every face of a cell in the collection is itself a cell of the collection, and (b) the intersection of any two closed cells in the collection is a face (see below for definition) of both these cells.

A geometric description of a cell may be given as follows: Let $x_1, \ldots, x_n$ be n linearly independent points in $\mathbb{R}^n$. The set σ_n of

$$x = \sum_{i=1}^{n} \lambda_i x_i, \qquad 0 < \lambda_i < 1 \text{ and } \sum_{i=1}^{n} \lambda_i = 1,$$

will be called an *open cell*. Its closure $\bar{\sigma}_n$ is a *closed cell* and σ_m $(m < n)$ is a *face* of σ_n.

Thus a polytope is a collection of such σ_n's satisfying (a) and (b) in (iv) of Definition 2. It is possible to give a geometric realization of a polytope by considering that each cell is a vertex in a linear space of appropriate dimension and thus the polytope P may be regarded as a simplex.

Definition 3. We endow a polytope P with the *Whitehead topology*: A subset U of P is open if and only if $U \cap \sigma_n$ for each closed cell $\bar{\sigma}_n$ in P is open in σ_n in the Euclidean topology. Thus a polytope P is a topological space. Sometimes a polytope with this topology is called a *CW-polytope* and its topology a *CW-topology*.

Proposition 1. Let X be a topological space and P a CW-polytope. A mapping $f: P \to X$ is continuous if and only if for each closed cell $\bar{\sigma}_n$ in P, $f \mid \bar{\sigma}_n: \bar{\sigma}_n \to X$ is continuous.

Proof. Obvious from the definition of topology on P (see Proposition 2, §21).

Proposition 2. A CW-polytope is a Hausdorff space.

Proof. Since each σ_n is Hausdorff and P has the so-called topological sum topology, it follows that P is Hausdorff. (Theorem 2, §21).

Definition 4. Let σ be a cell in a polytope. The set of all open cells or simplexes having a common face with σ is called the *star* of the cell.

Observe that the star of a cell is then an open set.

Definition 5. Let $\{P_\alpha: \alpha \in A\}$ be an open covering of a topological space X. To each P_α, let x_α denote a uniquely determined element of a real vector space E such that the set $\{x_\alpha: \alpha \in A\}$ is linearly independent and spans E. The points $\{x_{\alpha_i}: i = 1, \ldots, n\}$ determine an n-cell if and only if $\bigcap_{i=1}^n P_{\alpha_i} \neq \varnothing$. The polytope thus determined is called the *nerve* of the covering $\{P_\alpha\}$ and is denoted by $N(P)$. (It is assumed here that a cell is sitting in a finite-dimensional vector subspace of E which is homeomorphic with a Euclidean space and that E has the weak topology.)

Proposition 3. Let $\{P_\alpha : \alpha \in A\}$ be a locally (or neighborhood) finite open covering of a metric space X and $N(P)$ its nerve. Then there exists a continuous mapping $\varphi : X \to N(P)$ such that for each P_α, $\varphi^{-1}(\text{St } x_\alpha) \subset P_\alpha$, where St x_α is the star of x_α.

Proof. Let d denote the metric of X. For $x \in X$, define

$$\lambda_\alpha(x) = \frac{d(x, X \setminus P_\alpha)}{\sum\limits_{\alpha \in A} d(x, X \setminus P_\alpha)}, \qquad \text{where } \lambda_\alpha(x) = 0 \text{ if } P_\alpha = X.$$

Then we see that:

(a) $0 \leq \lambda_\alpha \leq 1$.

(b) $\sum_{\alpha \in A} d(x, X \setminus P_\alpha) < \infty$ because $d(x, X \setminus P_\alpha) \neq 0$ if and only if $x \in P_\alpha$ and since the covering is locally finite, x can belong to at most a finite number of P_α's.

(c) $\sum_\alpha d(x, X \setminus P_\alpha) \neq 0$ for all $x \in X$, because $\{P_\alpha\}$ is a covering.

Thus we conclude that λ_α is well-defined for each $x \in X$. Further,

(d) Each λ_α is continuous. It is because $x \to d(x, B)$ is continuous (see Exercise 19, Chapter III.)

(e) $\sum_{\alpha \in A} \lambda_\alpha(x) = 1$ for each $x \in X$, and in some neighborhood of x only a finite number of λ_α's are nonzero, by the same arguments as in (b).

(f) $\lambda_\alpha(x) \neq 0$ if and only if $x \in P_\alpha$.

Now we define:

$$\varphi(x) = \sum_{\alpha \in A} \lambda_\alpha(x) x_\alpha.$$

If $x \in \bigcap_{i=1}^n P_{\alpha_i}$, then (e) implies that $\varphi(x)$ is an interior point of the cell with vertices $\{x_{\alpha_i}\}$ and barycentric coordinates $\{\lambda_{\alpha_i}\}$ ($1 \leq i \leq n$). Also for each α,

$$\varphi^{-1}(\text{St } x_\alpha) \subset P_\alpha.$$

To show that φ is continuous, let $x \in \bigcap_{i=1}^n P_{\alpha_i}$. Then $\varphi(x)$ is in the interior of the cell $\bar{\sigma} = $ convex closure of $\{x_{\alpha_1}, \ldots, x_{\alpha_n}\}$. If V is an open set containing $\varphi(x)$, then $V \cap \bar{\sigma}$ is open in $\bar{\sigma}$ and hence there exists an open neighborhood W of x such that $\varphi(W) \subset V \cap \bar{\sigma} \subset V$, because each λ_α is continuous. This completes the proof.

Now we state and prove a proposition which will extend the mapping φ: $X \to N(P)$ obtained in Proposition 3 to a suitable topological space. More precisely, we have:

Proposition 4. Let (X, d) be a metric space and A a closed subspace of X. Then there is a Hausdorff topological space Y (not necessarily metrizable) and a continuous map $\mu\colon X \to Y$ satisfying:

(i) The restriction $\mu \mid A\colon A \to \mu(A)$ is a homeomorphism and hence $\mu(A)$ is closed;

(ii) $Y \backslash \mu(A)$ is an infinite polytope and $\mu(X \backslash A) \subset Y \backslash \mu(A)$;

(iii) each neighborhood of $a \in \mu(A) \backslash [\mu(A)]^{\circ}$ contains infinitely many cells of $Y \backslash \mu(A)$.

Proof. Observe $X \backslash A$ is a metric space. Let $\{\mathbf{P}_\alpha\}$ be a canonical covering of $X \backslash A$ and let $N(\mathbf{P})$ denote its nerve. Now let Y be the set which consists of points which are in $1 : 1$ correspondence with the points of the set A and the points of $N(\mathbf{P})$ with the topology as follows:

(a) $N(\mathbf{P})$ is endowed with the CW-topology (Definition 3);

(b) a subbase for the neighborhoods of $a \in A$ in Y is determined by selecting a neighborhood W of $a \in X$ and the set of points in Y, $W \cap A$ along with the star of every vertex of $N(P)$, corresponding to each P_α contained in W.

Then we observe that Y is a Hausdorff space, and A and $N(\mathbf{P})$ as subspaces preserve their topologies.

Now define

$$\mu(x) = \begin{cases} x & \text{if } x \in A; \\ \varphi(x) & \text{if } x \in X \backslash A. \end{cases}$$

To establish the continuity of μ, we observe that $\varphi\colon X \backslash A \to N(\mathbf{P})$ is continuous by Proposition 3. Obviously μ is continuous on A. Thus the continuity on the common points remains. Let $a \in \operatorname{Fr} A = A \cap \overline{X \backslash A}$. Let $\tilde{W}$ be a subbasic neighborhood of $\mu(a)$ in Y which is determined by a neighborhood W of a in X. Since the covering $\{\mathbf{P}_\alpha\}$ of $X \backslash A$ is canonical, there is a neighborhood $W' \subset W$ of a such that $P_\alpha \cap W' \neq \varnothing$ implies $\mathbf{P}_\alpha \subset W$. Clearly $\{\mathbf{P}_\alpha\colon \mathbf{P}_\alpha \subset W' \cap (X \backslash A)\}$ is not empty. We show $\mu(W') \subset \tilde{W}$.

If $x \in W' \cap X \backslash A$ and if $x \in \bigcap_{i=1}^{n} \mathbf{P}_{\alpha_i}$ (and only these $\mathbf{P}_{\alpha_i}$'s), then $\varphi(x)$ is an interior point of the cell with vertices $\{x_{\alpha_i}\colon 1 \leq i \leq n\}$ and hence $\varphi(x)$ belongs to the star of the vertex (say) x_α. Since $\mathbf{P}_{\alpha_1} \cap W' \neq \varnothing$, we have $\mathbf{P}_{\alpha_1} \subset W$ and so $\varphi(x) \in \tilde{W}$. The conditions (i)–(iii) now follow immediately. Also

$$\varphi\big(W' \cap (X \backslash A)\big) = \mu\big(W' \cap (X \backslash A)\big) \subset W'.$$

Finally, since $W' \subset W$, we have

$$\mu(W' \cap A) \subset W' \cap A \subset \tilde{W}$$

and so $\mu(W') \subset \tilde{W}$. This proves that μ is continuous.

Theorem 2. Let (X, d) be a metric space, A a closed subspace of X, E a locally convex Hausdorff linear space, and $f: A \to E$ a continuous mapping. Then there exists a continuous mapping $f\tilde{}: X \to E$ such that $f\tilde{} \mid A = f$ and $f\tilde{}(X) \subset$ [convex hull of $f(A)$, which is the smallest convex set in E containing $f(A)$], where E is assumed to be of sufficiently high dimension with weak topology.

Proof. Let Y be the topological space which exists in view of Proposition 4. Observe that Y contains a copy of A. We show that there exists a continuous map $f\tilde{}: Y \to E$ which extends $f: A \to E$.

Let $\mathbf{P} = \{\mathbf{P}_\alpha\}$ be a canonical covering of $X \backslash A$ and $N(\mathbf{P})$ its nerve. Let V_0 denote the vertices of $N(\mathbf{P})$. We first define an extension of f to f_0: $A \cup V_0 \to E$ as follows:

In each $\mathbf{P}_\alpha$, choose a point y_α and choose $a_\alpha \in A$ such that $d(y_\alpha, a_\alpha)$ $< 2d(y_\alpha, A)$. If x_α is the vertex of $N(\mathbf{P})$ corresponding to $\mathbf{P}_\alpha$, put $f_0(x_\alpha)$ $= f(a_\alpha)$, and $f_0(a) = f(a)$ for $a \in A$. This extends $f: A \to E$ to $f_0: A \cup V_0 \to E$. We show that f_0 is continuous. Since V_0 is discrete, f_0 is continuous on V_0.

Now let V be a neighborhood of $f_0(a) = f(a)$, $a \in A$. Since f is continuous on A by hypothesis, there exists $\delta > 0$ such that $d(a, a') < \delta$ implies $f(a') \in V$. Let W be a neighborhood of $a \in A$ with radius $< \delta/3$. If $\mathbf{P}_\alpha \in \{\mathbf{P}_\alpha\}$ and $\mathbf{P}_\alpha \subset W$, then $d(y_\alpha, a) < \delta/3$ and so

$$d(a_\alpha, a) \leq d(a_\alpha, y_\alpha) + d(y_\alpha, a)$$

$$\leq 2d(y_\alpha, A) + \frac{\delta}{3} < \delta.$$

Thus all vertices of $N(\mathbf{P})$ in $W\tilde{}$ have the property: $f_0(x_\alpha) = f(a_\alpha) \in V$. Hence for all $x\tilde{} \in W\tilde{} \cap [A \cup V_0]$, $f_0(x\tilde{}) \in V$ and this proves the continuity of f_0. By linearity we extend $f_0: A \cup V_0 \to E$ to $f\tilde{}: Y \to E$ by putting

$$f\tilde{}(x) = \sum_{\alpha \in A} \lambda_\alpha(x) f_0(x_\alpha) = \varphi(x).$$

In other words, we have defined $f\tilde{}$ as follows:

$$f\tilde{}(x) = \begin{cases} f(x), & x \in A; \\ \varphi(x), & x \in X \backslash A. \end{cases}$$

To show that $f^\sim$ is continuous let V be a convex neighborhood of $f(a)$
$= f^\sim(a)$. By continuity of f_0 at a there is a neighborhood $W^\sim$ of a with

$$f_0(W^\sim \cap [A \cup V_0]) \subset V,$$

where $W^\sim$ is determined by W. Choose a neighborhood $W' \subset W$ of a in X
such that $W' \cap \mathbf{P}_\alpha \neq \emptyset$ implies $\mathbf{P}_\alpha \subset W$. This means that the vertices corresponding to $\mathbf{P}_\alpha$ in W' have images lying in V. If $x_\alpha \in \bar{\mathrm{St}}\, x_\beta$ with $\mathbf{P}_\beta \subset W'$
then $\mathbf{P}_\alpha \cap W' \neq \emptyset$ and so $x_\alpha \in W^\sim$. Thus the vertices of any cell belonging
to the closure of the star of any x_β are sent into the convex set $V \subset E$ and
hence the linear extension over the cells has images lying in V. Hence
$f^\sim(W') \subset V$. This proves the continuity of $f^\sim$. The assertion that $f^\sim(X) \subset$
[convex hull of $f(A)$] follows from the definition of $f^\sim$.

§62. The Hahn–Banach Extension Theorem

In this section we deal with the extension of some special maps, viz.,
linear maps on linear spaces. We will see that for such special maps and
special spaces, the extension problem has a very satisfactory solution.
However, this has been achieved by the use of Zorn's lemma.

We need the following concepts:

Definition 6.   Let E be a real linear space. A map $p: E \to R$ is said to be
a *functional* and is:

  (a) *additive* if $p(x + y) = p(x) + p(y)$ for each $x, y \in E$;
  (b) *subadditive* if $p(x + y) \le p(x) + p(y)$ for $x, y \in E$;
  (c) *positive homogeneous* if $p(\lambda x) = \lambda p(x)$ for $x \in E$, $\lambda \ge 0$;
  (d) *homogeneous* if $p(\lambda x) = |\lambda| p(x)$ for $x \in E$ and $\lambda \in R$;
  (e) *linear* if it is additive and $p(\lambda x) = \lambda p(x)$ for $x \in E$ and $\lambda \in R$.

Similar definitions are given for functionals $p: E \to C$.

Theorem 3. (Hahn–Banach).   Let E be a real linear space and F a
subspace of E. Let p be a subadditive positive homogeneous functional on E.
If $f: F \to R$ is a linear mapping such that $f(x) \le p(x)$ for all $x \in F$, then there
is a linear mapping $f^\sim: E \to R$ such that $f^\sim(x) \le p(x)$ for all $x \in E$ and
$f^\sim(x) = f(x)$ for $x \in F$.

Proof.   Let $\mathscr{F}$ denote the family of pairs (G, g), where G is a linear subspace of E and g is a linear functional on G which extends f and $g(x) \le p(x)$
for all $x \in G$. Clearly $\mathscr{F}$ is nonempty, since the given pair (F, f) is in $\mathscr{F}$.

We define a relation "$\leq$" on $\mathscr{F}$ as follows: $(G_1, g_1) \leq (G_2, g_2)$ if G_1 is a subspace of G_2 and g_2 extends g_1. The relation "$\leq$" defines a partial ordering. (This is easy to check.) The family $\mathscr{F}$ under $\leq$ is inductive: Let $\{G_i, g_i\}$ be a linearly ordered subfamily of $\mathscr{F}$. Put $G = \cup\, G_i$. Then G is a linear subspace of E because G_i's are all linearly ordered. Define a linear map $g: G \to R$ by $g(x) = g_i(x)$ for $x \in G_i$. Then g is clearly a linear mapping. Also, since $g_i(x) \leq p(x)$ for $x \in G_i$, it follows that $g(x) \leq p(x)$ for $x \in G$. Hence (G, g) is in $\mathscr{F}$. Thus by Zorn's lemma, $\mathscr{F}$ has a maximal element $(H, f^\sim)$, which has the property that F is a linear subspace of H, $f^\sim(x) = f(x)$ for $x \in F$ and $f^\sim(x) \leq p(x)$ for $x \in H$.

The proof will be completed if we show that $H = E$. Suppose not, then there exists $x_0 \in E \backslash H$. Put $M = \{x + \lambda x_0 \colon x \in H,\ \lambda \in R\}$. Then M is clearly a linear subspace of E.

For $x, y \in H$, we see that $x - y \in H$ and hence $f^\sim(x - y) \leq p(x - y) = p(x + x_0 - x_0 - y)$. Using the linearity of $f^\sim$ and subadditivity of p, we have

$$f^\sim(x) - f^\sim(y) \leq p(x + x_0) + p(-x_0 - y)$$

or

$$-p(-x_0 - y) - f^\sim(y) \leq p(x + x_0) - f^\sim(x).$$

Since x, y are independent, we have

$$\alpha = \sup_{y \in H} \{-p(-x_0 - y) - f^\sim(y)\} \leq \inf_{x \in H} \{p(x + x_0) - f^\sim(x)\} = \beta.$$

Choose $\gamma \in R$ such that $\alpha \leq \gamma \leq \beta$. Now by definition each $m \in M$ has the representation: $m = x + \lambda x_0$, $x \in H$, $\lambda \in R$. We define

$$h(m) = f^\sim(x) + \lambda \gamma.$$

Then for $m_1, m_2 \in M$, $m_1 = x_1 + \lambda_1 x_0$, $m_2 = x_2 + \lambda_2 x_0$, where $\lambda_1, \lambda_2 \in R$ and $x_1, x_2 \in H$, and so

$$h(m_1 + m_2) = f^\sim(x_1 + x_2) + (\lambda_1 + \lambda_2)\gamma$$
$$= h(m_1) + h(m_2),$$

and also $h(\lambda m_1) = f^\sim(\lambda x_1) + \lambda \lambda_1 \gamma = \lambda(f^\sim(x_1) + \lambda_1 \gamma) = \lambda h(m_1)$, for $\lambda \in R$. This proves that $h: M \to R$ is a linear functional. Further, $m = x + \lambda \gamma \in M$ is in H if and only if $\lambda = 0$. Thus $h(m) = f^\sim(x)$ for $m \in H$ and hence h extends f because $f^\sim$ extends f. Now we have found a pair (M, h) such that $H \subset M \subset E$ and h extends $f^\sim$. We show that $h(m) \leq p(m)$ for all $m \in M$.

From the inequality shown above, viz.,

$$-p(-x_0 - y) - f^\sim(y) \le \gamma \le p(x + x_0) - f^\sim(x),$$

we obtain

$$-p\left(-x_0 - \frac{y}{\lambda}\right) - f^\sim\left(\frac{y}{\lambda}\right) \le \gamma \le p\left(\frac{x}{\lambda} + x_0\right) - f^\sim\left(\frac{x}{\lambda}\right)$$

for $\lambda \ne 0$ because if $x, y \in H$ then $x/\lambda, y/\lambda \in H$ for $\lambda \ne 0$. If $\lambda > 0$, then

$$\gamma\lambda \le \lambda p\left(\frac{x}{\lambda} + x_0\right) - \lambda f^\sim\left(\frac{x}{\lambda}\right)$$

and so (using the positive homogeneity of p)

$$h(m) = f^\sim(x) + \lambda\gamma \le p(x + \lambda x_0) = p(m).$$

And if $\lambda < 0$, then using the left-hand inequality at the top of this page (replacing y by x/λ) we have

$$-\lambda p\left(-x_0 - \frac{x}{\lambda}\right) - \lambda f^\sim\left(\frac{x}{\lambda}\right) \ge \lambda\gamma,$$

or

$$h(m) = f^\sim(x) + \lambda\gamma \le -\lambda p\left(-\frac{x}{\lambda} - x_0\right)$$

$$\le p(x + \lambda x_0) = p(m).$$

This establishes that (M, h) is also in $\mathscr{F}$ and $M \supset H$ contradicts the maximality of H. Hence $H = E$, and the theorem is proved.

Corollary 2. Let E be a topological linear space, F a linear subspace of E and p a continuous subadditive homogeneous functional on E. Let f be a continuous linear functional on F such that $|f(x)| \le p(x)$ for $x \in F$; then there exists a continuous linear functional $f^\sim$ on E such that $f^\sim(x) = f(x)$ for $x \in F$ and $|f^\sim(x)| \le p(x)$ for $x \in E$.

Proof. Let $f^\sim$ be the linear functional obtained in the theorem such that $f^\sim(x) = f(x)$ for $x \in E$ and $|f^\sim(x)| \le p(x)$ for $x \in E$. Since p is continuous, $p(x - x_0) \to 0$ as $x \to x_0$. Hence it follows that $f^\sim$ is also continuous, because

$$|f^\sim(x) - f^\sim(x_0)| \le p(x - x_0) \to 0 \quad \text{as} \quad x \to x_0.$$

Recall that the topology of a locally convex Hausdorff linear space E is defined by a family $\{p_\alpha\}$ of seminorms which are subadditive homogeneous functionals. Thus each p_α is continuous in the topology defining the locally convex topology of E. Suppose f is a continuous linear functional on E, then for $\varepsilon > 0$, $f^{-1}(-\varepsilon, \varepsilon)$ is a convex neighborhood of 0 in E. Now choose a seminorm p_α such that

$$|f(x)| \leq \varepsilon \, p_\alpha(x).$$

Thus we have:

Corollary 3. Let E be a locally convex Hausdorff linear space, F a linear subspace of E, and f a continuous linear functional on F. Then there exists a continuous linear functional $f^\sim$ on E such that $f^\sim(x) = f(x)$, $x \in F$.

A categorical interpretation of the above theorem is as follows:

Corollary 4. Let LCV denote the category of all locally convex Hausdorff real linear spaces and continuous linear maps as morphisms. Then the real line $\mathbb{R}$ is an injective in LCV.

In particular, Corollary 3 holds for normed spaces.

Corollary 5. Let E be a normed space and F a linear subspace of E. Let f be a linear continuous functional on F. Then there exists a continuous linear functional $f^\sim$ on E such that $f^\sim(x) = f(x)$ for all $x \in F$ and $\|f^\sim\| = \|f\|$, where $\|f\| = \sup\{|f(x)| : \|x\| \leq 1, \ x \in F\}$ and similarly $\|f^\sim\|$.

Proof. Since f is continuous, given $\varepsilon > 0$, there exists $\delta > 0$ such that for $\|x\| \leq \delta$, $x \in F$ we have $|f(x)| \leq \varepsilon$. But then for any $x \in F$,

$$\left\| \frac{x\delta}{\|x\|} \right\| \leq \delta$$

and so

$$\left\| \frac{f(x\delta)}{\|x\|} \right\| \leq \varepsilon \quad \text{or} \quad |f(x)| \leq (\varepsilon/\delta) \, \|x\|.$$

Put $\varepsilon/\delta = M > 0$ and we have $|f(x)| \leq M \|x\|$ for all $x \in F$. On the other hand, $|f(x)| \leq M \|x\|$, $x \in F$ easily implies that f is continuous. It is easy to verify that $\|f\| = \inf\{M > 0 : |f(x)| \leq M \|x\|, x \in F\}$. For each $x \in E$, define $p(x) = \|f\| \|x\|$. Then clearly p is a seminorm on E, and $|f(x)| \leq p(x)$ for $x \in F$. By Corollary 3 there exists a continuous linear functional $f^\sim$ on E such that $f^\sim(x) = f(x)$ for $x \in F$ and $|f^\sim(x)| \leq p(x) = \|f\| \|x\|$ for $x \in E$ and hence $\|f^\sim\| < \|f\|$. But the fact that $\|f\| \leq \|f^\sim\|$ is clear. Hence $\|f^\sim\| = \|f\|$.

Corollary 6. (Separation theorem)   Let E be a normed linear space, C a convex open subset of E, and $x_0 \notin C$. Then there exists a continuous linear functional f on E such that $f(x_0) \notin f(C)$.

Proof.   We may assume that $0 \in C$. Thus C is a convex neighborhood of 0. Let $p(x) = \inf\{\lambda > 0: x \in \lambda C\}$. Then it is easy to verify that p is a subadditive positive homogeneous functional on E and $\{x \in E: p(x) < 1\}$ $\subset C \subset \{x \in E: p(x) \leq 1\}$. Consider $F = \{\lambda x_0: \lambda \in R\}$ to be a linear subspace of E. Let $g(x) = \lambda p(x_0)$ for $x = \lambda x_0 \in F$. Then g is a continuous linear functional on F and $g(x_0) = p(x_0) > 1$ because $x_0 \notin C$ and $p(x) \leq 1$ for $x \in C$. We have: $g(x) = \lambda p(x_0) \leq p(x)$ for $x \in F$. Hence by Corollary 2 of the Hahn–Banach theorem there exists a continuous linear functional f on E such that $f(x) = g(x)$ for $x \in F$ and $|f(x)| \leq p(x)$ for $x \in E$. This shows that for $x \in C$, $|f(x)| \leq 1$ and $f(x_0) = g(x_0) > 1$. In other words, $f(x_0) \notin f(C)$.

Corollary 7.   (a)   In every locally convex Hausdorff space E for x, $y \in E$, $x \neq y$, there exists a continuous linear functional f such that $f(x) \neq f(y)$.

(b)   In each normed space $(E, \|\cdots\|)$, for any $x \neq 0$, there exists a continuous linear functional f on E such that $f(x) = \|x\|$.

Proof.   (a)   Let $z = x - y \neq 0$. Put $F = \{\lambda z: \lambda \in R\}$. Then F is a subspace of E. For each $u = \lambda z \in F$, define $g(u) = \lambda$. Then g is clearly a linear functional which is continuous on F. Then there exists a continuous linear functional f on E (Corollary 3) such that $f(u) = g(u)$ for all $u \in F$. For $\lambda = 1$, $u = z = x - y \neq 0$ and we have $1 = g(z) = g(x) - g(y) \neq 0$; hence $f(x) \neq f(y)$.

(b)   Since $x \neq 0$, the set $F = \{\lambda x: \lambda \text{ is a real scalar}\}$ is a linear subspace of E. For each $z = \lambda x \in F$, define $f_0(z) = \lambda \|x\|$. Then, as for (a), f_0 is a continuous linear functional on F and $f_0(x) = \|x\|$. Clearly $|f_0(x)| = \|x\|$ and so $\|f_0\| = \sup\{|f_0(x)|: \|x\| \leq 1\} = 1$. Hence there exists a continuous linear functional f on E such that $f(z) = f_0(z)$ for all $z \in F$ and so $f(x) = f_0(x)$ $= \|x\|$.

Remark.   Corollary 7 says that on a locally convex Hausdorff linear space there are lots of nonzero continuous linear functionals.

The set of all continuous linear functionals on E is denoted by E' and is called the (topological) *dual* of E. Clearly $E' \subset R^E$. We may endow E' with the pointwise convergence or product topology induced from R^E. We denote this topology by $\sigma(E', E)$ on E', which is called the weak* (or w*-) topology on E'.

Proposition 5. E' endowed with $\sigma(E', E)$ is a locally convex Hausdorff linear space.

Proof.   Observe that a subbase of the neighborhood system at $0 \in E'$ is given by the sets

$$T(x, \varepsilon) = \{f \in E': |f(x)| < \varepsilon\},$$

for each fixed $x \in E$, where ε runs over the positive reals. If $T(x, \varepsilon_1)$, $T(x, \varepsilon_2)$ are any two sets, then we see that

$$T(x, \varepsilon) \subset T(x, \varepsilon_1) \cap T(x, \varepsilon_2),$$

where $\varepsilon = \min(\varepsilon_1, \varepsilon_2)$.
   Also

$$T(x, \tfrac{1}{2}\varepsilon) + T(x, \tfrac{1}{2}\varepsilon) \subset T(x, \varepsilon),$$

and each $T(x, \varepsilon)$ is convex and circled, because for $f, g \in T(x, \varepsilon)$ and for $0 \leq \lambda \leq 1$, we see that $|f(x)| < \varepsilon$ and $|g(x)| < \varepsilon$ which imply that

$$|(\lambda f + (1 - \lambda)g)(x)| = |\lambda f(x) + (1 - \lambda)g(x)| < \varepsilon,$$

and for $|\lambda| \leq 1$,

$$|(\lambda f)(x)| = |\lambda f(x)| < \varepsilon.$$

In other words, $\lambda f + (1 - \lambda)g \in T(x, \varepsilon)$ and $\lambda f \in T(x, \varepsilon)$ for $|\lambda| \leq 1$. Hence the family $\{T(x, \varepsilon)\}$ forms a subbase of convex circled neighborhoods of 0 in the $\sigma(E', E)$-topology. Therefore E' is a locally convex space. Furthermore since R is Hausdorff, so is R^E and hence so is E'.
   Since E' is a uniform space (because each topological vector space is), one may describe the uniform structure by gauges which are derived from seminorms in linear spaces.
   Let E' be the dual of a locally convex Hausdorff space. Then the family $\{p_\alpha : \alpha \in A\}$ of seminorms defines a locally convex topology which coincides with $\sigma(E', E)$, where each p_α is defined as follows: Let A denote the set of all nonempty finite subsets of E. For each $\alpha \in A$, let there be $x_1, \ldots, x_\alpha \in E$ (α finite). For each $x' \in E'$, put

$$p_\alpha(x') = \max_{1 \leq i \leq \alpha} |x'(x_i)|.$$

A net $f_\beta \to f$ in E' with the $\sigma(E', E)$-topology if and only if for a given $\varepsilon > 0$ and for any seminorm p_α there exists β_0 such that $p_\alpha(f_\beta - f) < \varepsilon$ for $\beta \geq \beta_0$.

Consider now a normed space E. Then E' has lots of nonzero linear continuous functionals. E' is a Banach space in the norm topology defined by: For $x' \in E'$,

$$\| x' \| = \sup\{| x'(x) |: \| x \| \leq 1, \ x \in E\}.$$

Then we check that $\| x' \| \geq 0$ and $\| x' \| = 0$ iff $x'(x) = 0$ for all $\| x \| \leq 1$. But then for any

$$0 \neq x \in E, \qquad \left\| \frac{x}{\| x \|} \right\| \leq 1$$

implies

$$x'\left(\left\| \frac{x}{\| x \|} \right\| \right) = \frac{1}{\| x \|} x'(x) = 0 \Rightarrow x'(x) = 0.$$

Hence $x' = 0$ iff $\| x' \| = 0$. For λ real,

$$\| \lambda x' \| = \sup_{\|x\| \leq 1} | \lambda x'(x) | = | \lambda | \, \| x \|$$

and

$$\| x' + y' \| \leq \| x' \| + \| y' \|$$

follows easily. Thus $\| x' \|$ defines a norm on E'.

Proposition 6. E' is a Banach space, i.e., a complete normed space in the norm $\| x' \|$.

Proof. Let x_n' be a Cauchy sequence in E', i.e., for a given $\varepsilon > 0$ there exists n_0 such that for $m, n \geq n_0$, $\| x_n' - x_m' \| < \varepsilon$. But this shows that for any $x \in E$,

$$| x_n'(x) - x_m'(x) | \leq \| x_n' - x_m' \| \, \| x \| \leq \varepsilon \| x \|, \text{ for } m, n \geq n_0.$$

Thus for each $x \in E$, $\{x_n'(x)\}$ is a Cauchy sequence of real numbers, hence convergent. Put

$$x'(x) = \lim_{n \to \infty} x_n'(x), \qquad \text{for each } x \in E.$$

Then

$$\| x_n'(x) - x_m'(x) \| < \varepsilon \| x \| \qquad \text{for } n, m \geq n_0$$

implies (by letting $m \to \infty$)

$$| x_n'(x) - x'(x) | < \varepsilon \| x \|, \qquad \text{for all } n \geq n_0, \qquad (*)$$

for any x. Thus

$$\| x_n' - x' \| = \sup_{\|x\| \leq 1} | x_n'(x) - x'(x) | \leq \varepsilon, \qquad \text{for } n \geq n_0.$$

Since each x_n' is continuous, we have from $(*)$,

$$| x'(x) | < \varepsilon \| x \| + \| x_n' \| \, \| x \|$$

for any fixed $n \geq n_0$. It is easy to see that $\{\| x_n' \|\}$ is bounded because $\{x_n'\}$ is a Cauchy sequence in the norm of E'. Hence $| x'(x) \leq M | \, \| x \|$ for some $M > 0$. This shows that x' is continuous, i.e., $x' \in E'$ and so E' is a Banach space.

Theorem 4. (Alaoglu). Let E be a normed space and E' its dual. Then the unit ball $B' = \{x' \in E' : \| x \| \leq 1\}$ of E' is $\sigma(E', E)$-compact and equicontinuous.

Proof. For $x \in E$ and $x' \in B'$, we have

$$| x'(x) | \leq \| x' \| \, \| x \| \leq \| x \|.$$

Thus

$$- \| x \| \leq | x'(x) | \leq \| x \|.$$

Put $I_x = [- \| x \|, \| x \|]$, which is a compact interval of the real line. Hence by Tychonoff's theorem (§35, Theorem 3), the product $\prod_{x \in E} I_x$ is compact. We show that B' is a closed subset of $\prod_{x \in E} I_x$, if we identify each $x' \in B'$ by a point in $\prod_{x \in E} I_x$ via the mapping: $x' \to (x'(x))_{x \in E}$. Let (x_α') be a net in B' converging to x', which is clearly linear. Then for each $x \in E$, we have $| x_\alpha'(x) | \leq \| x \|$. Hence $| x'(x) | \leq \| x \|$ and this shows that $x' \in B'$. In other words, B' is $\sigma(E', E)$-closed and hence compact in the induced topology, which is $\sigma(E', E)$.

Example. Let X be a compact Hausdorff space. The set $C(X)$ is a real Banach space (Corollary 10, §59). Hence its dual $[C(X)]'$ is also a Banach space by Proposition 6, with the sup norm. $[C(X)]'$ is usually identified with the set $M(X)$ of all bounded, regular, Borel measures on X (cf. Halmos [42]) and the norm on $[C(X)]'$ is the total variation of measures. The identification of $M(X)$ and $[C(X)]'$ is given by the following:

For each continuous linear functional T on $C(X)$ there exists a unique measure $\mu \in M(X)$ such that for all $f \in C(X)$

$$T(f) = \int_X f \, d\mu,$$

where the integration is the Radon functional (see Royden [107]).

The unit ball of $[C(X)]'$ is compact in the topology $\sigma(C(X)', C(X))$ and is clearly convex. We say that $\mu \in C(X)'$ is nonnegative or $\mu \geq 0$ if for each $f \in C(X), f \geq 0$ implies $\int f \, d\mu \geq 0$.

The upper half-ball

$$B'^+ = \{\mu \in C(X)': \mu \geq 0, \, \|\mu\| = 1\}$$

of the unit ball B' is called the set of *probability measures*. B'^+ is clearly convex and $\sigma(C(X)', C(X))$-compact.

In general, it is not true that $B' = \{x' \in E': \|x'\| \leq 1\}$ is norm compact in E'. This is true if and only if E is a finite-dimensional normed space (cf. Schaeffer [99]).

For each normed space E, we have seen that E' is a Banach space. If we repeat the operation, we see that E'', the set of all continuous linear functionals on E' (also called the *bidual* of E), is also a Banach space with the norm:

$$\|x''\| = \sup_{\|x\| \leq 1} |x''(x')|, \qquad \text{where } x'' \in E''.$$

For each $x \in E$, we define $f_x(x') = x'(x)$, where $x' \in E'$. Since x' is linear and continuous, it is clear that f_x is linear and continuous. Hence $f_x \in E''$. Thus to each $x \in E$ there is an element $f_x \in E''$. The correspondence $x \to f_x$ defines a mapping of E into E''. We prove:

Proposition 7. The mapping $\varphi: x \to f_x$ is an isometry of the Banach space E into the Banach space E''.

Proof. To show φ is one-to-one, suppose $\varphi(x) = \varphi(y)$. Then we have $f_x(x') = f_y(x')$ for all $x' \in E$, i.e., $x'(x) = x'(y)$ for all $x' \in E'$. But we have seen (Corollary 6) that if $x \neq y$ in a normed space E then there exists $f \in E'$ such that $f(x) \neq f(y)$. Thus $x'(x) = x'(y)$ for all $x' \in E'$ implies $x = y$. Now we show that $\|\varphi(x)\| = \|f_x\| = \|x\|$. By definition of f_x, $|f_x(x')| = |x'(x)| \leq \|x'\| \, \|x\|$ for $x' \in E'$. Hence

$$\|f_x\| = \sup_{\|x'\| \leq 1} |f_x(x')| \leq \|x\|.$$

On the other hand, by Corollary 7, for each $x \neq 0$ there exists $x' \in E'$ such that $x'(x) = \| x' \|$. This shows that $\| f_x \| = \| x \|$.

§63. A General Extension Theorem

In this section, we wish to generalize Tietze's extension theorem for function spaces. This result is due to E. Bishop [12]. It includes the special cases for continuous functions on the unit disk in the complex plane or closed unit interval. Recall that if X is a compact Hausdorff space, $C(X)$ is the Banach space of all continuous real- or complex-valued functions. The set of all continuous linear functionals on $C(X)$, i.e., the dual $[C(X)]'$ of $C(X)$, is usually identified with the Banach space of all bounded regular Borel measures $M(X)$ on X. The duality between $C(X)$ and $M(X)$ is usually described by the so-called Riesz representation theorem (see the example before Proposition 7, §62), viz., each continuous linear functional T on $C(X)$ can be given by

$$T(f) = \int_X f(x) \, d\mu(x),$$

for $f \in C(X)$. Sometimes $T(f)$ is written as $\langle T, f \rangle$. Since T and μ can be identified, the duality between $C(X)$ and $M(X)$ is expressed by

$$\langle \mu, f \rangle = \int_X f(x) \, d\mu(x).$$

For any subset $A \subset C(X)$, we denote by $A^\perp = \{\mu \in M(X): \langle \mu, f \rangle = 0,$ for all $f \in A\}$. $A^\perp$ is called the *annihilator* of A.

It is well known in measure theory that every Baire measure can be extended to a regular Borel measure (cf. Halmos [42]). With these preliminaries, we have a general extension theorem:

Theorem 5. (Bishop [12]) Let X be a compact Hausdorff space, $C(X)$ the Banach space of all continuous real- or complex-valued functions on X, and B a closed linear subspace of $C(X)$. Let $B^\perp$ denote the set of all Baire measures μ such that $\int f \, d\mu = 0$ for $f \in B$ and let $\hat\mu$ be the regular Borel extension of $\mu \in B^\perp$. Let S be a closed subset of X such that $\hat\mu(T) = \int \chi_T \, d\hat\mu = 0$ for all Borel subsets $T \subset S$ and $\mu \in B^\perp$. Let $\Delta \in C(X)$ such that $\Delta > 0$ and for a fixed $f \in C(X)$, $|f(x)| < \Delta(s)$ for $s \in S$. Then there exists a $f^\sim \in C(X)$ such that $|f^\sim(x)| < \Delta(x)$ for all $x \in X$, and $f^\sim(x) = f(x)$ for $x \in S$.

Proof. (a) First assume that $\Delta(s) = r < 1$. Then $|f(x)| < r < 1$ for all $x \in S$. We show that there is a $g \in C(X)$ with $\| g \| < 1$ and $g(x) = f(x)$, $x \in S$. Let $U_r = \{h \in B : \| h \|_S < r\}$ and let φ denote the restriction mapping of B into $C(S)$, i.e., $\varphi(h) = h \mid S$, $h \in B$. Then by assumption $f \in \varphi(U_1)$. Put $V_r = \overline{\varphi(U_r)}$. Then $f \in V_r$. By the Hahn–Banach theorem (Theorem 3, Corollary 6, §62) there exists $T_{\mu_1} \in [C(S)]' = M(S)$ such that

$$T_{\mu_1}(f) = \langle T_{\mu_1}, f \rangle > 1 \quad \text{and} \quad |T_{\mu_1}(h)| = |\langle T_{\mu_1}, h \rangle| \le 1 \qquad (*)$$

for $h \in V_r$. In other words, by the duality theorem for $C(S)$ and $M(S)$, we can write

$$T_{\mu_1}(h) = \int_S h(x)\, d\mu_1(x)$$

for all $h \in C(S)$.

Now we define a linear functional on B by $T(h) = T_{\mu_1}(\varphi(h))$. Since for $h \in U_r$, $\varphi(h) \in \varphi(U_r) \subset V_r$, by (*) we have

$$|T(h)| = |T_{\mu_1}(\varphi(h))| < 1.$$

Hence

$$\| T \| = \sup_{\substack{\|h\| \le 1 \\ h \in B}} |T(h)| = \sup_{\substack{\|h\| \le 1 \\ h \in B}} |T_{\mu_1}(\varphi(h))|$$

$$= r^{-1} \sup_{\substack{\|h\| \le r \\ h \in B}} |T_{\mu_1}(\varphi(h))| = r^{-1} \sup_{\substack{\|k\| \le r \\ k \in V_r}} |T_{\mu_1}(k)|$$

$$\le r^{-1} \text{ [where } k = \varphi(h)],$$

because $\sup_{k \in V_r} |T_{\mu_1}(k)| \le 1$ by (*).

Thus again by the Riesz representation theorem and the Hahn–Banach theorem there is a measure μ_2 on X with $\| \mu_2 \| \le r^{-1}$ and

$$T(h) = \int_X h(x)\, d\mu_2(x) \qquad \text{for all } h \in B.$$

Put $\mu = \mu_1 - \mu_2$. Then for $h \in B$,

$$\int_X h\, d\mu(x) = \int_X h\, d\mu_1 - \int_X h\, d\mu_2$$

$$= T_{\mu_1}(\varphi(h)) - T(h) = 0.$$

Hence $\mu \in B^\perp$. Also

$$\left| \int_S f\, d\mu \right| \ge \left| \int_S f\, d\mu_1 \right| - \left| \int_S f\, d\mu_2 \right|$$

$$\ge T_{\mu_1}(f) - \left| \int_S f\, d\mu_2 \right|.$$

Since

$$\| \mu_2 \| \leq \overset{\iota}{r}^{-1}, \quad \left| \int_S f \, d\mu_2 \right| \leq \| f \|_S \| \mu_2 \| < r(r^{-1}) = 1,$$

because $|f(x)| < r$ on S and $T_{\mu_1}(f) > 1$, we get $| \int_S f \, d\mu | > 1 - 1 = 0$. This yields a contradiction because otherwise

$$\hat{\mu}(S) = \sup_{|f| < \chi_S} \left| \int f \chi_S \, d\mu \right| \geq \left| \int_S f \, d\mu \right| > 0$$

contradicts the hypothesis for S, where χ_S is the characteristic function of S. Hence $f \in V_r = \overline{\varphi(U_r)}$.

Thus there exists $g_1 \in B$, $\| g_1 \| < r$ and $|f(x) - g_1(x)| < \lambda/2$ for all $x \in S$, where $\lambda = 1 - r > 0$. Put $f_1 = f - g_1$. Then $f_1 \in C(S)$ and $|f_1(x)| = |f(x) - g_1(x)| < \frac{1}{2}\lambda$. By induction there is a sequence $\{g_n\}$,

$$g_n \in B, \qquad \| g_n \| < \frac{\lambda}{2^{n-1}}, \quad n \geq 2$$

and

$$\left| f(x) - \sum_{k=1}^n g_k(x) \right| < \frac{\lambda}{2^n},$$

for all $n \geq 2$ and $x \in S$.

Put $g(x) = \sum_{n=1}^{\infty} g_n(x)$. Then from $\| g_n \| < \lambda/2^{n-1}$ $(n \geq 2)$, it follows that $g \in C(X)$. Since $g_n \in B$ (which is a closed subspace), we conclude that $g \in B$ and

$$\| g \| < \| g_1 \| + \sum_{n=2}^{\infty} \| g_n \| < r + \sum_{n=2}^{\infty} \frac{\lambda}{2^{n-1}}$$

$$< r + \lambda = 1.$$

Clearly $g(x) = f(x)$ for all $x \in S$.

(b) Now let $\Delta > 0$, $\Delta \in C(X)$. Let $B_0 = \{h: \Delta h \in B\}$. Then B_0 is a closed subspace of $C(X)$ and $B_0{}^\perp = \{\Delta^{-1}\mu : \mu \in B^\perp\}$. Also $\hat{\mu}(T) = 0$ for all Borel subsets $T \subset S$ and for all $\hat{\mu} \in B_0{}^\perp$. By case (a) there exists $g_0 \in B_0$ with $\| g_0 \| < \| \Delta \| < 1$ and $g_0(x) = \Delta^{-1}(x) f(x)$ for all $x \in S$. Put $g(x) = \Delta(x)g_0(x)$. Then $g \in B$ and $|g(x)| = \Delta(x)|g_0(x)| < \Delta(x)$ for all $x \in X$ and $g(x) = \Delta(x)g_0(x) = f(x)$ for all $x \in S$. This completes the proof.

Corollary 8 (Rudin [100]). Let $X = \{$complex numbers $z: |z| = 1\}$; $B = \{f \in C(X)$: which are restrictions of the functions g, which are holomorphic on the open disk $|z| < 1$ and continuous on $|z| \leq 1\}$. Let S be any measurable set of Lebesgue measure 0 in X. Then the same conclusion as in Theorem 5 holds.

Proof. Immediate.

Examples and Exercises

1. Let $f: E \to [0, 1]$ be a continuous function. Then $f^{-1}(0) = Z_f$ is called the *zero set* of f. Show that the countable intersection of zero sets is a zero set.

2. Let f be a uniformly continuous map of a metric space E into a complete metric space F. Show that there exists a unique extension $f^{\sim}$ of f from the completion $E^{\sim}$ of E to F and $f^{\sim}$ is uniformly continuous.

3. Show that the image of a complete uniform space under a uniformly continuous map need not be complete. [*Hint*: consider $E = R$, $f(x) = x/(1 + |x|)$].

4. Can the continuous map $f(x) = x/|x|$ of $E = \{x \in R, x \neq 0\}$ be extended to R? (If not, why not?).

5. If f is a continuous map of a complete metric space E into any Hausdorff topological space F, and if $\{A_n\}$ is a monotonically decreasing sequence in E (i.e., $A_n \supset A_{n+1}$ for $n \geq 1$) such that the diameter $\delta(A_n) \to 0$, then

$$f\left(\bigcap_{n=1}^{\infty} A_n\right) = \bigcap_{n=1}^{\infty} f(A_n).$$

6. Show that the identity map of $S = \{(x, y): x^2 + y^2 = 1\} \subset R^2$ into itself cannot be extended to a continuous map of $D = \{(x, y): x^2 + y^2 \leq 1\}$ into S. (Hence S is not an AR-space or absolute-retract space (cf. Exercise 28, Chapter III).

7. Let $S^n = \{(x_1, \ldots, x_{n+1}) \in R^{n+1}: x_1^2 + \cdots + x_{n+1}^2 = 1\}$ and let F be a closed subset of a normal space E. Let $f: F \to S^n$ be a continuous map; then there exists an open set $U \supset F$ and a continuous extension $f^{\sim}: U \to S^n$ of f. [*Hint*: By Example 28, Chapter III, there exists an extension of f from E to $I^{n+1} \supset S^n$. Use the radial projection of $I^{n+1} \setminus \{\frac{1}{2}, \ldots, \frac{1}{2})\}$ to S^n, which is continuous, to obtain an open set of E.]

8. Let $1 \leq p < \infty$. For each p, let ℓ_p denote the set of all real or complex sequences $x = \{x_i\}$ such that $\sum_{i=1}^{\infty} |x_i|^p < \infty$ with the norm:

$$\|x\| = \left\{\sum_{i=1}^{\infty} |x_i|^p\right\}^{1/p},$$

and let ℓ_∞ be the space of all bounded sequences, i.e., $\| x \|_\infty = \sup| x_i |$ $< \infty$. Define q by $1/p + 1/q = 1$ if $1 < p < \infty$ and $q = \infty$ if $p = 1$. Show that the dual ℓ_p' of the Banach space ℓ_p $(1 \leq p \leq \infty)$ can be identified with ℓ_q.

9. Let X be a compact Hausdorff space and $C(X)$ the Banach space of all continuous real- or complex-valued functions on X. Show that every bounded regular Borel measure μ on X can be identified with a functional:

$$T(f) = \int f(x) \, d\mu(x),$$

$f \in C(X)$. (This is called a Riesz representation theorem.)

10. (Hilbert space Riesz representation theorem) Let H be a Hilbert space. Then the dual H' of H can be identified with H via:

$$f(x) = \langle x, y_f \rangle,$$

where $f \in H'$ and $y_f \in H$ depending upon f and $\langle , \rangle$ denotes the scalar product. Hence show that $l_2' \approx l_2$.

11. Let (X, m, μ) be a measure space (cf. Halmos [42]) and let $L_p(X)$, $1 \leq p < \infty$, denote the space of equivalence classes of measurable functions f on X such that $\int |f(x)|^p \, d\mu < \infty$. Then $L_p(X)$ with

$$\| f \|_p = \left\{ \int_X |f(x)|^p \, du(x) \right\}^{1/p}$$

is a Banach space. Show that for $1 < p < \infty$, $L_p' = L_q$, where $1/p + 1/q = 1$. Let $L_\infty(X)$ denote the space of equivalence classes of all essentially bounded functions with essential sup norm (cf. Halmos [42]). Show that $L_1' = L_\infty$ and that L_∞ is a nonseparable Banach space.

12. Let E, F be two normed spaces and $f: E \rightarrow F$ a continuous linear map. Let E', F' denote the topological duals of E and F, respectively. For each $x' \in F'$, define $f^*(x') = x' \circ f$, a linear functional on E. Show that f^* is also continuous. If $\| f \|$ and $\| f^* \|$ are their norms, show that $\| f \| = \| f^* \|$ (f^* is called the *transpose* or *adjoint* map of f).

13. Let E be a locally convex Hausdorff space and E' its dual. Show that:
    (a) A subset $M \subset E'$ is equicontinuous iff $M^\circ = \{x \in E: |f(x)| \leq 1 \text{ for all } f \in M\}$ (the polar set of M) is a neighborhood of 0 in E iff there exists a neighborhood U of 0 in E such that $M \subset U^\circ = \{f \in E': |f(x)| \leq 1 \text{ for all } x \in U\}$.
    (b) A subset $B \subset E$ is absorbing iff B° is $\sigma(E', E)$-bounded in E'.

(c)   A subset $B \subset E$ absorbs all bounded subsets of E iff B° is strong-ly bounded or $\mathcal{E}_\beta$-bounded in E'.

(d)   Show that closed convex subsets of E and $\sigma(E, E')$-closed convex subsets of E are the same.

(e)   If B is a convex, circled, and $\sigma(E, E')$-closed subset of E, then $B^{\circ\circ} = \{x \in E: |f(x)| \leq 1 \text{ for all } f \in B^\circ\} = B$ (see [45], [13], or [55b]).

X

$C(X)$ Spaces

The space $C(X)$ of all continuous complex- or real-valued functions on a topological space X plays an important role. This space, with pointwise multiplication, turns out to be an algebra. With a suitable topology, it is even a topological algebra. Thus the Stone–Weierstrass theorem can be formulated and proved. Furthermore, because of the lattice structure of the real line, there is a lattice structure on $C(X)$. This is turn enables one to study some other, deeper properties of $C(X)$. In particular, we prove the Banach–Stone theorem and other results which exploit the algebraic structure on $C(X)$, which may not be available for $C(X, Y)$, if X and Y are topological spaces and have no richer structure than that of the set of real numbers.

Further, it will be possible for us to characterize $C(X)$ for some X. That is, given suitable properties of X, we find $C(X)$ and with suitable properties on $C(X)$ we recover X.

If we regard $C(X)$ as a linear topological space, then the interrelation between X and $C(X)$ is actually functorial in categorical language. Thus the category of certain linear topological spaces and continuous linear maps and the category of Hausdorff compact spaces and continuous maps are dual to each other via an adjoint functor. The precise formulation will follow in the text.

§64. Stone–Weierstrass Theorem

One of the basic theorems in approximation theory is the Stone–Weierstrass theorem. It deals with the conditions under which a subalgebra of $C(X)$ is either dense in $C(X)$ or coincides with it.

Observe that $C(X) = C(X, R)$, where the real line R is given the Euclidean topology. If we want to regard $C(X)$ as a set of all complex-valued functions on X, we shall mention it explicitly or write $C(X) = C(X, C)$.

It is possible to consider equivalent bounded metrics on the real line. In that case the nature of the space $C(X)$ is different. However, we will have no occasion to consider this situation. Indeed in some cases, e.g., when X is compact, it does not matter what equivalent metric on R is chosen because $C(X)$ remains invariant. In the sequel, it is understood that $C(X)$ is nonempty, i.e., X is nonempty.

Theorem 1. $C(X)$, endowed with the compact–open topology $\mathcal{C}_c$ and with pointwise-addition, multiplication, and scalar multiplication is a locally convex commutative topological algebra with a unit.

Proof. Let $f, g \in C(X)$ and α a real scalar. Then

$$(f + g)(x) = f(x) + g(x),$$
$$(\alpha f)(x) = \alpha f(x),$$
$$(fg)(x) = f(x)g(x)$$

for $x \in X$ define continuous real functions. It is thus clear that $C(X)$ is a linear space and indeed a commutative algebra with identity 1, the constant function equal to 1. To show that the maps

$$\left. \begin{array}{l} (f, g) \to f + g \\ (f, g) \to fg \end{array} \right\} \quad \text{of } C(X) \times C(X) \quad \text{into } C(X),$$

$$(\alpha, f) \to \alpha f \qquad \text{of } R \times C(X) \qquad \text{into } C(X)$$

are continuous, let $T(K, U)$ be a $\mathcal{C}_c$-neighborhood of O in $C(X)$, where K is a compact subset of X, and U is an open neighborhood of O in the real line R. Choose neighborhoods V and W of $O \in R$ such that $V + W \subset U$. But then $f + g + T(K, U)$, $f + T(K, V)$, and $g + T(K, W)$ are $\mathcal{C}_c$-neighborhoods of $f + g$, f, and g, respectively. For each $h \in f + T(K, V)$ and $k \in g + T(K, W)$, we have $h(x) - f(x) \in V$ and $k(x) - g(x) \in W$ for all $x \in K$. Therefore

$$h(x) + k(x) - f(x) - g(x) \in V + W \subset U$$

for all $x \in K$. But this shows that $h + k \in f + g + T(K, U)$ for all $h \in f + T(K, V)$ and $k \in g + T(K, W)$. This proves the continuity of the map:

$(f, g) \to f + g$. Similar proofs work for the other maps, viz., (f, g): $\to fg$ and (α, f): $\to \alpha f$. Thus $C(X)$ is a topological algebra.

If we take U to be a symmetric convex neighborhood of $0 \in R$, then $T(K, U)$ is symmetric and convex. Hence $C(X)$ is a locally convex topological algebra.

Remark. We have not made any assumptions about the topological space X. It is possible that without restrictions, $C(X)$ may consist of only constant functions. If we require that $C(X)$ consist of more than constant functions, then we assume that X is a completely regular space, which we will do throughout in the sequel.

Corollary 1. Let X be a topological space and let $BC(X)$ denote the set of all bounded continuous real- or complex-valued functions. Then $BC(X)$ is a Banach algebra with the norm $\| f \| = \sup_{x \in X} | f(x) |, f \in BC(X)$, i.e., a Banach space in which $\| fg \| \leq \| f \| \, \| g \|$.

Proof. We have seen that for each $f \in BC(X)$, $f(X) = \{f(x): x \in X\}$ is bounded and hence the metric d^+ defined in Proposition 14, §59 is finite. We put $\| f \| = \sup\{| f(x) |: x \in X\}$. Then $\| f \| \geq 0$ for all $f \in BC(X)$ and $\| f \| = 0$ iff $f = 0$. Also for any scalar λ,

$$\| \lambda f \| = \sup_{x \in X} | \lambda f(x) | = | \lambda | \sup_{x \in X} | f(x) | = | \lambda | \, \| f \|.$$

Finally, from

$$| f(x) + g(x) | \leq | f(x) | + | g(x) |,$$

we obtain

$$\| f + g \| \leq \| f \| + \| g \|.$$

Hence $BC(X)$ is a normed space. As shown before it is complete (Proposition 14, §59), and for $f, g \in C(X)$, $| fg(x) | \leq | f(x) | \, | g(x) |$ implies $\| fg \| \leq \| f \| \, \| g \|$. This proves that $C(X)$ is a Banach algebra with identity which is the constant function equal to 1.

Definition 1. Let X be a topological space. Let $C_\infty(X)$ denote the set of all continuous functions $f: X \to R$ or C such that for each $\varepsilon > 0$ there is a compact set $F \subset X$ such that $| f(x) | < \varepsilon$ whenever $x \in X \setminus E$. Then each f is called a *function vanishing at infinity*.

Proposition 1. Every $f \in C_\infty(X)$ is bounded.

Proof.   Let $\varepsilon = 1$. Then there is a compact set $F \subset X$ such that for all $x \in X \setminus F$, $|f(x)| \leq 1$. Since F is compact, $f(F)$ is compact and hence bounded in R or C, i.e., there exists an $M > 0$ such that $|f(x)| < M$ for all $x \in F$. Hence $|f(x)| \leq \max(1, M)$ for all $x \in X$. Thus f is bounded on X.

Corollary 2.   Let X be a locally compact space and let $C_\infty(X)$ be as defined above. Then $C_\infty(X)$ is a closed proper subalgebra of $BC(X)$, the space of all bounded continuous functions

Proof.   It follows by combining Proposition 1 and Theorem 1.

Corollary 3.   Let X be a compact Hausdorff space. Then $C(X) = BC(X) = C_\infty(X)$ is a Banach algebra.

Proof.   Obvious.

Remark.   Recall that if A is any topological algebra and B is a subalgebra of A, then $\bar{B}$ is also a subalgebra.

We are now heading toward the Stone–Weierstrass theorem. For this we need some preparatory results.

Proposition 2.   Let A be a closed subalgebra with identity of the topological algebra $C(X)$, endowed with the compact–open topology. Then:

(i)   For all $f \in A$, $|f| \in A$, where $|f|(x) = |f(x)|$ for all $x \in X$.

(ii)   For $f, g \in A$ if $(f \vee g)(x) = \max(f(x), g(x))$ and $(f \wedge g)(x) = \min(f(x), g(x))$, $x \in X$, then $f \vee g \in A$ and $f \wedge g \in A$.

Proof.   (i)   By the binomial theorem for $x \in [0, 1)$, we have:

$$(1 - x)^{1/2} = 1 + \tfrac{1}{2}(- x) + \frac{\tfrac{1}{2}(\tfrac{1}{2} - 1)}{2!}(- x)^2 + \cdots$$

or

$$1 - (1 - x)^{1/2} = \tfrac{1}{2}x + \binom{\tfrac{1}{2}}{2}(- x)^2 + \cdots = \sum_{n=1}^{\infty} \left| \binom{\tfrac{1}{2}}{n} \right| x^n.$$

Since all the coefficients of powers of x are positive, we have a series of positive terms, the partial sums of which form a monotonic sequence which is bounded because $1 - (1 - x)^{1/2} \leq 1$ on $[0, 1]$. Thus the series $\sum_{n=1}^{\infty} \left| \binom{1/2}{n} \right| x^n$ converges at $x = 1$ and converges uniformly to $1 - (1 - x)^{1/2}$. Replacing

x by $1 - t$, we obtain

$$p_m(t) = \sum_{n=1}^{m} \left| \binom{\frac{1}{2}}{n} \right| (1 - t)^n \to 1 - t^{1/2}$$

uniformly on $[0, 1]$. Observe that the p_m's are polynomials.

Now let $f \in A$ and let $\bigcap_{i=1}^{n} T(K_i, U_i)$ be a basic neighborhood of $|f|$, where the K_i's are compact in X and the U_i's open in $\mathbb{R}$ for $i = 1, \ldots, n$. We may assume that the U_i's are given by ε_i, i.e., $T(K_i, U_i) = \{f \in C(X): |f(x) - |f|(x)| < \varepsilon_i, x \in K_i\}$, $i = 1, \ldots, n$. By taking $\varepsilon = \min \varepsilon_i$, we are required to find a $g \in A$ such that $|g(x) - |f|(x)| < \varepsilon$, for $x \in K = \bigcup_{i=1}^{n} K_i$. Since K is compact, $|f(x)| \leq M < \infty$ for all $x \in K$ and some M. Hence,

$$1 - p_m\left(\frac{f^2}{M^2}\right) = 1 - \sum_{n=1}^{m} \left| \binom{\frac{1}{2}}{n} \right| \left(1 - \frac{f^2}{M^2}\right)^n$$

converges to $(f^2/M^2)^{1/2} = |f|/M$ uniformly on K. Since A is an algebra with identity, $1 - p_m(f^2/M^2) \in A$ for each m. Hence its limit $|f|/M \in \bar{A}$, and so $|f| \in \bar{A} = A$.

(ii) For $f, g \in A$, $f + g$ and $f - g \in A$, because A is an algebra. By (i), $|f + g|, |f - g| \in A$. Hence

$$f \vee g = 2^{-1}(f + g + |f - g|) \in A$$

and

$$f \wedge g = 2^{-1}(f + g - |f - g|) \in A.$$

This proves (ii).

Now we state and prove the Stone–Weierstrass theorem. There are two aspects of this theorem, viz., real and complex. The real case is proved with fewer conditions on the algebra while the complex case requires one additional condition.

Theorem 2 (Stone–Weierstrass theorem: real case). Let X be a topological space and let A be a subalgebra of $C(X)$ of all real continuous functions on X, with the compact–open topology such that:

(i) For $x, y \in X$, $x \neq y$, there exists $f \in A$ such that $f(x) \neq f(y)$ (i.e., A is separating or separates points of X)

(ii) for $x \in X$, there exists $f \in A$ such that $f(x) \neq 0$.

Then $\bar{A} = C(X)$, i.e., A is dense in $C(X)$. Thus if A is already closed, then $A = C(X)$.

Proof. From Proposition 2, we observe that for $f, g \in A$, $|f|, f \vee g$, $f \wedge g \in \bar{A}$. If $X = \{x\}$, then $C(X) = R$ and (ii) implies that $A = C(X) = R$. Now for $x, y \in X$, $x \neq y$, and any real numbers r and s, by hypothesis (i), there exists $g \in A$ such that $g(x) \neq g(y)$. If for $z \in X$, we put

$$g'(z) = r \frac{g(z) - g(y)}{g(x) - g(y)} - s \frac{g(z) - g(x)}{g(x) - g(y)},$$

then $g' \in C(X)$ and $g'(x) = r, g'(y) = s$. Since A is an algebra, it is clear that $g' \in A$. We use this fact below.

To complete the proof, let $f \in C(X)$ and let $\varepsilon > 0$. We want to show that there is a $g \in \bar{A}$ such that $g - f \in \bigcap_{i=1}^{n} T(K_i, \varepsilon_i) = T$, where the K_i's are compact subsets of X, and T is a $\mathscr{C}_c$-neighborhood of 0. By taking $\varepsilon = \min \varepsilon_i$ and $K = \bigcup_{i=1}^{n} K_i$, it is sufficient to show the existence of $g \in A$ such that $|f(k) - g(k)| < \varepsilon$ for all $k \in K$, which is compact. For any fixed $x, y \in K$, $x \neq y$, from the opening paragraph there exists $g_{x,y} \in A$ such that $g_{x,y}(x) = f(x), g_{x,y}(y) = f(y)$. Put

$$U_{x,y} = \{z \in K: g_{x,y}(z) < f(z) + \varepsilon\}.$$

Since $g_{x,y}$ is continuous, $U_{x,y}$ is an open neighborhood of x such that $g_{x,y}(z) < f(z) + \varepsilon$ for all $z \in U_{x,y}$. Clearly $\{U_{x,y}\}_x$ is an open covering of the compact set K, and so only a finite subcovering $\{U_{x_i,y}\}_{i=1}^{n}$ covers K. Let $g_{x_i,y} \in \bar{A}$, to which corresponds $U_{x_i,y}$. Put $g_y = \wedge_i g_{x_i,y}$. Then by (ii) of Proposition 2, $g_y \in \bar{A}$ and if $z \in K$, $z \in U_{x_i,y}$ for some i and so $g_y(z) \leq g_{x_i,y}(z) < f(z) + \varepsilon$ for all $z \in K$. As before, by continuity there is an open neighborhood V_y of y such that for all $z \in V_y$, $g_y(z) > f(z) - \varepsilon$. Again $\{V_y\}$ is an open covering of the compact set K as y runs over K and by choosing a finite open covering $\{V_{y_1}, \ldots, V_{y_m}\}$ we obtain $g_{y_1}, \ldots, g_{y_m}$, to which correspond $V_{y_1}, \ldots, V_{y_m}$. Put $g = \vee_{i=1}^{m} g_{y_i}$. As before $g \in \bar{A}$ and if $z \in K$ then $z \in V_{y_i}$ for some i and so $g(z) \geq g_{y_i}(z) > f(z) - \varepsilon$ for all $z \in K$. By Proposition 2, $g \in \bar{A}$ and $g(z) \geq g_{x_i}(z) > f(z) - \varepsilon$ for all $z \in K$. Thus we have obtained, by combining the two inequalities, that $|f(z) - g(z)| < \varepsilon$ for all $z \in K$, which means that $f \in \bar{A}$. This proves that A is dense in $C(X)$. Now if A is $\mathscr{C}_c$-closed, then we have $A = \bar{A} = C(X)$.

Theorem 3 (Stone–Weierstrass theorem: complex case). Let X be a topological space and let A be a subalgebra of $C(X)$ of all complex-valued continuous functions on X with the compact–open topology such that:

(i) For $x, y \in X$, $x \neq y$, there exists $f \in A$ such that $f(x) \neq f(y)$, i.e., A is separating;

(ii) for $x \in X$ there exists $f \in A$ with $f(x) \neq 0$;

(iii) for each $f \in A, \tilde{f} \in A$, where $\tilde{f}(x) = \overline{f(x)}$, the complex conjugate of $f(x)$.

Then $\bar{A} = C(X)$. If A is $\mathscr{C}_c$-closed, then $A = C(X)$.

Proof. We use the real case. Let A_R denote the set of all real-valued functions in A. More precisely, we know that for each $f \in A, f = f_1 + if_2$, $i = \sqrt{-1}$, where f_1, f_2 are real-valued continuous functions on X and $\mathscr{R}(f)$ = real part of $f = f_1 \in A_R$ and also $f_2 \in A_R$.

We denote by $C_R(X)$ the set of all real functions in $C(X)$. We show that $\overline{A_R} = C_R(X)$, from which it is immediate that $\bar{A} = C(X)$.

To show that $\overline{A_R} = C_R(X)$, we verify the conditions of the Stone–Weierstrass theorem for the real case. Since $\mathscr{R}(f) = 2^{-1}(f + \tilde{f})$ and $\tilde{f} \in A$ by (iii), $\mathscr{R}(f) \in A_R$. Further, if $x \neq x, y \in X$, by (i) there exists $f \in A$ such that $f(x) \neq f(y)$, which implies that $\mathscr{R}(f)(x) \neq \mathscr{R}(f)$ or $\mathscr{I}(f)(x) \neq \mathscr{I}(f)(y)$ and $\mathscr{R}(f)$ or $\mathscr{I}(f) \in A_R$. Also if there is $f \in A$ such that $f(x) \neq 0$, then $f\tilde{f} = |f|^2 \in A_R$ and $|f|^2(x) \neq 0$. Hence by the real case of the Stone–Weierstrass theorem, we have $\overline{A_R} = C_R(X)$ and hence $\bar{A} = C(X)$.

Remark. For the complex case, condition (iii) is necessary because otherwise the continuous complex-valued functions on the unit open disk $\{z: |z| < 1\}$, being the uniform limit on compact sets of polynomials $\{1, z^n\}_{n \geq 1}$, have to be analytic, which is not true.

Corollary 4. Let X be a topological space and let $BC(X)$ denote the set of all real- or complex-valued bounded continuous functions on X with the sup–norm topology. Let A be a subalgebra of $BC(X)$ satisfying conditions (i) and (ii) for the real case or (i)–(iii) for the complex case. Then $\bar{A} = BC(X)$.

Proof. Immediate from the above theorem with the observation that on $BC(X)$ the compact–open topology coincides with the norm topology on $BC(X)$.

Corollary 5. Let X be a locally compact Hausdorff space and let $C_\infty(X)$ denote the set of all real or complex continuous functions on X which vanish at infinity, with the sup–norm topology. Let A be a subalgebra of $C_\infty(X)$ satisfying (i) and (ii) for the real case and (i)–(iii) for the complex case. Then $\bar{A} = C_\infty(X)$.

Proof.   Immediate from Corollary 4, since $C_\infty(X)$ with the sup–norm topology is a closed subspace of $BC(X)$.

Corollary 6.   Let X be a compact Hausdorff space and let $C(X)$ denote the set of all real- or complex-valued continuous functions on X. Let A be a subalgebra of $C(X)$ satisfying conditions (i), (ii) of Theorem 2 for the real case or (i)–(iii) of Theorem 3 for the complex case. Then $\bar{A} = C(X)$.

Proof.   We see that when X is compact, $C(X) = BC(X) = C_\infty(X)$ and the corollary follows from the above.

Corollary 7 (Classical Weierstrass theorem).   Let $I = [a, b]$, $-\infty < a < b < +\infty$, and let $C(I)$ be the set of all continuous real-valued functions with the sup–norm: $\| f \| = \sup_{x \in I} | f(x) |$. Then each $f \in C(I)$ can be approximated uniformly by real polynomials.

Proof.   Let A denote the set of all real polynomials, i.e., $p_m(x) = \sum_{n=0}^{m} a_n x^n$, where the a_n's are real. Then certainly A separates points because if $t_1 \neq t_2$ then the polynomial $p(x) = x \in A$ and $p(t_1) \neq p(t_2)$. Also condition (ii) is satisfied. Hence $\bar{A} = C(X)$ by the real case of the Stone–Weierstrass theorem.

Remark.   In the last corollary, the compact interval I can be replaced by any n-dimensional compact subset of R^n, e.g., $I^n = \{(x_1, \ldots, x_n) : a_i \leq x_i \leq b_i$ for $i = 1, \ldots, n\}$ to obtain the Weierstrass approximation theorem in the n-dimensional cube.

§65. Embeddings of X into $C(X)$

We have seen that each completely regular space can be embedded into a compact space (Proposition 51, §44), every metric space can be embedded into a complete metric space (Corollary 8, §30), and every uniform Hausdorff space can be embedded into a complete uniform Hausdorff space (Theorem 8, §30).

In this section we show that certain topological spaces can be embedded into their function space. Indeed, if X is a completely regular space than we shall see that X can be embedded into the dual of $C(X)$.

Recall, if X is any space, then $BC(X)$, the set of all bounded continuous real-valued functions, is a metric space with the metric d^+ defined, for f, $g \in BC(X)$, by

$$d^+(f, g) = \sup_{x \in X} |f(x) - g(x)|.$$

By embedding generally one means that the topological space which is embedded in a topological space carries the same topological structure as the space into which it is embedded. Now, if we want to embed a metric space into another metric space, we want to have their metrics unchanged. For this, we have:

Definition 2. Let (X, d), (Y, d') be two metric spaces and f a mapping of X into Y. Then f is said to be an *isometry* if for all $x_1, x_2 \in X$,

$$d'(f(x_1), f(x_2)) = d(x_1, x_2).$$

Proposition 3. Let f be an isometry of a metric space (X, d) into another metric space (Y, d'). Then

   (i) f is $1:1$;

  (ii) f is continuous;

 (iii) f is uniformly continuous;

 (iv) f is open, i.e., f^{-1} is continuous.

Thus an isometry is a homeomorphism if and only if it is onto.

Proof. Since f is an isometry, for $x_1, x_2 \in X$, we have

$$d'(f(x_1), f(x_2)) = d(x_1, x_2).$$

Then (i) is obvious and so are (ii) and (iii). Since f^{-1} exists and maps the range $R(f) \subset Y$ onto X, we obtain

$$d(f^{-1}(y_1), f^{-1}(y_2)) = d'(y_1, y_2) \qquad \text{for all } y_1, y_2 \in R(f).$$

This shows that f^{-1} is continuous. Thus an isometry is always a homeomorphism *into*. Therefore an isometry f is a homeomorphism if and only if f is onto.

Corollary 8. If the metric spaces (X, d) and (Y, d') are isometric onto, then the topological vector space $C(X)$ is linearly isomorphic and homeomorphic with $C(Y)$.

Proof. Let h be an isometry between X and Y. Then for each $g \in C(Y)$, we define $h^*(g) = g \circ h$, given by $h^*(g)(x) = g(h(x))$ for $x \in X$. $h^*(g)$ is a continuous real-valued map in $C(X)$ (Proposition 1, §16). Thus h^* is easily seen to be continuous when $C(X)$ and $C(Y)$ are given the $\mathscr{C}_c$-topology. Clearly h^* is a linear map from the real vector space $C(Y)$ to $C(X)$. It is not difficult to see that h^* is an isomorphism and a homeomorphism because the compact sets of X are mapped onto compact sets of Y and vice versa, because h is an isometry onto.

Theorem 4. Let (X, d) be a metric space and $BC(X)$ the linear space of all bounded continuous real-valued functions on X with metric d^+. Then (X, d) can be isometrically embedded into $(BC(X), d^+)$.

Proof. Let x_0 be a fixed point in X. For $a \in X$, define $f_a(x) = d(x, a) - d(x, x_0)$. By Exercise 19, Chapter III, f_a is continuous and by the triangular inequality of d, we have

$$|f_a(x)| = |d(x, a) - d(x, x_0)| \leq d(a, x_0) < \infty,$$

for all $x \in X$. Hence for each $a \in X$, f_a is a continuous bounded real-valued function and so $f_a \in BC(X)$. Now we define the map $\varphi : a \to f_a$ of X into $BC(X)$ and show that φ is an isometry. For $a, b \in X$, consider

$$
\begin{aligned}
d^+(\varphi(a), \varphi(b)) &= d^+(f_a, f_b) \\
&= \sup_{x \in X} |f_a(x) - f_b(x)| \\
&= \sup_{x \in X} |[d(x, a) - d(x, x_0)] - [d(x, b) - d(x, x_0)]| \\
&= \sup_{x \in X} |d(x, a) - d(x, b)| \leq d(a, b).
\end{aligned}
$$

But for $x = b$, we have

$$d(x, a) - d(x, b) = d(b, a) - 0 = d(a, b).$$

Thus

$$\sup_{x \in X} |d(x, a) - d(x, b)| = d(a, b)$$

and so

$$d^+(\varphi(a), \varphi(b)) = d^+(f_a, f_b) = d(a, b)$$

implies that $\varphi : a \to f_a$ is an isometry. This completes the proof.

Since $BC(X)$ is a linear space, it is possible to determine the size of the image of X under the isometry φ. We indeed have:

Proposition 4. Let (X, d), $BC(X)$, and φ be as indicated in Theorem 4. Then $\varphi(X)$ is a closed subset of its convex hull in $BC(X)$.

Proof. Let Y denote the convex hull of $\varphi(X)$ in $BC(X)$. This means that

$$Y = \left\{ \sum_i \lambda_i f_{a_i} \text{ (finite sum)}: \sum \lambda_i = 1,\ 0 < \lambda_i \leq 1,\ f_{a_i} \in \varphi(X) \right\}.$$

Let $f \in Y \backslash \varphi(X)$. Then $f = \sum_i \lambda_i f_{a_i}$, where $a_i \in X$, $1 \leq i \leq n$, $\sum \lambda_i = 1$, $\lambda_i > 0$. Now for each $a \in X$, we have

$$f(x) - f_a(x) = \sum_{i=1}^{n} \lambda_i f_{a_i}(x) - f_a(x)$$

$$= \sum_{i=1}^{n} \lambda_i [d(x, a_i) - d(x, x_0)] - [d(x, a) - d(x, x_0)]$$

$$= \sum_{i=1}^{n} \lambda_i d(x, a_i) - d(x, a)$$

(because $\sum \lambda_i = 1$).

Thus for $n \geq 2$ we have

$$f(x) - f_a(x) \geq \min[d(x, a_1)\lambda_1 + \lambda_2 d(x, a_2)] - d(x, a)$$

$$\geq \min_{1 \leq i \leq 2} (\lambda_i)[d(a_1, a_2)] - d(x, a).$$

Put $x = a$. Then we have

$$f(a) - f_a(a) \geq \min_{1 \leq i \leq 2} \lambda_i d(a_1, a_2) > 0$$

for any $a \in X$. This shows that f is not in the closure of $\varphi(X)$. In other words, $\varphi(X)$ is closed in Y.

Corollary 9. Let X be a compact metric space. Then X can be embedded isometrically into $\left(C(X), d^+\right)$ as a closed subset of its convex hull.

Proof. We observe that $C(X) = BC(X)$, because X is compact and hence the corollary follows from Theorem 4 and Proposition 4.

Now we assume that X is a topological group (§31). We denote a topological group by G. Then we know that the map $f_a: x \to ax$ of G into G

is called the *left translation* and the map f^a: $x \to xa$ of G into G is called the *right translation*.

Proposition 5.   The mappings f_a and f^a, as defined above, are homeomorphisms of G onto itself.

Proof.   It is enough to prove the proposition for f_a, since the proof of the other part is similar. Clearly f_a is one-to-one, because if $ax = ay$ for x, $y \in G$ then by multiplying both sides by the inverse a^{-1} of a, we obtain $x = y$. The map f_a is onto, because for any $y \in G$, $y = ax$ implies $a^{-1}y = x$ and so $f_a(a^{-1}y) = aa^{-1}y = y$. Also, f_a is continuous, because the map: $(x, y) \to xy$ of $G \times G$ into G is continuous. Next, f_a is open iff f_a^{-1} is continuous. But f_a^{-1} is the map: $x \to a^{-1}x$ and hence continuous by the above argument.

This proposition shows that each f_a or f^a is in $C(G, G)$. It should be noted that f_a and f^a may not be homomorphisms $(a \neq e)$ because $f_a(xy) = f_a(x)f_a(y)$ may not hold. Nevertheless, we have a map: $a \to f_a$ of G into $C(G, G)$.

Theorem 5.   The mapping: $a \to f_a$ of G into $C(G, G)$ is one-to-one and continuous when $C(G, G)$ is endowed with the compact–open topology.

Proof.   If $f_a = f_b$, then $ax = bx$ for all $x \in G$ and hence in particular for $x = e$ (identity), we have $a = b$. Thus $a \to f_a$ is $1 : 1$. Let $\{a_\alpha : \alpha \in A\}$ be a net in G such that $a_\alpha \to a$. Then for any $x \in G$, $a_\alpha x \to ax$, because the multiplication $(x, y) \to xy$ is continuous. This immediately shows that the map: $a \to f_a$ is continuous if $C(G, G)$ is endowed with the pointwise convergence topology. Now to show that the map: $a \to f_a$ is continuous when $C(G, G)$ has the $\mathscr{C}_c$-topology, let $T(K, U)$ be an open subbase neighborhood of f_a. This means that $f_a(K) = aK \subset U$, where K is any compact subset and U an open subset of G. Hence we have $K \subset a^{-1}U$. Clearly $a^{-1}U$ is open, hence $F = G \setminus a^{-1}U$ is closed and $F \cap K = \varnothing$. By Exercise 15 (iv), Chapter V, there is an open neighborhood V of the identity such that $VK \subset G \setminus F = a^{-1}U$ or $aVK \subset U$. But then aV is an open neighborhood of a and for $b \in aV$, $f_b(K) = bK \subset U$ shows that the mapping $a \to f_a$ is continuous, when $C(G, G)$ is given the compact–open topology.

Corollary 10.   Let G be a compact Hausdorff topological group. Then the mapping $a \to f_a$ of G into $C(G, G)$ is an embedding, i.e., a homeomorphism into.

Proof. Since G is a compact Hausdorff group, the continuous image G' of G into $C(G, G)$ under the map: $a \to f_a$ is also a compact Hausdorff space. Since G is compact, the mapping: $a \to f_a$ is a homeomorphism into (Proposition 12, §35) and so an embedding.

§66. C(X) Spaces for Compact Spaces X

In this section, we consider compact Hausdorff spaces X. Then we know that $C(X) = BC(X) = C_\infty(X)$ and the compact–open topology is the uniform metric topology:

$$d^+(f, g) = \sup_{x \in X} |f(x) - g(x)|.$$

Also sometimes we write $\|f\| = \sup_{x \in X} |f(x)|$ and so $d^+(f, g) = \|f - g\|$. Then $\|f\|$ is called the sup–norm on the linear space $C(X)$. We shall show that the Banach space $C(X)$ is completely determined by X. This means that two compact Hausdorff spaces X and Y are homeomorphic if and only if their associated Banach function algebras $C(X)$ and $C(Y)$ are isomorphic and homeomorphic.

In order to establish this, we have to exploit some algebraic properties of $C(X)$. We have seen that $C(X)$ is a locally convex algebra with identity 1 and being a complete normed space is a Banach algebra with identity 1. We see that the set of all nonzero real or complex homomorphisms on $C(X)$ is a nonempty subset because, for each $x \in X$, the evaluation map

$$\varphi_x(f) = f(x), \qquad f \in C(X)$$

clearly defines a nonzero real or complex algebra homomorphism. We denote the set of all nonzero homomorphisms by M and the map: $x \to \varphi_x$ embeds X into M when the latter is endowed with the pointwise convergence topology because

$$|\varphi_x(f) - \varphi_x(f_0)| = |f(x) - f_0(x)| \to 0$$

as $f \to f_0$ for each fixed x. Thus each φ_x is continuous on $(C(X), \mathcal{E}_c)$.

We show that every element of M is determined by a point $x \in X$ and therefore each $\varphi \in M$ is continuous. First we have:

Proposition 6. Let X be a completely regular space. Then every nonzero real (or complex) homomorphism of $C(X)$ is onto and $\varphi(1) = 1$.

Proof. Since $1^2 = 1$, we have $[\varphi(1)]^2 = \varphi(1)$. Furthermore, for each $f \in C(X), f = f \cdot 1$ and so $\varphi(f) = \varphi(f)\varphi(1)$. These observations show that $\varphi(1) = 1$. Since $C(X)$ contains constant functions, $C(X)$ contains a copy of the real or complex numbers depending upon whether $C(X)$ is considered to be the set of real- or complex-valued continuous functions. Since φ is linear multiplicative and $\varphi(1) = 1$, it follows that φ is onto.

Proposition 7. Let x_0 be a fixed point in a completely regular space X. The set $\mathfrak{I}$ of all $f \in C(X)$ such that $f(x_0) = 0$ forms a proper ideal. When X is a compact Hausdorff space, each proper ideal of $C(X)$ is contained in an ideal of functions which vanish at some point x_0.

Proof. Let $\mathfrak{I}_{x_0} = \{f \in C(X): f(x_0) = 0\}$. Then $f(x_0) = 0$ and $g \in C(X)$ imply $fg(x_0) = f(x_0)g(x_0) = 0$. Hence $fg \in \mathfrak{I}_{x_0}$. Thus it is clear that $\mathfrak{I}_{x_0}$ forms an ideal. Since X is completely regular, there exists $f \in C(X)$ such that $f(x_0) \neq 0$ and so $f \notin \mathfrak{I}_{x_0}$, i.e., $\mathfrak{I}_{x_0}$ is a proper ideal.

Now for the second part, suppose $\mathfrak{I}$ is a proper ideal of $C(X)$. Let $f \in \mathfrak{I}$. Then $\{x \in X: f(x) = 0\}$ is nonempty because if it were empty, then $f(x) \neq 0$ for $x \in X$ and so $1/f \in C(X)$. But since $\mathfrak{I}$ is an ideal, $1 = f(1/f) \in \mathfrak{I}$ and so $C(X) = \mathfrak{I}$, a contradiction to the assumption that $\mathfrak{I}$ is proper. Let $Z(f) = \{x \in X: f(x) = 0\} = f^{-1}(0)$. Then $Z(f)$ is clearly closed because f is continuous and $\{0\}$ is a closed subset of the real or complex numbers. Let $f_1, \ldots, f_n \in \mathfrak{I}$. Then $f = \sum_{i=1}^{n} |f_i|^2 \in \mathfrak{I}$ and so $Z(f) = \bigcap_{i=1}^{n} Z(f_i) \neq \varnothing$. Thus the family of zero-sets $\{Z(f), f \in \mathfrak{I}\}$ has the finite-intersection property. Since X is compact, $\cap \{Z(f): f \in \mathfrak{I}\} \neq \varnothing$. Now let $x_0 \in \cap \{Z(f): f \in \mathfrak{I}\}$; then we have $f(x_0) = 0$ for all $f \in \mathfrak{I}$, i.e., $\mathfrak{I} \subset \mathfrak{I}_{x_0}$.

The next proposition characterizes the maximal proper ideals.

Proposition 8. Let X be a compact Hausdorff space and let $\mathfrak{I}$ be a proper maximal ideal of $C(X)$; then there exists $x_0 \in X$ such that $\mathfrak{I} = \mathfrak{I}_{x_0}$. Hence every maximal ideal in $C(X)$ is closed.

Proof. Since $\mathfrak{I}$ is a proper ideal, by the previous proposition there exists $x_0 \in X$ such that $\mathfrak{I} \subset \mathfrak{I}_{x_0} \subset C(X)$. Since $\mathfrak{I}$ is a proper maximal ideal, $\mathfrak{I} = \mathfrak{I}_{x_0}$. Since $\mathfrak{I}_{x_0} = \varphi_{x_0}^{-1}(0)$ and since each φ_{x_0} is continuous, it follows that $\mathfrak{I}_{x_0}$ is closed. (This fact is also easy to verify directly.)

Now we show that every nonzero real or complex homomorphism of $C(X)$ arises from a point of X.

Theorem 6. Let X be a compact Hausdorff space. Then every nonzero real or complex homomorphism φ of the Banach algebra $C(X)$ is φ_x for some

$x \in X$, where φ_x is the evaluation map, i.e., $\varphi_x(f) = f(x)$ for $f \in C(X)$. Thus every homomorphism of $C(X)$ is continuous.

Proof. For each homomorphism $\varphi: C(X) \to R$ or C, we see that $\varphi^{-1}(0)$ is a maximal ideal of $C(X)$. Hence there exists $x_0 \in X$, by Proposition 8, such that $\varphi^{-1}(0) = \mathfrak{I}_{x_0}$. Now to show that $\varphi = \varphi_{x_0}$, let $f \in C(X)$. Assume $f(x_0) = a$. Then clearly $(f - a)$ vanishes at x_0 and so $f - a \in \mathfrak{I}_{x_0} = \varphi^{-1}(0)$. This means that $\varphi(f) = \varphi(a) = a\varphi(1) = a = f(x_0) = \varphi_{x_0}(f)$ for $f \in C(X)$. This completes the proof.

Thus we have shown that every $\varphi \in M$ is continuous and $\varphi = \varphi_{x_0}$ for some $x_0 \in X$. Now we prove:

Theorem 7 (Gelfand–Kolmogoroff). Let X be a compact Hausdorff space and let M be the set of all nonzero real or complex homomorphisms of $C(X)$. Then the identification map: $x \to \varphi_x$ of X onto M is a homeomorphism, where M is given the product topology induced from $R^{C(X)}$ or $C^{C(X)}$ and φ_x is the evaluation homomorphism. (Observe that each element in M is given by φ_x for some x.)

Proof. Let $\eta: x \to \varphi_x$ denote the mapping. Since X is compact, it is enough (Proposition 12, §35) to show that η is one-to-one, onto, and continuous. Let $x_1 \neq x_2$; then because X is compact and so completely regular, there exists $f \in C(X)$ such that $f(x_1) \neq f(x_2)$. But this means that $\varphi_{x_1}(f) \neq \varphi_{x_2}(f)$. Hence η is one-to-one. Clearly η is onto. To show the continuity at $x_0 \in X$, let $T = \{\varphi_x \in M: |\varphi_x(f) - \varphi_{x_0}(f)| < \varepsilon\}$ (for a fixed f) be a subbase neighborhood of $\eta(x_0) = \varphi_{x_0}$. Then

$$\eta^{-1}(T) = \{x \in X: |f(x) - f(x_0)| < \varepsilon\} = f^{-1}(U),$$

where

$$U = \{z \in R \text{ or } C: |z - f(x_0)| < \delta\}$$

and δ is determined by the continuity of f. Hence $f^{-1}(U)$ is a neighborhood of x_0 and so is $\eta^{-1}(T)$. Thus η is continuous and hence it is a homeomorphism (Proposition 12, §35).

Now we describe the topological isomorphisms of $C(X)$ onto $C(Y)$ with the homeomorphisms of X and Y.

Theorem 8. Let X, Y be two compact Hausdorff spaces and let $h: X \to Y$ be a homeomorphism of X onto Y. Then h induces an algebra homomorphism h^* from $C(Y)$ onto $C(X)$ which is a homeomorphism and isomorphism.

Conversely, let g be any continuous nonzero algebra homomorphism of $C(Y)$ onto $C(X)$; then there exists a unique continuous map $h: X \rightarrow Y$ such that $h^* = g$, and if g is a continuous isomorphism onto then h is a homeomorphism.

Proof. For each $f \in C(Y)$, define $h^*(f) = f \circ h$, where h is a homeomorphism of X onto Y. Then it is clear that h^* is a continuous algebra homomorphism of $C(Y)$ into $C(X)$. Since h is one-to-one and onto, so is h^*. Hence h^* is also a homeomorphism because of the open mapping theorem (§52, Theorem 15), since $C(X)$, $C(Y)$ are Banach spaces. For the converse, let g be an algebra homomorphism of $C(Y)$ into $C(X)$. Then $g(1_Y) = 1_X$, where 1_X (or 1_Y) is the identity of $C(X)$ [or $C(Y)$]. Let M_X and M_Y denote the sets of all nonzero complex homomorphisms of $C(X)$ and $C(Y)$ respectively. We know that $M_X \subset [C(X)]'$ and $M_Y \subset [C(Y)]'$. We define a map g^* from M_X to M_Y by: $g^*(\varphi) = \varphi \circ g$, for $\varphi \in M_X$. If $f_1, f_2 \in C(X)$, then

$$g^*(\varphi)(f_1 f_2) = \varphi\big(g(f_1 f_2)\big) = \varphi\big(g(f_1)\big)\varphi\big(g(f_2)\big) = g^*(\varphi)(f_1)(g^*(\varphi)(f_2)$$

shows that $g^*(\varphi) \in M_Y$ if g is continuous. Thus g^* maps M_X into M_Y. It is clear that $g^*: M_X \rightarrow M_Y$ is the restriction of the adjoint map (Exercise 12, Chapter IX). $g^*: [C(X)]'^\sigma \rightarrow [C(Y)]'^\sigma$, which is an isomorphism and homeomorphism if $g: C(Y) \rightarrow C(X)$ is a continuous isomorphism. Thus its restriction $g^*: M_X \rightarrow M_Y$ is $1:1$, onto, and continuous. Since M_X, M_Y are w*-compact, g^* is a homeomorphism of M_X onto M_Y. Thus the required homeomorphism $h: X \rightarrow Y$ is obtained by looking at the following diagram:

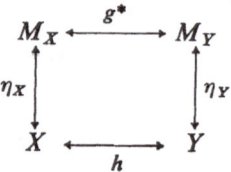

viz., $h = \eta_Y^{-1} \circ g^* \circ \eta_X$. It is clear that the induced (by h) algebra homomorphism $h^* = g$.

Corollary 11. Let X, Y be compact Hausdorff spaces and $C(X)$, $C(Y)$ their respective Banach algebras of continuous functions. Then $X \simeq Y$ (homeomorphic) iff $C(X)$ and $C(Y)$ are algebraically isomorphic and topologically homeomorphic.

Proof. The proof follows immediately from Theorem 8.

§67. Separability in $C(X)$

Let X be a completely regular topological space and $C(X)$ the set of real or complex-valued continuous functions on X with the compact–open topology. It is often of interest to know when $C(X)$ is separable, i.e., it contains a countable dense subset, and reflexive, i.e., its bidual $[C(X)]''$ coincides with $C(X)$ pointwise and topologically, where $[C(X)]''$ is given the compact–open topology if we regard $[C(x)]'' \subset R^{[C(x)]'}$.

Proposition 9. Let $\{f_n\}$ be a countable subset of $C(X)$ such that for any pair x, $y \in X$, $x \neq y$, there exists $f_n \in \{f_n\}$ such that $f_n(x) \neq f_n(y)$. Let $\mathscr{C}$ be the coarsest topology on X which makes each f_n continuous as a map from $(X, \mathscr{C})$ into R or C. Then $(X, \mathscr{C})$ is a separable metrizable space.

Proof. For each $x \in X$, we consider the map $\sigma \colon x \to (f_n(x))_{n \geq 1}$ of X into R^N or C^N. Then σ is a homeomorphism into: because σ is one-to-one (since f_n is separating), continuous (because each f_n is continuous), and open (because $\mathscr{C}$ is the coarsest topology which makes each f_n continuous). Since the countable product R^N (or C^N) is a metric space, and $(X, \mathscr{C})$ is homeomorphic with a separable subspace of R^N, the proposition follows.

Theorem 9. Let X be a compact Hausdorff space. Then $C(X)$ with the norm topology is separable if and only if X is metrizable.

Proof. *"Only if" part*: Suppose $C(X)$ is separable. Let $\{f_i\}$ be a countable dense subset of $C(X)$. Then for x, $y \in X$, if $f_i(x) = f_i(y)$ for all $i \geq 1$, then by denseness we have $f(x) = f(y)$ for all $f \in C(X)$. Since X is a compact Hausdorff space (hence completely regular), $C(X)$ separates points and therefore $x = y$. Now let $\mathscr{C}'$ be the coarsest topology on X which makes each f_i continuous. Then clearly $\mathscr{C}'$ is Hausdorff and metrizable because $\{f_i\}$ is a countable dense subset. Since each f_i is continuous with respect to the initial topology $\mathscr{C}$ on X, we have $\mathscr{C} \supset \mathscr{C}'$. Since X with $\mathscr{C}$ is compact, we have $\mathscr{C} = \mathscr{C}'$ and so X is metrizable.

"If" part: Suppose X is metrizable. Then X is a compact metric space. We denote the metric of X by d. For each n, the open balls of diameter $1/n$ form an open covering of X. Since X is compact, only a finite number of open balls cover X. We denote this finite open covering of X by $\mathscr{U}_n$ for each $n \geq 1$. Let $\{f_{n,i}\}$ denote the partition of unity subordinated to $\mathscr{U}_n$ for each $n \geq 1$ (§43, Definition 17, Theorem 23). This means that each $f_{n,i}$ is a contin-

uous map from X into $[0, 1]$ and $\sum_i f_{n,i}(x) = 1$ for each n, $x \in X$. Since X is compact each $f_{n,i}$ is uniformly continuous. We show that the linear combinations of $f_{n,i}$ with rational coefficients are dense in $C(X)$. This will prove that $C(X)$ is separable, because the set of linear combinations of $\{f_{n,i}\}$ with rational coefficients is countable. Let $f \in C(X)$ and $\varepsilon > 0$ be given. Since f is continuous and X compact, f is uniformly continuous. Thus there is a member $\mathscr{U}_n$ of the open coverings in $\{\mathscr{U}_n\}$ such that for all $x, y \in X$, $d(x, y) < 1/n$, we have $|f(x) - f(y)| < \varepsilon$. We choose $x_{n,i} \in U_{n,i} \in \mathscr{U}_n$ (i finite); then $|f(x) - f(x_{n,i})| < \varepsilon$ for each i. Put $a_{n,i} = f(x_{n,i})$ and let $g_n = \sum_i a_{n,i} f_{n,i}$. Then for $x \in X$, we have

$$|f(x) - g_n(x)| = \left| \sum_i f(x) f_{n,i}(x) - \sum_i a_{n,i} f_{n,i}(x) \right|$$
$$\leq \sum_i |f(x) - f(x_{n,i})| \, f_{n,i}(x) < \varepsilon$$

(because $\sum_i f_{n,i}(x) = 1$). This shows that finite linear combinations of $\{f_{n,i}\}$ with real coefficients are dense in $C(X)$. Since the rational numbers are dense in the set of real members, we have shown that the linear combinations of $\{f_{n,i}\}$ with rational coefficients are dense in $C(X)$ and so $C(X)$ is separable.

Corollary 12. Let $X = [0, 1]$; then $C(X)$ is a separable Banach algebra.

Proof. This follows from Theorem 9, since $[0, 1]$ is a compact metric space.

Theorem 10. Let X be a completely regular space. Then $(C(X), \mathscr{E}_c)$ is separable if and only if the topology of X is finer than a separable metrizable topology on X.

Proof. Suppose $C(X)$ is separable. Let $\{f_n\}$ be a countable dense subset of $C(X)$. Let $\mathscr{E}'$ be the coarsest topology on X which makes each f_n continuous and let $\mathscr{E}$ be the initial topology on X. Since for any pair $x, y \in X$, $x \neq y$, there exists a $f_n \in \{f_n\}$ such that $f_n(x) \neq f_n(y)$ [because otherwise $f_n(x) = f_n(y)$ for all $n \geq 1$ implies $f(x) = f(y)$ for all $f \in C(X)$ since $\{f_n\}$ is dense, which is a contradiction due to the complete regularity of X] by Proposition 9, $\mathscr{E}'$ is a metrizable topology such that $(X, \mathscr{E}')$ is separable and $\mathscr{E} \supset \mathscr{E}'$. Thus the "only if" part follows.

For the "if part", assume that $\mathscr{C}'$ is a separable metrizable topology on X which is coarser than the initial topology $\mathscr{C}$ on X. Since $\mathscr{C}'$ is a separable metrizable topology on X, $(X, \mathscr{C}')$ is homeomorphic with a subspace M of a countable product I^N of closed bounded unit intervals (Theorem 9, §19). Since I^N is a compact metric separable space, by Theorem 9, $C(I^N)$ is a separable metric space with a countable dense subset, $\{h_n\}$.

First we show that the set of the restriction maps $\{h_n \mid M\}$ is dense in $C(M)$. Let $h \in C(M)$ and K a compact subset of M. Then $h \mid K$ has an extension $h' \in C(I^N)$ (Theorem 1, §61). Since $\{h_n\}$ is dense in $C(I^N)$, given $\varepsilon > 0$ there exists h_n such that $\mid h'(z) - h_n(z) \mid < \varepsilon$ for all $z \in K$. Since $h'(z) = h(z)$ for all $z \in K \subset M$, it follows that $C(M) \simeq C(X')$ is separable, where $X' = (X, \mathscr{C}')$.

Now we show that $C(X)$ is separable. Let $\{f_n\}$ be a countable dense subset of $C(X')$. We show that $\{f_n\}$ is dense in $C(X)$. Let $f \in C(X)$, K a compact subset of X, and $\varepsilon > 0$. Since $\mathscr{C} \supset \mathscr{C}'$, K is $\mathscr{C}'$-compact and $\mathscr{C}$, $\mathscr{C}'$ coincide on K. Clearly $f \mid K$ is continuous with respect to $\mathscr{C}$ and $\mathscr{C}'$. Let g be a function in $C(X')$ which extends $f \mid K$ (Theorem 1, §61). But since there exists $f_n \in \{f_n\}$ such that $\mid g(x) - f_n(x) \mid < \varepsilon$ for $x \in K$ implies $\mid f \mid K(x) - f_n(x) \mid < \varepsilon$ for all $x \in K$ and so $\mid f(x) - f_n(x) \mid < \varepsilon$ for all $x \in K$, which shows that $\{f_n\}$ is dense in $C(X)$. This completes the proof.

Theorem 11. Let X be a completely regular space which is the union of a countable family of compact metrizable subsets; then $(C(X), \mathscr{C}_c)$ is metrizable.

Proof. Let $\{K_n, n \geq 1\}$ be a countable family of compact metrizable subsets of X such that $X = \bigcup_{n=1}^{\infty} K_n$. Since the finite union of compact metrizable spaces is also metrizable, we may assume that $\{K_n\}$, in addition, is an increasing sequence. By Theorem 9, $C(K_n)$ is separable for each n. Let $\{g_{m,n}\}_{m \geq 1}$ be a countable dense subset of $C(K_n)$ for each $n \geq 1$. Let $h_{m,n}$ denote the extension of $g_{m,n}$ from K_n to X by the extension theorem (Theorem 1, §61). Let $\mathscr{C}'$ be the coarsest topology which makes $h_{m,n}$ continuous. Suppose $x, y \in X$, $x \neq y$; then $x, y \in K_n$ for some $n \geq 1$. Since $\{g_{m,n}\}_{m \geq 1}$ is dense in $C(K_n)$, there exists m such that $g_{m,n}(x) = h_{m,n}(x) \neq h_{m,n}(y) = g_{m,n}(y)$. Hence by Proposition 9, $\mathscr{C}'$ is a separable metrizable topology which is coarser than the initial topology on X. Thus by Theorem 10, $(C(X), \mathscr{C}_c)$ is separable.

Remark. In particular, $(C(R), \mathscr{C}_c)$ is a metrizable separable complete locally convex algebra.

§68. $C(X)$ Spaces for Completely Regular Spaces X

Let X be a Hausdorff topological space and $C(X)$ the set of all real- or complex-valued continuous functions on X, endowed with the compact–open topology. We have seen that $C(X)$ is a locally convex Hausdorff commutative topological algebra, where the family of seminorms, $\{p_K\}$, defining the locally convex topology, is given by: for $f \in C(X)$,

$$p_K(f) = \sup \{|f(x)|: \ x \in K\},$$

in which K runs over all compact subsets of X. We first determine closed ideals of the algebra $C(X)$.

Notation. Let A be a subset of X. We denote

$$\mathfrak{J}_A = \{f \in C(X): f(A) = 0\} \subset C(X).$$

Proposition 10. If A is a closed subset of X, then $\mathfrak{J}_A$ is a closed ideal of $C(X)$.

Proof. We have to show first that $\mathfrak{J}_A \times C(X) \subset \mathfrak{J}_A$. Let $f \in \mathfrak{J}_A$ and $g \in C(X)$; then $fg(A) = f(A)g(A) = 0$ implies $fg \in \mathfrak{J}_A$. To show that $\mathfrak{J}_A$ is closed, let $f \in \bar{\mathfrak{J}}_A$. Then there is a net $f_\alpha \in \mathfrak{J}_A$ such that $f_\alpha(K) \to f(K)$ for any compact subset K of X. Since A is closed, $A \cap K$ is also compact and therefore $f_\alpha(A \cap K) \to f(A \cap K)$ shows that $f(A \cap K) = 0$, since $f_\alpha(A \cap K) = 0$. From this we conclude that $f(A) = 0$, because if not, then there exists $x_0 \in A$ such that $f(x_0) \neq 0$. But $f_\alpha(x_0) = 0$ implies that $f_\alpha(x_0) \not\to f(x_0)$ on the compact subset $\{x_0\}$ a contradiction. Hence $f \in \mathfrak{J}_A$ and $\mathfrak{J}_A$ is closed.

Recall that if X is compact, then there is a one-to-one correspondence between maximal ideals of $C(X)$ and singletons of X, as shown in Proposition 8, §66. Actually we can improve upon this. But first we have the following:

Proposition 11. The correspondence of closed subsets A of X and closed ideals $\mathfrak{J}_A$ is one-to-one if and only if X is completely regular. [Hence for a completely regular space X, there is a one-to-one correspondence between closed subsets of X and closed ideals of $C(X)$.]

Proof. Suppose X is completely regular and let $\mathfrak{J}$ be a closed ideal of $C(X)$. Put $A = \{x \in X: f(x) = 0 \text{ for all } f \in \mathfrak{J}\} = \bigcap_{f \in \mathfrak{J}} Z(f)$. Then A is a closed subset of X because each $Z(f)$ is closed by continuity of f. (A is possibly empty.) Clearly $\mathfrak{J} \subset \mathfrak{J}_A$. Observe that $\mathfrak{J}_A$ is a closed algebra. Let K be any compact subset of X. Let $\mathfrak{J}' = \{f' \in C(K): f' = f \mid K, f \in \mathfrak{J}\}$ and $\mathfrak{J}_A' = \{f' \in C(K): f' = f \mid K, f \in \mathfrak{J}_A\}$. Since K is compact, $\mathrm{Cl}_c \mathfrak{J}'$ in $C(K)$ is an ideal of functions which vanish on $A \cap K$ and hence $\mathfrak{J}_A' \subset \mathrm{Cl}_c \mathfrak{J}'$ (Corollary 6, §64). This shows that $\mathfrak{J}$ is dense in $\mathfrak{J}_A$. Since $\mathfrak{J}$ and $\mathfrak{J}_A$ are closed, we have $\mathfrak{J} = \mathfrak{J}_A$. Further, $\mathfrak{J}_A \neq \mathfrak{J}_B$ if and only if $A \neq B$, because X is completely regular.

For the converse, suppose $A \to \mathfrak{J}_A$ is one-to-one, i.e., $\mathfrak{J}_A = \mathfrak{J}_B \Leftrightarrow A = B$. If X is not completely regular, then there is a closed subset $A \subset X$ and a point $x \in X \setminus A$ such that for each $f \in C(X)$, $f(A) = 0$, $f(x) = 0$. Put $B = A \cup \{x\}$. Then $\mathfrak{J}_A = \mathfrak{J}_B$ but $A \neq B$. This completes the proof of the converse.

Corollary 13. Let X be a completely regular space. Then the closed proper maximal ideals are in one-to-one correspondence with $\mathfrak{J}_{\{x\}}$, where $\{x\}$ is a singleton of X.

Proof. Since $\{x\}$ is closed, $\mathfrak{J}_{\{x\}}$ is a closed maximal ideal. Suppose $\mathfrak{J}$ is a closed proper maximal ideal of $C(X)$ such that $f(x) = 0$ for all $f \in \mathfrak{J}$ and for some $x \in X$. Then clearly $\mathfrak{J} \subset \mathfrak{J}_{\{x\}} \subset C(X)$. Since $\mathfrak{J}$ is a proper maximal ideal, we have $\mathfrak{J} = \mathfrak{J}_{\{x\}}$.

Corollary 14. Let X be a completely regular space and $C(X)$ the locally convex algebra with the compact–open topology. Let M denote the set of all nonzero continuous multiplicative linear functionals. Then there is a one-to-one correspondence between the points of X and M.

Proof. For each $x \in X$, $\varphi_x(f) = f(x)$ $[f \in C(X)]$ is an element of M. The map: $x \to \varphi_x$ of X into M is one-to-one. By Corollary 13, the map $x \to \mathfrak{J}_{\{x\}}$ is one-to-one. Now let $\varphi \in M$. Then it is easy to check that $\varphi^{-1}(0)$ is a proper closed maximal ideal in $C(X)$, and hence given by an element of X by Corollary 13. The proof is thus completed.

More specifically, Corollary 14 tells us that each real or complex continuous homomorphism of $C(X)$, when X is completely regular, arises from the points of X. For future reference, we state this fact in the following:

Proposition 12. Let X be a completely regular space. Then each real or complex continuous algebra homomorphism on $C(X)$ is defined by $\varphi_x(f) = f(x)$ for all $f \in C(X)$ and some $x \in X$.

§69. Characterization of Banach and Fréchet Spaces $C(X)$

We assume throughout this section that X is a completely regular space and $C(X)$ denotes the space of all continuous real- or complex-valued functions on X with the compact–open topology. We wish to establish the necessary and sufficient conditions on X for $C(X)$ to be a specified topological vector space. First of all, we give a necessary and sufficient condition for $C(X)$ to be a Banach space, and then a complete topological vector space.

Theorem 12. $(C(X), \mathcal{C}_c)$ is a Banach algebra if and only if X is compact.

Proof. If X is compact, we have already shown that $C(X)$ is a Banach algebra (§64, Corollary 3). (Observe that $\mathcal{C}_c$ coincides here with the norm topology).

Assume $C(X)$ is a Banach algebra with the identity 1 (the constant function equal to 1 on X). We have seen that there is a one-to-one correspondence between X and the set M of all nonzero real or complex homomorphisms of $C(X)$ (§66, Theorem 6) which are continuous on $C(X)$ and are given by elements $x \in X$, i.e., $x \to \varphi_x = \varphi$ is one-to-one. It is easy to verify that $\| \varphi_x \| = 1$. Since M is a w*-closed subset of the unit ball of the Banach space $[C(X)]'$, which is w*-compact by Alaoglu's theorem (§62, Theorem 4), it follows that M is w*-compact. Since the map: $x \to \varphi_x$, where $\varphi_x(f) = f(x), f \in C(X)$, $\varphi_x \in M$, and $x \in X$, is a homeomorphism, when M is given the w*-topology induced from the dual $[C(X)]'$ of $C(X)$, it follows that X is compact. This completes the proof.

Remark. For the "only if" part, one may appeal to Gelfand's theory (see, for example, [46] or [59]).

Theorem 13. $(C(X), \mathcal{C}_c)$ is complete if and only if X is a k_r-space (Definition 10, §40).

Proof. Assume that X is a k_r-space. Observe that R^X (or C^X) with the compact–open topology is complete (cf. Theorem 7, §57), where R or C is the field of real or complex numbers, and $C(X) \subset R^X$ (or C^X). It is therefore enough to show that $C(X)$ is a closed subspace of R^X. Let (f_α) be a net in $C(X)$ such that $f_\alpha \to f \in R^X$, where the convergence is taken in the $\mathcal{C}_c$-topology. Let K be a compact subset of X; then the restriction map $\varrho_K : R^X \to R^K$ on K is clearly continuous. Hence $\varrho_K(f_\alpha) \to \varrho_K(f)$ uniformly on K. Since $\varrho_K(f_\alpha) \in C(K)$, $\varrho_K(f) \in C(K)$ because $C(K)$ is a Banach space. Since K is an arbitrary compact set, $f \mid K$ being continuous implies $f \in C(X)$ because

X is a k_r-space. Hence $C(X)$ is complete. Conversely, assume $(C(X), \mathscr{C}_c)$ is complete. Then each Cauchy net in $C(X)$ is convergent. Also each closed bounded set is complete. Let K be a compact subset of X and let $f \in R^X$ such that $f_K = f \mid K \in C(K)$. First assume that f is bounded on X. Clearly f_K is a bounded real (or complex) function with $\| f_K \|_K = \sup_{x \in K} | f_K(x) | < \infty$. But then there exists (Theorem 1, §61) an extension $\tilde{f}_K$ of f_K to the whole of X such that $\tilde{f}_K(x) = f_K(x)$ for $x \in K$ and $\| f_K \|_K \leq \| \tilde{f}_K \|_X$. If we partially order the family of compact subsets of X by inclusion, then $\{f_K\}$ is clearly a bounded net in $C(X)$ with f as a limit point. Since $C(X)$ is complete, we have $f \in C(X)$. If f is not bounded then for each positive interber n, we define $f_n(x) = f(x)$, if $f(x) \leq n$ and $f_n(x) = n$, if $f(x) > n$. Then $| f_n(x) | \leq n$ for all $x \in X$, i.e., each f_n is a bounded function whose restriction to each compact subset is continuous because f has this property. Clearly $f_n \to f$ in R^X in the $\mathscr{C}_c$-topology and therefore $\{f_n\}$ is a Cauchy sequence in $C(X)$ and hence its closure is in $C(X)$. This proves that $f \in C(X)$ and X is thus a k_r-space.

Recall. Given a completely regular space X, for $x, y \in X$, $x \neq y$, there exists, by definition, a continuous real-valued function $f \in C(X)$ such that $f(x) \neq f(y)$. We show that for any compact subset $K \subset X$ and $x \notin K$, there exists $f \in C(X)$ such that $f(x) \notin f(K)$, in order to indicate the method of proof of the next result.

Proposition 13. Let X be a Hausdorff space such that $C(X)$ is separating and K a compact subset of X such that $x_0 \notin K$. Then there exists $f \in C(X)$ such that $f(x) \notin f(K)$.

Proof. For each $x \in K$, $x \neq x_0$ (Since $C(X)$ is separating), there is a continuous function $f_x: X \to [0, 1]$ such that $f_x(x) = 0$ and $f_x(x_0) = 1$. Let $\varepsilon > 0$ be given with $\varepsilon < 1$. Then $U_x = \{y \in X: | f_x(y) | < \varepsilon\}$ is an open neighborhood of x. As x runs over K, we have an open covering of K. Since K is compact, only a finite subcovering $\{U_{x_i}\}_{i=1}^n$ covers K. Let $f_{x_i}(1 \leq i \leq n)$ be the continuous functions corresponding to x_i. Then we have $f_{x_i}(x_i) = 0$, $f_{x_i}(x_0) = 1$, and $| f_{x_i}(y) | < \varepsilon$ for all $y \in U_{x_i}$. Put $f = \bigwedge_i f_{x_i}$. Then clearly f is a continuous function on X and for each $x \in K \subset \bigcup_{i=1}^n U_{x_i}$ there exists x_i such that $x \in U_{x_i}$ and so

$$0 \leq f(x) \leq f_{x_i}(x) < \varepsilon < 1$$

and

$$f(x_0) = \min_{1 \leq i \leq n} f_{x_i}(x_0) = 1.$$

Thus we have shown that $f(x_0) \notin f(K)$.

Remark. If X is completely regular, then Proposition 13 holds trivially by the definition of complete regularity.

Theorem 14. Let X be a completely regular space and let K_1, K_2 be two compact subsets of X such that $K_1 \cap K_2 = \varnothing$. Then there exists $f \in C(X)$ such that $f(K_1) \neq f(K_2)$.

Proof. By the definition of complete regularity of X, $C(X)$ is separating. Hence for each $x \in K_1$, there exists $f_x \in C(X)$ such that $f_x(x) \notin f_x(K)$ by Proposition 13. We may assume that $f_x(x) = 0$ and $f_x(K) = 1$. Let $0 < \varepsilon < 1$ be given. Then $V_x = \{y \in X : f_x(y) < \varepsilon\}$ is an open neighborhood of x and so $\{V_x : x \in K_1\}$ is an open covering of K_1. Since K_1 is compact, there is a finite subcovering $\{V_{x_i} : x_i \in K_1, \ 1 \le i \le n\}$ which covers K_1. Let f_{x_i} be the function which defines V_{x_i}, $1 \le i \le n$. Put $f = \wedge f_{x_i}$. Then $f \in C(X)$ and for each $x \in K_1 \subset \bigcup_{i=1}^{n} V_{x_i}$, $f(x) \le f_{x_i}(x) < \varepsilon < 1$ and for $x \in K_2$, $f(x) = \min_i f_{x_i}(x) = 1$. This proves that $f(x) \notin f(K_2)$ for all $x \in K_1$ or, in other words, $f(K_1) \neq f(K_2)$.

Definition 3. A topological space X is said to be *hemicompact* if there exists a countable family $\{K_n\}$ of compact subsets of X such that $X = \bigcup_{n=1}^{\infty} K_n$ and each compact subset of X is contained in some K_n. (Sometimes, the family $\{K_n\}$ is said to form a *fundamental system of compact subsets* of X.)

Clearly each locally compact hemicompact space is σ-compact.

Theorem 15. Let X be a completely regular space. Then $\big(C(X), \mathscr{T}_c\big)$ is metrizable if and only if X is hemicompact.

Proof. Suppose X is hemicompact. Let $\{K_n\}$ be a countable family of a fundamental system of compact subsets of X. Let $\mathscr{T}$ denote the $\mathfrak{S}_0$-topology (Definition 5, §57) defined by $\mathfrak{S}_0 = \{K_n\}$. Then $(C(X), \mathscr{T})$ has a countable base of neighborhoods of 0 in $C(X)$. Hence by Corollary 14(b), §32, $(C(X), \mathscr{T})$ is metrizable. Thus to prove that $(C(X), \mathscr{T}_c)$ is metrizable, we show that $\mathscr{T}_c = \mathscr{T}$. First let $\mathfrak{S}$ denote the collection of all compact subsets of X. Then $\mathscr{T}_c$ is an $\mathfrak{S}$-topology. Since $\mathfrak{S} \supset \mathfrak{S}_0$, by the remark following Definition 5, §57, $\mathscr{T}_c \supset \mathscr{T}$. For the reverse inclusion, let

$$T(K, \varepsilon) = \{f \in C(X) : |f(x)| < \varepsilon, \ x \in K\}$$

be a basic $\mathscr{T}_c$-neighborhood of 0 in $C(X)$, in which K is compact and $\varepsilon > 0$.

Since X is hemicompact, there exists $K_n \in \{K_n\}$ such that $K \subset K_n$. Clearly $T(K_n, \varepsilon) \subset T(K, \varepsilon)$ and so $\mathscr{C} \supset \mathscr{C}_c$. [Observe that since $C(X)$ is a topological linear space under $\mathscr{C}$ as well as $\mathscr{C}_c$, it is enough to compare the neighborhoods at 0 in order to compare the corresponding topologies.] Thus we have $\mathscr{C}_c = \mathscr{C}$ and so $(C(X), \mathscr{C}_c)$ is metrizable.

Conversely, assume $(C(X), \mathscr{C}_c)$ is metrizable. We show that X is a hemi-compact space. Clearly $C(X)$ is first countable. Let 0 be the zero function and let $T(K_n, W_n)$ be a countable base of neighborhoods of $0 \in C(X)$, where K_n is compact in X and W_n is an open neighborhood of $0 \in R$ or C. Since $\{T(K_n, W_n)\}$ is a base of neighborhoods of 0, each $T(K, W)$ [for K compact in X and $W = (-\varepsilon, \varepsilon)$, $0 < \varepsilon < 1$, open in R] contains some $T(K_n, W_n)$, i.e., $T(K_n, W_n) \subset T(K, W)$. But this means that $W_n \subset W$ and $K \subset K_n$: For otherwise if $x \in K \backslash K_n$ there is a $f \in C(K)$ by Proposition 13 such that $f(K_n) = 0$ and $f(x) \notin W$, i.e., $f \in T(K_n, W_n)$ but $f \notin T(K, W)$. Further, since $K = \{x\}$ is a compact set, by the previous argument we have $T(K_n, W_n) \subset T(\{x\}, W)$, which implies that $x \in K_n$ and so $X = \bigcup_{n=1}^{\infty} K_n$. This proves that X is hemicompact.

Corollary 15. Let X be a completely regular space. $C(X)$ is a Fréchet space (§31) if and only if X is a hemicompact k_r-space.

Proof. This follows by combining Theorems 13 and 15.

Corollary 16. Let X be a completely regular space. Then $C(X)$ with the compact–open topoogy $\mathscr{C}_c$ is separable and metrizable if and only if X is a countable union of compact sets $\{K_n\}$, each of which is metrizable and each compact subset of X is contained in some K_n.

Proof. $C(X)$ is metrizable if and only if X is hemicompact (Theorem 15). Hence by Theorem 11, §67, it follows that $C(X)$ is separable, because the weak topology (§21, Chapter IV) on X defined by a countable family of compact metric spaces $\{K_n\}$ is metrizable, separable, and coarser than the initial topology of X. For the "only if" part, we know that X is hemi-compact by Theorem 15 and there exists a separable metrizable topology $\mathscr{C}'$ on X coarser than the initial topology $\mathscr{C}_c$ by Theorem 10, §67. But then $\mathscr{C}$ and $\mathscr{C}'$ coincide on each compact subset K of X and hence each compact subset of X is $\mathscr{C}$-metrizable because $\mathscr{C}'$ is metrizable.

Corollary 17. Let X be a locally compact Hausdorff space with the topoloyg $\mathscr{C}$. Then the following statements are equivalent:

(a) $\mathscr{C}$ is separable and metrizable;

(b) X is the union of a countable family of compact metrizable subsets;

(c) $C(X)$ is separable and metrizable.

Proof.   (a) $\Rightarrow$ (b)   Since X is locally compact and second countable, it is paracompact (Exercise 21, Chapter VI) and therefore it is the topological sum of a family of subsets $\{B_\alpha : \alpha \in A\}$ such that B_α is compact and countable at infinity (Exercise 21, Chapter VI). Since X is separable, A must be countable and (b) follows.

(b) $\Rightarrow$ (a)   Let $X^\sim = X \cup \{\infty\}$ denote the one-point compactification of X. Let K be a compact metrizable subset of X; then $K^\sim = K \cup \{\infty\}$ is a compact metrizable subset of $X^\sim$ and hence $X^\sim$ is the union of a countable family of compact metrizable sets. Hence by Corollary 16, $C(X^\sim)$ is separable and so $X^\sim$ is metrizable (Theorem 9, §67). Thus $X^\sim$, being a compact metrizable space, is separable and therefore X is also separable and metrizable.

(b) $\Leftrightarrow$ (c)   follows by Theorems 10 and 11, §67.

§70. Characterization of Locally Convex Spaces $C(X)$

If X is a completely regular space, then $C(X)$ is a locally convex Hausdorff topological vector space in the compact–open topology. As noted before, we may describe this locally convex topology by seminorms: For each compact set $K \subset X$ and $f \in C(X)$, $p_K(f) = \sup\{|f(x)| : x \in K\}$. Then as K runs over the family of all compact subsets of X, we obtain a family of seminorms (which give pseudo-metrics) defining the topology of $C(X)$. Thus the dual $[C(X)]'$ of $C(X)$ consists of many nonzero continuous linear functionals (Corollary 7, §62). First we need a criterion of equicontinuous subsets of $[C(X)]'$ to be used later. But first we define:

Definition 4.   Let $\ell \in [C(X)]'$. The smallest compact subset A of X such that for all $f \in C(X)$ vanishing on A implies $\ell(f) = 0$ is called the *support of ℓ* and is written as *Supp ℓ*. If B is a subset of $[C(X)]'$, by Supp B we mean the set $\overline{\bigcup\{\text{Supp } \ell : \ell \in B\}}$.

Remark.   The fact that Supp ℓ for each $\ell \in C(X)'$ exists is left for the reader to verify.

Now we give the criterion promised above.

Proposition 14. Let X be a completely regular space. A subset $B \subset [C(X)]'$ is equicontinuous if and only if Supp $B = K$ is compact and

$$\sup\{|\ell(f)| : \ell \in B, \ p_K(f) \le 1, \ f \in C(X)\} < \infty.$$

Proof. *"Only if"* part: Since B is equicontinuous, there is a neighborhood U of 0 in $C(X)$ such that $B \subset U^\circ$ (see the proof of Theorem 13, §59), i.e., for some $n > 0$ and some compact subset $K' \subset X$, $B \subset n\{\ell \in C(X)' : |\ell(f)| \le 1, f \in C(X)$ with $p_{K'}(f) \le 1\}$. Let $\ell \in B$ and suppose $f \in C(X)$ vanishes on K'. Then $p_{K'}(\lambda f) = 0 \le 1$ for all $\lambda > 0$. Hence $\lambda |\ell(f)| \le 1$ for $\lambda > 0$ and so $\ell(f) = 0$. This proves that Supp $\ell \subset K'$ and hence Supp $B \subset K'$. Supp B is, therefore, compact. Let $K = $ Supp B and let $f \in C(X)$ with $p_K(f) \le 1$. There is an extension (Theorem 1, §61) $f' \in C(X)$ of the restriction $f_{|K}$ with $p_K(f') \le 1$. Then $|\ell(f')| \le n$ for some $n > 0$ and all $\ell \in B$. On the other hand, $f - f'$ vanishes on K so that $\ell(f) = \ell(f')$ for all $\ell \in B$. Hence

$$\sup\{|\ell(f)| : \ell \in B, \ f \in C(X), \ p_K(f) \le 1\} \le n < \infty.$$

Conversely, suppose

$$r = \sup\{|\ell(f)| : \ell \in B, \ f \in C(X), \ p_K(f) \le 1\}.$$

Then

$$B \subset r\{\ell \in C(X)' : |\ell(f)| \le 1 \text{ for } f \in C(X) \text{ with } p_K(f) \le 1\}.$$

Hence B is contained in the polar of a neighborhood of 0 in $C(X)$ and is therefore equicontinuous (Exercise 13, Chapter IX).

Proposition 15. Let X be a completely regular space and $[C(X)]'$ the dual of $C(X)$. A subset M' of $[C(X)]'$ is w*-bounded if and only if

$$M'^\circ = \{f \in C(X) : \sup\{|\ell(f)| : \ell \in M'\} \le 1\}$$

is absorbing, i.e., for each $f \in C(X)$ there exists $\alpha > 0$ such that $f \in \alpha M'^\circ$.

Proof. Let M' be a subset of $[C(X)]'$. Then for each $f \in C(X)$,

$\sup\{|\ell(f)| : \ell \in M'\} \le \alpha < \infty, \ \alpha \ne 0$ iff $\sup\{|\alpha^{-1}\ell(f)| : \ell \in M'\}$
$\qquad = \sup\{|\ell(\alpha^{-1}f)| : \ell \in M'\} \le 1$ iff $\alpha^{-1}f \in M'^\circ$ or $f \in \alpha M'^\circ$.

This proves that M'° is absorbing iff M' is w*-bounded.

Definition 5. A subset A of X is said to be $C(X)$-*pseudocompact* if each $f \in C(X)$ is bounded on A, i.e., $\sup_{x \in A} |f(x)| < \infty$.

Proposition 16. If a subset M' of $[C(X)]'$ is w*-bounded then Supp M' is $C(X)$-pseudocompact.

Proof. Assume M' is a w*-bounded subset of $[C(X)]'$ and let $A = \operatorname{Supp} M'$. We show that A is $C(X)$-pseudocompact. Since M' is w*-bounded, M'° is absorbing, i.e., for each $f \in C(X)$ there exists $\alpha > 0$ such that $f \in \alpha M'^\circ$. If A is not $C(X)$-pseudocompact there exists $f \in C(X)$ with $\sup_{x \in A} |f(x)| = \infty$. Let $\{U_n\}$ be a decreasing sequence of open sets in X such that $\bigcap_{n=1}^\infty \bar{U}_n = \varnothing$ and $U_n \cap A \neq \varnothing$, where $U_n = \{x \in X : f(x) > n\}$. Choose $\ell_n \in M'$ such that $U_n \cap \operatorname{Supp} \ell_n \neq \varnothing$, for each $n \geq 1$. By taking a subsequence, we may assume that $U_n \cap \operatorname{Supp} \ell_m = \varnothing$ if $n > m$. Due to complete regularity of X and the definition of Supp, there exists $f_n \in C(X)$ with $f_n(X \setminus U_n) = 0$ and $\ell_n(f_n) = 1$. Put $\alpha_1 = 1$ and by induction $\alpha_n = n - \sum_{i=1}^{n-1} \alpha_i \ell_n(f_i)$ for $n \geq 2$. Set $g_n = \sum_{i=1}^n \alpha_i f_i$. Then $\ell_n(g_m) = n$ for $m > n$. Since $\bigcap_{n=1}^\infty \bar{U}_n = \varnothing$, we see that $g_n \to f \in C(X)$. But $\ell_n(g_m) = n$ for all $m \geq n$ implies that $\{\ell_n(f') : n \geq 1\}$ is not a bounded sequence of real or complex numbers. Since $\{\ell_n : n \geq 1\} \subset M'$, it follows that M' is not w*-bounded, which is a contradiction in view of the assumption. Hence A is pseudocompact.

Recall a locally convex Hausdorff space E is said to be a *Mackey space* if each w*-compact, convex, circled subset of its dual E' is equicontinuous. (Exercise 17, Chapter VIII).

Every barreled space, in particular, each Baire, complete metrizable, or Banach space is a Mackey space (see [46]).

Theorem 16. Let X be a completely regular space. Then $C(X)$ in the compact–open topology is a Mackey space if and only if for each w*-compact, convex, circled subset M' of $[C(X)]'$, Supp M' is compact.

Proof. Suppose $C(X)$ is a Mackey space. Let M' be a w*-compact, convex, circled subset of $[C(X)]'$. Then M' is equicontinuous. Hence by Proposition 14, Supp M' is compact.

Conversely, assume that for each w*-compact, convex, circled subset M' of $[C(X)]'$, Supp M' is compact. We show that M' is equicontinuous. Put Supp $M' = K$. Since M' is w*-compact and hence w*-bounded, $\sup\{|\ell(f)| : \ell \in M'\} < \infty$ for each $f \in C(X)$. For each $\ell \in M'$, we define a linear functional ℓ^* on $C(K)$ by $\ell^*(h) = \ell(h')$, where h' is a continuous extension (Theorem 1, §61) of h from K to X. ℓ^* is well-defined because if h'' is another extension of h, then $h'' - h' = 0$ on K and so $\ell(h'' - h') = 0$ or $\ell(h'') = \ell(h')$ because Supp $\ell \subset K$. We show that $\ell^* \in [C(K)]'$. Since $C(K)$ is a Banach space because K is compact, it will follow that ℓ^* is bounded if it is bounded on a neighborhood of 0 in $C(K)$. Clearly

$$U = \{f \in C(X) : p_K(f) = \sup_{x \in K} |f(x)| \leq 1\} \subset C(X)$$

induces a neighborhood of 0 in $C(K)$. Since ℓ is continuous on $C(X)$ and Supp $\ell \subset K$, it follows that $\sup\{|\ell(f)| : f \in U\} \leq 1$. This means that $M^* = \{\ell^* : \ell \in M'\}$ is a w*-bounded subset of $[C(K)]'$ and hence norm-bounded by the Banach–Steinhous theorem (Theorem 17, §52) because $C(K)$ is a Banach space. In other words,

$$\sup\{|\ell(f)| : \ell \in M', f \in U\} < \infty.$$

This shows that M' is equicontinuous by Proposition 14, and so $C(X)$ is a Mackey space.

Definition 6. A completely regular space X is said to be a *μ-space* if for each w*-bounded subset M' of $[C(X)]'$, Supp M' is compact.

Theorem 17. Let X be a completely regular space. Then $(C(X), \mathscr{C}_c)$ is barreled (§52) if and only if X is a μ-space.

Proof. Suppose $C(X)$ is barreled. Let M' be a w*-bounded subset of $[C(X)]'$. Then by Proposition 16, Supp M' is $C(X)$-pseudocompact. Since $C(X)$ is barreled, it follows that M' is equicontinuous and so by Proposition 14, Supp M' is compact and so X is a μ-space. For the "if" part, suppose X is a μ-space, i.e., for each w*-bounded subset M' of $[C(X)]'$, Supp M' is compact. If B is a barrel (Definition 16, § 52) in $C(X)$ then by Proposition 15, $B^\circ = \{\ell \in [C(X)]' : \sup_{f \in B} |\ell(f)| \leq 1\}$ is w*-bounded and hence Supp B° is compact, i.e., B° is equicontinuous by Proposition 14, and so $B = B^{\circ\circ}$ (Exercise 13, Chapter IX) is a neighborhood of 0 in $C(X)$. This proves that $C(X)$ is barreled.

We give another characterization for barreledness of $C(X)$ in the following:

Theorem 18. Let X be a completely regular space. Then $(C(X), \mathscr{C}_c)$ is barreled if and only if each $C(X)$-pseudocompact subset of X is relatively compact.

Proof. Let $C(X)$ be barreled. If A is a $C(X)$-pseudocompact subset of X, put

$$B' = \{\ell \in [C(X)]': \text{Supp } \ell \subset A \text{ and } |\ell(f)| \leq 1 \text{ for all } f \in C(X) \text{ with } \sup_{x \in A} |f(x)| \leq 1\}.$$

If $B = \{f \in C(X): \sup_{x \in A} |f(x)| \leq 1\}$, then B is convex, circled, and closed. Also A being $C(X)$-pseudocompact implies B is absorbing (Propositions 15, 16). Hence B is a barrel in $C(X)$ and so a neighborhood of 0 because $C(X)$ is barreled. Hence B' is equicontinuous and therefore Supp $B' = \bar{A}$ is compact by Proposition 14.

Conversely, assume each $C(X)$-pseudocompact subset of X is relatively compact. If B is a barrel in $C(X)$, then $B^\circ = \{\ell \in [C(X)]': |\ell(f)| \leq 1$ for $f \in B\}$ is w*-bounded in $[C(X)]'$ (Exercise 13, Chap. IX). Hence Supp B° is $C(X)$-pseudocompact by Proposition 16. But then by assumption, Supp B° is compact and so B is equicontinuous (Proposition 14). This shows that $B = B^{\circ\circ}$ is a neighborhood of 0. In other words, $C(X)$ is barreled.

Definition 7. A locally convex Hausdorff space E is said to be *countably barreled* if each w*-bounded subset of its dual E' which is a countable union of equicontinuous sets is equicontinuous.

This class of locally convex spaces was introduced by Husain [101]. Clearly each barreled space is countably barreled. But the converse does not hold [102].

Theorem 19 (Morris and Wulbert [102]). Let X be a completely regular space. $(C(X), \mathscr{C}_c)$ is countably barreled if and only if every $C(X)$-pseudocompact subset A of X which is the closure of a countable union of compact sets is actually compact.

Proof. Suppose $C(X)$ is countably barreled. Let A be a $C(X)$-pseudocompact subset of X such that $A = \text{Cl } \bigcup_{n=1}^{\infty} K_n$, where each K_n is compact. Put

$$B_n' = \{\ell \in [C(X)]': \text{Supp } \ell \subset K_n \text{ and }$$

$$|\ell(f)| \leq 1 \text{ for } f \in C(X) \text{ with } \sup_{x \in K_n} |f(x)| \leq 1\}.$$

Since K_n is compact, $B_n = \{f \in C(X): \sup_{x \in K_n} |f(x)| \leq 1\}$ is a neighborhood of 0 in $C(X)$ and so each B_n' is equicontinuous (Proposition 14). Put $B' = \bigcup_{n=1}^{\infty} B_n'$. Since A is $C(X)$-pseudocompact, B' is w*-bounded (Proposition 16). Since $C(X)$ is countably barreled, B' is equicontinuous and therefore A is compact (Proposition 14).

Conversely, assume the condition holds and suppose $C(X)$ is not countably barreled. Then there is a w*-bounded set $B' = \bigcup_{n=1}^{\infty} B_n'$ where each B_n' is equicontinuous but B' is not equicontinuous. Since B_n' is equicontinuous Supp B_n' is compact (Proposition 14) and Supp B' is $C(X)$-pseudocompact (because B' is w*-bounded) but not compact (Proposition 14). Clearly Supp $B' = \text{Cl} \bigcup_{n=1}^{\infty} \text{Supp } B_n' \subset X$ contradicts the condition assumed on X. Hence $C(X)$ is countably barreled.

Recall that a locally convex Hausdorff space is said to be quasibarreled if each barrel which absorbs every bounded subset is a neighborhood of 0 (See Definition 16, §52).

Theorem 20. Let X be a completely regular space. Then $(C(X), \mathscr{C}_c)$ is quasibarreled if and only if for every closed noncompact subset A of X there is a real-valued lower semicontinuous function $g \geq 0$ on X which is bounded on every compact subset of X but unbounded on A.

Proof. Assume $C(X)$ is quasibarreled. Let A be a closed noncompact subset of X. Let $\hat{A}$ denote the image of A under the homeomorphism: $x \to \varphi_x$ of X into $[C(X)]'$, where $\varphi_x(f) = f(x)$ for $f \in C(X)$ (Corollary 14, §68). Since A is not compact, $\hat{A}$ is not a relatively compact subset of $\hat{X}$ and hence not equicontinuous and therefore not strongly bounded, because $C(X)$ is quasibarreled (Theorem 14, §59). Let B be a circled, bounded subset of $C(X)$ such that $\sup\{|\ell(f)| : f \in B, \ell \in \hat{A}\} = \infty$. Set $g(x) = \sup_{f \in B} |f(x)|$. Since B is bounded and circled, $g \geq 0$ is a real-valued lower semicontinuous function such that

$$\sup\{g(x) : x \in A\} = \sup\{|\ell(f)| : f \in B, \ell \in \hat{A}\} = \infty.$$

Thus g is unbounded on A. But if K is a compact subset of X, then, since B is bounded in $C(X)$, we have

$$\|g\|_K = \sup\{\|f\|_K : f \in B\} < \infty.$$

For the converse, let B' be a strongly bounded subset of $[C(X)]'$. Then B' is w*-bounded. To show that B' is equicontinuous, it is enough to show (in view of Proposition 14) that Supp $B' = A$ is compact. Suppose Supp B' is not compact. Let $g \geq 0$ be a real-valued lower semicontinuous function on X which is bounded on every compact subset of X but unbounded on Supp B'. For each positive integer n, let $U_n = \{x \in X : g(x) > n\}$. Then U_n is open and $U_n \cap$ Supp $B' \neq \emptyset$ for all $n \geq 1$. Let K be any compact subset of X. Then g is bounded on K and so $U_n \cap K = \emptyset$ for all but a finite number of n. Since U_n is open, there exists $\ell_n \in B'$ so that Supp $\ell_n \cap U_n \neq \emptyset$. By the definition of Supp, there exists $g_n \in C(X)$, vanishing on $X \setminus U_n$ with $\ell_n(g_n) = 1$. We put $\lambda_1 = 1$ and by induction $\lambda_n = n - \sum_{i=1}^{n-1} \lambda_i \ell_n(g_i)$. Set $h_n = \sum_{i=1}^{n} \lambda_i g_i$. Then $\ell_n(h_n) = n$ and let $M = \{h_n\} n \geq 1$. Suppose for $n > n_0$, $U_n \cap K = \emptyset$; $i = 1$ then for $x \in K$,

$$| h_n(x) | < \sum_{i=1}^{n_0} | \lambda_i | \, | g_i(x) | < \sum_{i=1}^{n_0} | \lambda_i | \sup_{x \in K} | g_i(x) | < \infty.$$

This proves that M is a bounded set in $C(X)$. But since $\ell_i \in B'$ and $\ell_n(h_n) = n$ for all $n \geq 1$, B' is not strongly bounded in $[C(X)]'$, which is a contradiction.

Recall a locally convex Hausdorff space is said to be bornological if every convex set which absorbs each bounded subset is a neighborhood of 0 (Definition 15, §52).

Each metrizable locally convex space is bornological and so is a normed space, in particular. Every bornological space is quasibarreled [45] but the converse need not hold (cf. §59, Chapter VIII). Also recall that every completely regular space is uniformizable (Corollary 1, §27).

Definition 8. A completely regular space X is called a Q-space if for $x \in \beta X \setminus X$ (βX is the Stone–Čech compactification of X) there is a countable set of neighborhoods of x in βX whose intersection is empty.

Proposition 17. The following statements are equivalent for any completely regular space X:

(a) X is a Q-space;

(b) for each $x \in \beta X \setminus X$, there is a sequence of continuous functions on βX into $[-\infty, \infty]$ whose values at x are equal to $+\infty$, but whose restriction to X is finite.

Proof. Easy.

Theorem 21 (Nachbin [67b] and Shirota [75]). Let X be a completely regular space. Then $\big(C(X), \mathscr{E}_c\big)$ is a bornological space if and only if X is a Q-space.

Proof. Assume X is not a Q-space. Then by Proposition 17, there exists $x \in \beta X \setminus X$ such that for each $f \in C(X)$, its unique continuous extension $\beta(f)$ to βX is finite at x, i.e., $\beta(f)(x)$ is finite for every $f \in C(X)$. The mapping $f \to \beta(f)(x)$ is a linear functional which is discontinuous on $C(X)$. But this maps bounded sets into bounded sets because if not, there is a sequence $f_n \in B$ (bounded) in $C(X)$ such that $\beta(f_n)(x) \to \infty$ as $n \to \infty$. Putting

$$U_n = \{y \in \beta X: \; |\beta(f_n)(y)| > |\beta(f_n)(x)| - 1\},$$

we see that U_n is a neighborhood of x in βX and the assumption on X implies that there exists $x_0 \in \bigcap_{n=1}^{\infty} U_n \cap X$. Hence $f_n(x_0) \to \infty$ as $n \to \infty$, contrary to the assumption that B is bounded. Hence the existence of a discontinuous linear functional on $C(X)$ which maps bounded sets into bounded sets implies that $C(X)$ is not bornological (Definition 16′, §52, and Exercise 26(c), Chapter VII). Thus if $C(X)$ is bornological, then X must be a Q-space.

For the converse, assume X is a Q-space. We have shown that $C(X)$ is a vector lattice (Proposition 2, §64). A subset B of $C(X)$ is called order-bounded if there exist $f, g \in C(X)$ such that $f \leq h \leq g$ for all $h \in B$. Every order-bounded subset is bounded in the vector topology but not conversely. Thus to prove that $C(X)$ is bornological it is sufficient to prove that every convex circled subset B which absorbs each order-bounded subset of $C(X)$ is a neighborhood of 0 in $C(X)$.

Let K be a closed subset of X (hence compact in βX) such that for each $f \in C(X)$, if $\beta(f)$ vanishes on K then $f \in B \subset C(X)$. (If so, then we call K a *supporting* set for B.) Indeed, βX is a supporting set for B. Since B absorbs the order-bounded segment $[-e, e]$, where e is the identity function on X, there exists $\lambda > 0$ such that $[-e, e] \subset \lambda B$. Now take $\delta = 1/\lambda > 0$; then for each $f \in C(X)$, $\sup_{x \in K} |f(x)| \leq \delta$ implies that $f \in B$ as above. Now we show that B has a supporting set K contained in X. This will prove that B is a $\mathscr{C}_e$-neighborhood of 0, since K is compact.

We show that K is a supporting set for B if and only if for $f \in C(X)$, $\beta(f) = 0$ in a neighborhood of K in βX implies $f \in B$. If K is a supporting set then it is clear. Now assume that K has the property that $f \in C(X)$ implies $\beta(f)$ vanishes on K. Put $g = \sup(f, \frac{1}{2}\delta e) + \inf(f, -\frac{1}{2}\delta e)$ with $\delta = 1/\lambda > 0$. Since $\beta(2g)$ vanishes on the set $V = \{x \in \beta X: \; |\beta V(x)| < 2\delta\}$, which is a neighborhood of K in βX, we have $2g \in B$. Also $2(f - g) \in \delta[-e, e] \subset B$. Thus $f = \frac{1}{2}\{2(f - g) + 2g\} \in B$.

If K_1, K_2 are two supporting sets for B, then so is $K = K_1 \cap K_2$. Assume $f \in C(X)$ and that $\beta(f)$ vanishes in an open neighborhood W of K in βX. Since K_1 and $K_2 \setminus W$ are disjoint closed sets in βX, we can find disjoint open sets $W_1, W_2, K_1 \subset W_1, K_2 \backslash W \subset W_2, \beta(g) = 1$ on W, and $= 0$ on W_2. Clearly $\beta(g)$ is the extension of some $g \in C(X)$. Now $2gf$ vanishes on $(W \cap W_2) \cap X$ and $W \cap W_2$ is open in βX whereas X is dense in βX. Therefore $2fg \in B$. Similarly $\beta(2f(e - g))$ vanishes on W_1 and so it is in B. Finally $f = \frac{1}{2}[2fg + 2f(e - g)] \in B$ and so K is a supporting set. The intersection of all supporting sets of B is again a supporting set.

The intersection $K(B)$ of all supporting sets of B is a supporting set of B. We show that $K(B) \subset X$. Let $x \in \beta X \backslash X$ and W_n be a decreasing sequence of open neighborhoods of $x \in \beta X$ such that $\bigcap_{n=1}^{\infty} \overline{W}_n \cap X = \varnothing$. Since X is a Q-space one of the sets $\beta X \backslash W_n$ is a supporting set of B, because otherwise there exists $f_n \in C(X)$, such that $f_n \in B$ and $\beta(f_n)$ vanishes on $\beta X \backslash W_n$, for $n \geq 1$. Put $g = \sup_n(n \, |f_n|)$. As $X \backslash \overline{W}_n \subset \beta X \backslash W_m$, so f_m vanishes on $E \backslash \overline{W}_n$ for $m \geq n$, and so g and $\sup\{|f_1|, \ldots, n \, |f_n|\}$ coincide on $E \backslash \overline{W}_n$. Since the sets $\{X \backslash W_n\}$ cover X, we have $g \in C(X)$. Since B absorbs the order-bounded subset $[-g, g]$, there exists $\lambda > 0$ such that $f_n \in \lambda B$, which implies $f_n \in B$, for $n \geq \lambda$, contrary to the assumption. Thus $\beta X \backslash W_n$ is a supporting set for some n. Hence $x \notin K(B)$ and so $K(B) \subset X$. This completes the proof.

Epilogue

In the theory of topological vector spaces and topological algebras there are some important classes of locally convex spaces which form a sub-class of complete locally convex spaces and which include Fréchet spaces. These include, for example, hypercomplete, B-complete, and B_r-complete spaces, each of which includes the previous class (see [45]). But so far there are no theorems asserting that $C(X)$ is hypercomplete, B-complete under the compact–open topology for specific conditions on X. There are still some open questions. However, some partial answers are known, e.g., see Rosa [103] and Summers [104].

Examples and Exercises

1. Show that Theorem 2, §64 is not true for complex-valued functions, using the example given at the end of Theorem 3, §64.

2. Show that every continuous periodic function on $[0, 2\pi]$ is the uniform limit of trigonometric polynomials.

3. Show that each metric space can be embedded into a complete metric space as a dense isometric space and as a subspace of the bounded functions of Lipschitz' kind.

4. Show that each Hausdorff abelian topological group can be embedded into a complete Hausdorff topological group as a dense subgroup.

5. Show that the set $\mathscr{E}$ of all entire functions on the complex plane is a Fréchet space with countable seminorms $\{p_n\}$ given by

$$p_n(f) = \sup_{\|z\| \leq n} |\sum_{n=1}^{\infty} a_i z^i|, \text{ where } f(z) = \sum_{i=1}^{\infty} a_n z^n.$$

6. Show that $\mathscr{E}$ of Example 5 is *not* a Banach space but is a Fréchet locally *m*-convex algebra, i.e., $p_n(fg) \leq p_n(f)p_n(g)$.

7. What are all the continuous complex homomorphisms of $\mathscr{E}$?

8. Let $I = [0, 1]$. Show that I^I is a compact Hausdorff space. Is it metrizable? Is it separable? Is it a second countable space?

9. Let (X, d) be a compact metric space. Show that the set of all isometries of (X, d) into itself is a compact topological group with the point–open topology induced from X^X, which is identical with the metric topology defined by

$$d^*(f_1, f_2) = \sup\{d(f_1(x), f_2(x)) : x \in X\},$$

where f_1, f_2 are isometries.

10. Show that for some completely regular space X, $C(\beta X)$ has complex homomorphisms which are discontinuous on $C(X)$. Also show that the map: $X \to C(X)$ is a contravariant functor from the category of completely regular spaces and continuous maps to the category of TVS's and continuous linear maps.

11. A topological vector space is said to be *quasicomplete* if each closed and bounded subset of it is complete in the induced uniformity. Let X be a completely regular space. Then $(C(X), \mathscr{C}_c)$ is complete iff it is quasicomplete.

12. A topological vector space is said to be *semicomplete* if each Cauchy sequence in it converges. Show that if X is a hemicompact space, then $C(X)$ is complete iff quasicomplete iff semicomplete, where $C(X)$ is endowed with the $\mathscr{C}_c$-topology.

13. Let X be a completely regular space and let $C(X)$ be endowed with the $\mathscr{C}_c$-topology. The following statements are equivalent:

  (a) $C(X)$ is a *Schwartz space* (see [55b]);

  (b) every bounded subset of $C(X)$ is precompact;

  (c) every compact subset of X is finite;

  (d) $C(X)$ has the weak topology $\sigma(C(X), [C(X)]')$;

  (e) $C(X)$ is a dense subset of R^X.

14. For any completely regular space X, the following statements for $C(X)$ are equivalent:

  (i) $C(X)$ is a *Montel* space (i.e., a barreled locally convex space in which each closed bounded set is compact);

  (ii) $C(X)$ is reflexive (i.e., the bidual $[C(X)]''$ of $C(X)$ is iso-morphic and homeomorphic with $C(X)$);

  (iii) $C(X)$ is semireflexive [i.e., $C(X)''$ is only isomorphic with $C(X)$];

  (iv) X is discrete.

15. (a) Let X be a Q-space. Then X has a countable fundamental system of bounded sets (i.e., any bounded set is contained in any one of these countable sets) in $C(X)$ under the embedding $X \rightarrow C(X)$ iff X is compact.

  (b) Show that a completely regular space X is a Q-space if there exists a coarsest uniformity making X complete and each $f \in C(X)$ uniformly continuous (see [105]).

16. Let X be a Q-space. Then X is compact iff $C(X)$ is sequentially complete, quasibarreled, and X is pseudocompact.

17. Let X be a completely regular space. Then $C(X)$ is a *DF-space* (viz., a locally convex space E with a countable fundamental system of bounded sets and such that every strongly bounded subset of E' which is the countable union of equicontinuous sets is itself equicontinuous) iff the union of any countable family of compact subsets of X is relatively compact.

18. A completely regular space X is called *real compact* if every real homomorphism $\varphi: C(X) \rightarrow R$ of the algebra $C(X)$ of all continuous real-valued functions is given by a point of X, i.e., there exists $x \in X$ such that $\varphi(f) = f(x)$ for all $f \in C(X)$.

  (a) Show that X is real compact iff it is a Q-space;

  (b) show that each compact Hausdorff space is real compact;

  (c) is the space $[0, \Omega)$ real compact (see Exercise 38, Chapter VI)?

  (d) show that R^A is real compact, where A is any cardinal number;

  (e) show that each closed subset of a real-compact space is real compact;

(f) let X be a completely regular space and $C(X)$ the space of all continuous real-valued functions on X. Let X be embedded in $\mathbb{R}^{C(X)}$. The closure $v(X)$ of X in $\mathbb{R}^{C(X)}$ is real compact. [$v(X)$ is called the *real compactification* of X.]

Unsolved problems

(a) $C(X)$ is hypercomplete iff X is?

(b) $C(X)$ is B-complete iff X is?

(c) $C(X)$ is B_r-complete iff X is?

(d) $C(X)$ is a B_r-$(\mathscr{C})$-space iff X is?

(e) $C(X)$ is a $B(\mathscr{C})$-space iff X is?

(f) $C(X)$ is countably quasibarreled iff X is?

(For the definition of terms used here see Husain [45]. For some partial answers, see [103] and [104].)

Bibliography

1. A. Abian, *The Theory of Sets and Transfinite Arithmetic*, W. B. Saunders, Philadelphia (1965).
2. P. Alexandroff and H. Hoff, *Topologie I*, Berlin (1935).
3. P. Alexandroff and P. Urysohn, Mémoire sur les espaces topologiques compacts, *Verh. Akad. Wetensch. Amsterdam* **14**, 1–96 (1929).
4. R. Arens, Note on convergence in topology, *Math. Magazine* **23**, 229–234 (1950).
5. R. Arens, Topologies for homeomorphism groups, *Amer. J. Math.* **68**, 593–610 (1946).
6. R. Arens and J. Dugundji, Remark on the concept of compactness, *Portugaliae Math.* **9**, 141–143 (1950).
7. R. Arens and J. Dugundji, Topologies for function spaces, *Pacific J. Math.* **1**, 5–31 (1951).
8. R. H. Bing, Metrization of topological spaces, *Canadian J. Math.* **3**, 175–186 (1951).
9. G. Birkhoff, *Lattice Theory* (revised ed.), *AMS* Colloquium Publ. XXV, New York (1948).
10. G. Birkhoff, Moore–Smith convergence in general topology, *Ann. of Math.* (2) **38**, 39–56 (1937).
11. G. Birkhoff and S. Mac Lane, *A Survey of Modern Algebra* (3rd ed.), Collier–Macmillan, Canada Ltd. (1965).
12. E. Bishop, A general Rudin–Carlson theorem, *Proc. Amer. Math. Soc.* **13**, 140–143 (1962).
13. N. Bourbaki, Topologie générale, *Actualités Sci. Indust.* (Paris), 858(1940); 1084 (1948).
14. N. Bourbaki, Espaces vectoriels topologiques, *Actualités Sci. Indust.* (Paris), 1189 (1953); 1229(1955).
15. N. Bourbaki and J. Dieudonné, Note de tératopologie II, *Revue Scientifique* **77**, 180–181 (1939).
16. K. Borsuk, Über Isomorphie der Funktionalräume, *Bull. Acad. Polonaise*, 1–10 (1933).
17. J. B. Brown, Connectivity, semicontinuity, and the Darboux property, *Duke Math. Journal* **36**, 559–562 (1969).

327

18. A. M. Bruckner and J. G. Ceder, Darboux continuity, *Jahresbericht d. Deutschen Mathem., Vereingung* **67**, 93–117 (1965).

19. C. Brueanu and I. Tevy, Asupra Functiieor Aproape continue in sensue lui H. Blumberg si V. Pták, *Studii Sr. Cer. Math.* **23**, 17–26 (1971).

20. E. Čech, On bicompact spaces, *Ann. of Math.* (2), **38**, 823–844 (1937).

21. C. Chevalley, *Theory of Lie groups*, Princeton (1946).

22. E. W. Chittenden, On the metrization problem and related problems in the theory of abstract sets, *Bull. AMS* **33**, 13–34 (1937).

23. H. J. Cohen, Sur un probléme de M. Dieudonné, *C-R Acad. Sci. Paris*, 234 (1952).

24. L. W. Cohen and C. Goffman, On the metrization of uniform spaces, *Proc. AMS* **1**, 750–753 (1950).

25(a) M. M. Day, Convergence, closure, and neighborhoods, *Duke Math. Journal* **11**, 81–100 (1944).

25(b) M. M. Day, *Normed Linear Spaces*, Academic Press, New York (1962).

26. J. Dieudonné, Une généralization des espaces compacts, *J. Math. Pures App.* **23**, 65–76 (1944).

27. J. Dieudonné, Sur un espace localement compact non métrisable, *Anais de Acad. Bras. Ci.* **19**, 67–69 (1947).

28. J. Dieudonné, Un example d'espace normal non susceptible d'une structure uniform d'espace complet, *C. R. Acad. Sci. Paris* **209**, 145–147 (1939).

29. J. Dieudonné, Sur les espaces uniformes complets, *Ann. Sci. Ecole Norm. Sup.* **56**, 227–291 (1939).

30. J. Dixmier, Sur certains espaces considérés par M. H. Stone, *Summa Brasil. Math.* **2**, 151–182 (1951).

31. C. H. Dowker, An embedding theorem for paracompact metric spaces, *Duke Math. Journal* **14**, 639–645 (1947).

32. C. H. Dowker, On countably paracompact spaces, *Canadian J. Math.* **3**, 219–244 (1951).

33. C. H. Dowker, An estension of Alexandroff's mapping theorem, *Bull. Amer. Math. Soc.* **54**, 386–391 (1945).

34(a) J. Dugundji, An extension of Tietze's theorem, *Pacific J. Math.* **1**, 353–367 (1951).

34(b) J. Dugundji, *Topology*, (2nd ed.), Allyn and Bacon (1965).

35. W. T. (Van) Est and H. Freudenthal, Trennung durch stetige Functionen in topologischen Räumen, *Indagationes Math.* **13**, 359–368 (1951).

36(a) M. K. Fort, Category theorems, *Funda Math* **XLII**, 267–288 (1955).

36(b) M. K. Fort, Points of continuity of semicontinuous functions, *Publ. Math.* **2**, 100–102 (1951).

37. A. A. Fraenkel and Y. Bar-Hillel, *Foundations of Set Theory*, North Holland, Amsterdam (1958).

38. P. Freyd, *Abelian Categories*, Harper and Row, New York (1964).

39. D. Gale, Compact sets of functions and function rings, *Proc. Amer. Math. Soc.* **1**, 303–308 (1950).

40. C. Goffman, On a theorem of Henry Blumberg, *Canadian Math. Bull.* 21–22 (1953).

41. A. Grothendieck, Critères de compacité dans les espaces fonctionnels généraux, *Amer. J. Math.* **74**, 168–186 (1952).

42. P. R. Halmos, *Measure Theory*, Van Nostrand, New York (1950).

43. E. Hewitt, On two problems of Urysohn, *Ann. of Math.* (2) **47**, 503–509 (1946).

44. E. Hewitt, Rings of real-valued continuous functions, I, *Trans. Amer. Math. Soc.* **64**, 45–99 (1948).

45. T. Husain, The open mapping and closed graph theorems in topological vector spaces, *Oxford Math. Monographs* (1965).

46. T. Husain, *Introduction to Topological Groups*, W. B. Saunders, Philadelphia (1966).

47. T. Husain, Almost continuous mappings, *Prace Matematyezne Series 1*, X, 1–7 (1966).

48. T. Husain, $B(\mathscr{C})$-spaces and the closed graph theorem, *Math. Ann.* **153**, 293–298 (1964).

49. T. Husain, Some remarks about real almost continuous functions, *Math. Magazine* **40**, 250–254 (1967).

50. K. Iseki, On definitions of topological spaces, *J. Osaka Inst. Sci. Tech.* **1**, 97–98 (1949).

51. S. Kakutani, Topological properties of unit sphere of a Hilbert space, *Proc. Imp. Acad., Tokyo* **19**, 269–271 (1943).

52. M. Katetov, On *H*-closed extensions of topological spaces, *Casopis Pest. Mat. Fys.* **72**, 17–32 (1947).

53. J. L. Kelley, *General Topology*. Van Nostrand, New York (1955).

54. V. L. Klee, Invariant metrics in groups (solution of a problem of Banach), *Proc. Amer. Math. Soc.* **3**, 483–487 (1953).

55(a) G. Köthe, Über die Vollständigkeit einer Klasse lokal konvexer Räume, *Math. Zeit.* **52**, 627–630 (1950).

55(b). G. Köthe, *Topologische Lineare Räume*, Springer Verlag, Berlin (1966).

56. A. Kolmogoroff, Zur Normierbarkeit eines allgemeinen topologischen linearen Räumes, *Studia Math.* **5**, 29–33 (1934).

57. C. Kuratowski, Topologie I (2nd ed.), Warsaw (1948).

58(a). N. Levine, A decomposition of continuity in topological spaces, *Amer. Math. Monthly* **68**, 44–46 (1961).

58(b). N. Levine, Semiopen sets and semicontinuity in topological spaces, *Amer. Math. Monthly* **70**, 36–41 (1963).

59. L. H. Loomis, *Abstract Harmonic Analysis*, Van Nostrand, New York (1953).

60. Shwu-Yeng T. Lin, Almost continuity of mappings, *Canadian Math. Bull.* **11**, 453–455 (1968).

61. Paul E. Long and E. E. McGehee, Jr., Properties of almost continuous functions, *Proc. Amer. Math. Soc.* **24**, 175–180 (1970).

62. E. Michael, A note on paracompact spaces, *Proc. Amer. Math. Soc.* **4**, 831–838 (1953).

63. E. Michael, Topologies on spaces of subsets, *Trans. Amer. Math. Soc.* **71**, 151–182 (1951).

64. E. H. Moore, Definition of limit in general integral analysis, *Proc. Nat. Acad. Sci., USA* **1**, 628 (1915).

65. E. H. Moore and H. L. Smith, A general theory of limits, *Amer. J. Math.* **44**, 102–121 (1922).

66. L. Nachbin, *Topology and Order* (Math. Studies ♯ 4), Van Nostrand, New York (1965).

67(a). L. Nachbin, *Topological Vector Spaces*, Rio de Janeiro (1948).

67(b). L. Nachbin, Topological vector spaces of continuous functions, *Proc. Nat. Acad. Sci, USA* **40**, 471–474 (1954).

68. J. Nagata, On a necessary and sufficient condition of metrizability, *J. Inst. Polytech. Osaka City Univ.* **1**, 93–100 (1950).

69. J. Novak, Regular space on which every continuous function is constant, *Casopis Pest. Mat. Fys.* **73**, 58–68 (1948).

70. W. J. Pervin and N. Levine, Connected mappings of Hausdorff spaces, *Proc. AMS* **9**, 488–495 (1958).

71. Anand Prakas and P. Srivastava, A note on weak continuity, *J. of Math. Soc., Banaras Hindu Univ.*, **2**, 23–24 (1969).

72. H. Ribeiro, Une extension de la notion de convergence, *Portugaliae Math.* **2**, 153–161 (1941).

73. H. Ribeiro, Caractérisations des espaces réguliers normaux et complément normaux au moyen de l'opération de dérivation, *Portugaliae Math.* **2**, 1–7 (1940).

74. P. Samuel, Ultrafilters and compactifications of uniform spaces, *Trans. Amer. Math. Soc.* **64**, 100–132 (1948).

75. T. Shirota, On systems of structures of a completely regular space, *Osaka Math. Journal* **2**, 131–143 (1950).

76. W. Sierpiński, *General Topology* (2nd ed.), Toronto (1952).

77. M. K. Singal and A. R. Singal, Almost continuous mapping, *Yokohama Math. Journal* **2**, 63–73 (1968).

78. Yu. M. Smirnov, A necessary and sufficient condition for metrizability of a topological space, *Doklady Akad. Nauk SSSR* N.S. **77**, 197–200 (1951).

79. Yu. M. Smirnov, On metrization of topological spaces, *Uspekhi Matem. Nauk* **6**, 100–111 (1957).

80. B. D. Smith, An alternate characterization of continuity, *Proc. AMS* **39**, 318–320 (1973).

81. R. H. Sorgenfrey, On the topological product of paracompact spaces, *Bull. Amer. Math. Soc.* **53**, 631–632 (1947).

82. J. Stallings, Fixed-point theorems for connectivity maps, *Funda Math.* **47**, 249–263 (1959).

83. A. H. Stone, Paracompactness and product spaces, *Bull. Amer. Math. Soc.* **54**, 977–982 (1948).

84. M. H. Stone, The generalized Weierstrass approximation theorem, *Math. Magazine* **21**, 167–184 (1948).

85. M. H. Stone, Applications of the theory of Boolean rings to general topology, *Trans. Amer. Math. Soc.* **41**, 375–481 (1937).

86. J. W. Tukey, Convergence and uniformity in topology, *Ann. of Math. Studies* **2** (1940).

87. A. Tychonoff, Über die topologische Erweiterung von Räumen, *Math. Ann.* **102**, 544–561 (1929).

88. H. Umegaki, On the uniform space, *Tôhôku Math. Journal* (2), 57–63 (1950).

89. P. Urysohn, Über die Machtigkeit der zusammenhagen Mengen, *Math. Ann.* **94**, 262–295 (1925).

90. A. D. Wallace, Separation spaces, *Ann. of Math.* (2) **42**, 687–697 (1941).

91. A. Weil, Sur les espaces a structure uniforme et sur la topologie générale, *Actualités Sci. Industr.* (Paris), 551 (1937).

92. G. T. Whyburn, *Analytic Topology*, AMS Colloquium Publ. XXVII, New York (1942).

93. A. Wilansky, *Functional Analysis*, Blaisdell (1964).

94. E. Zermelo, Neuer Beweis für die Wohlordnung, *Math. Ann.* **65**, 107–128 (1908).

95. R. E. Zink, On semicontinuous functions and Baire functions, *Tran. Amer. Math. Soc.* **117**, 1–9 (1965).

96. H. Blumberg, New properties of all real functions, *Trans. Amer. Math. Soc.* **24**, 113–128 (1922).

97(a). V. Pták, Completeness and the open mapping theorem; *Bull. Soc. Math. France*, **86**, 41–74 (1958).

97(b). V. Pták, On complete topological linear spaces, *Čech. Math. Journal* **78**, 301–364 (1953).

98. J. L. Kelley, I. Namioka *et al.*, *Linear Topological Spaces*, Van Nostrand, Princeton (1963).

99. H. H. Schaefer, *Topological Vector Spaces*, MacMillan, New York (1966).

100. W. Rudin, *Principles of Mathematical Analysis*, McGraw-Hill, New York (1953).

101. T. Husain, Two new classes of locally convex spaces, *Math. Ann.* **166**, 289–299 (1966).

102. P. D. Morris and D. E. Wulbert, Functional representation of topological algebras, *Pacific J. of Math.* **22**, 323–337 (1967).

103. D. Rosa, *On Locally m-convex Function Algebras*, Ph. D. Thesis, McMaster University, Hamilton, Ontario (1974).

104. W. H. Summers, Full completeness in weighted spaces, *Canadian J. Math.* **22**, 1196–1207 (1970).

105. V. A. Efremovic, The geometry of proximity, *Mat. Slovnik*, N. S. 31 (**73**), 189–200 (1952).

106. M. D. Weir, *Hewitt–Nachbin Spaces*, North Holland–American Elsevier, Amsterdam (1975).

107. H. L. Royden, *Real Analysis* (2nd edition), MacMillan, New York (1968).

Index

Absolute complement, 2
Absolute retract, 76
Absorbing set, 129
Accumulation point, 26, 44
Adjoint (or transpose) map, 287
Alaoglu's theorem, 234, 281
Alexander theorem, 151
Alexander—Hoff Trennungsaxioms, 60
Algebra, 16
Algebraic dual, 16, 235
Annihilator, 283
Ascoli's theorem, 252
Associative law, 3
Axioms, separation, 60

Baire-category theorem, 176
Baire space, 175
Balls
  closed, 34
  open, 34
Banach—Schauder fixed-point theorem, 138
Banach—Steinhaus theorem, 228
Barrel, 227
Base
  of filter, 43
  of topology, 30
Bidual, 282
Bishop theorem, 283
Boundary, 27
Bounded set, 159, 227
Brouwer fixed-point theorem, 138

Canonical covering, 269

Canonical map, 93
Cantor set, 101
Cantor's diagonalization process, 40
Cardinal number, 7
Cartesian product, 5, 10
Category, 17
Cauchy filter, 117
Cauchy net, 117
Cauchy sequence, 117
Cell, 269
Closed graph, 199
  theorem, 225
Closed set, 22
Closure, 23
Cluster point, 26
Cofinal net, 37
Commutative law, 3
Compactifications, 186
  real, 324
Compatible proximity, 142
Complete metric space, 119
Complete uniform space, 117
Completion of uniform spaces, 121
Composition of maps, 17, 54
Composition of relations, 5
Continuous extension, 70
Continuous functions, 53
Continuous map, 53
Contraction map, 136
Converge simply, 238
Convergence
  of a filter, 44, 117
  of a net, 37, 117
  of a sequence, 117

Coproduct, 80
Coset, 14
Countable at infinity, 181
Covering, 30
CW-polytope, 270
CW-topology, 270

Derived set, 27
DF-space, 324
Diagonal, 5, 103
Diameter, 159
Differential equations, 140
Direct limit, 97, 98
Direct product, 15
Direct sum of groups, 15
Direct system, 97, 98
Directed set, 36
Distance between sets, 162
Distributive law, 3
Dual
   algebraic, 16, 235
   topological, 234, 278
Dual ideal, 12

Edges, 269
Elements of a set, 1
Equicontinuous maps, 250
Equipotent sets, 7
Equivalence class, 5
Equivalence relation, 5
Evaluation map, 238
Eventually constant net, 38
Extension of maps, 5, 70
Extension theorem, 267
Extremally disconnected spaces, 193

Family of mutually pairwise disjoint sets, 3
Filter, 43
Filter base, 43
Finite intersection property, 149
Fixed-point, 136
Fixed-point theorem, 137
Free union, 83
F-space, 129
Function, 5
Functional, 16
   additive, 274
   homogeneous, 274
   linear, 16, 274
   positive homogeneous, 274
   Radon, 282
   subaddIitive, 274

Functions vanishing at infinity, 291
Functor, 18
   contravariant, 18
   covariant, 18
Fundamental system
   of compact sets, 312
   of neighborhoods, 32

Gauges, 111
Gelfand—Kolmogoroff theorem, 303
General extension theorem, 283
Generalized inductive limit, 99
Graph of map, 5
Group, 14
   abelian or commutative, 14
G_δ-set, 67

Hahn—Banach extension theorem, 274
Heine—Borel theorem, 159
Hilbert cube, 190
Homeomorphism, 58
Homomorphism, 15, 16

Ideal (of an algebra), 15, 17
   left, 15, 17
   right, 15, 17
   two-sided, 15, 17
Ideal of a lattice, 12
Implicit function theorem, 138
Inductive limit, 97
Injective objects, 19
Interior of a set, 23
Intersection of sets, 1
Inverse system, 95
Isometry, 297

Join, 10

Kernel of a homomorphism, 15
k-Extension of a topology, 48
k-Spaces, 171
k_r-Spaces, 175
Kuratowski axioms, 24

Lattice, 10
   complete, 11
   distributive, 11
Lattice-operations, 10
Lebesgue measure, 220
Lebesgue number of coverings, 163
Left translations, 300
Limit inferior, 4

Limit points, 26, 39, 44
Limit superior, 4
Linear functional, 16
Linear map, 16
Linear space, 16
Linearly ordered set, 9
Lipschitz's condition, 138
Locally finite covering, 31
Locally σ-finite covering, 31
Lower bound, 9

Map or mapping, 5
  adjoint, 287
  almost continuous, 196
  almost open, 223
  approximately continuous, 220
  bijective, 6
  Cesaro type, 215
  closed, 57
  continuous, 55
  contraction, 136
  evaluation, 238
  finitely closed, 207
  graphically continuous, 207
  inclusion, 77
  injection, 6
  injective, 6
  inversely compact, 235
  lower semicontinuous, 217, 218
  nearly continuous, 210
  one-to-one (or injective), 6
  onto (or surjective), 6
  open, 57
  perfect or proper, 235
  projection, 10, 83
  quasicontinuous, 232
  semicontinuous, 218
  σ-continuous, 233
  uniformly continuous, 114
  upper semicontinuous, 217, 218
  w-continuous, 213
  w^{-1}-continuous, 215
  weakly continuous, 213
  with closed graph, 199
Maximal element, 9
Maximal ideal, 12
MB-space, 168
Meager set, 175
Meet, 10
Members of a set, 1
Metric, 34
Metric density, 220

Metrically topologically complete, 145
Metrizability
  of groups, 134
  of spaces, 131, 132
  of vector or linear spaces, 134
Minimal Cauchy filter, 122
Monotonic sequence, 4
Morphisms, 17
Morris and Wulbert theorem, 318

Nachbin–Shirota theorem, 320
Neighborhood system, 25
Neighborhoods, 25
Nerve of covering, 270
Net, 36
  Cauchy, 117
  convergent, 117
Norm, 130
Normal subgroup, 14

Objects, 17
One-point compactification, 187
Open map, 57
Open mapping theorem, 225
Operator, 5
Order or ordering, 8, 9
Order-complete, 9

Partial ordering, 8
Partition or decomposition, 93
Partition of unity, 184
Partly ordered set, 9
Peano curve, 101
Polar set, 287
Polytope, 269
Power of a set, 7
Power set, 7
Probability measure, 282
Projection map, 10, 83
Projective limit, 95, 96
Proximity map, 144
Proximity relation, 141
Proximity space, 141
Pseudobase, 30
Pseudometric, 34
P-space, 22

Q-space, 320
Quasicomplete space, 323
Quasimetric, 34
Quasimetrizable space, 34
Quasi-uniform space, 104

Quotient or factor group, 15

Real-compact space, 324
Reducible covering, 30
Refinement of a covering, 30
Relations, 5
  identity, 5
  inverse, 5
  reflexive, 5
  symmetric, 5
  transitive, 5
Relative complement, 2
Restriction of a map, 6
Riesz representation theorem, 283, 287
Right ideal, 15
Right translations, 300
Ring, 15
Rudin's theorem, 285

Schroeder—Bernstein theorem, 7
Semicomplete space, 323
Semigroup, 14
Semimetric, 34
Seminorm, 130
Separating family of pseudometrics, 211
Separation axioms, 60, 61
Separation theorem, 278
Sequence, 37
Sequential continuity, 57
Sequential convergence, 261
Sequential compact space, 157
Sequentially complete space, 118
$\mathfrak{S}$-equicontinuous set, 253
Set
  absorbing, 129
  of all maps, 237
  bounded, 159, 227
  Cantor, 101
  circled, 129
  closed, 22
  compact, 154
  convex, 129
  countable, 7
  dense, 33
  disjoint, 2
  empty, null, or void, 2
  finite, 7
  of first category or meager set, 175
  linearly ordered, 9
  nondense, 175
  open, 21
  relatively compact, 154

Set (cont'd)
  residual, 175
  of second category, 175
  supporting, 321
  symmetric, 103
  uncountable, 8
Sième, 175
Simplex, 269
Singleton, 1
Smirnov compactification, 146
$\mathfrak{S}$-topology, 248
$\mathfrak{S}$-uniformity, 248
Subgroup, 14
Supremum, 9
Symmetric difference, 2
Space
  Baire, 175
  Banach, 130
  barreled, 227
  bornological, 227
  Cantor, 240
  compact, 30, 149
  complete metric, 119
  complete uniform 117
  completely normal, 61
  completely regular, 64
  countably barreled, 318
  countably compact, 157
  countably paracompact, 183
  countably ultratopological, 22
  $C(X)$-pseudocompact, 316
  discrete, 22
  first countable, 32
  Fréchet, 130
  gauge, 111
  Hausdorff, 61
  hemicompact, 312
  Hilbert, 130
  infratopological, 21
  locally compact, 163, 164
  locally convex, 130
  Lindelöf, 30
  MacKey, 265
  MB-, 168
  metacompact, 191
  metric, 34
  metrizable, 34, 131
  minimal, 202
  Montel, 235, 323
  μ-, 317
  normal, 61
  normed, 130

Space (*cont'd*)
  paracompact, 180
  perfectly normal, 67
  precompact, 120
  pre-Hilbert, 130
  proximity, 141
  pseudocompact, 177
  pseudometric, 34
  Q-, 320
  quasibarreled, 227
  quasicomplete, 323
  quotient, 90
  real compact, 324
  reflexive, 323
  regular, 61
  Schwartz, 323
  second countable, 32
  semicomplete, 323
  semimetric, 34
  semireflexive, 323
  separable, 33
  sequentially compact, 157
  Sierpiński, 91
  σ-compact, 166
  supratopological, 21
  T_i- ($i = 1, ..., 6$), 61, 65
  topological, 21
  totally bounded, 120
  Tychonoff, 65
  uniform, 103, 104
  uniformizable, 111
Star of cell, 270
Stone–Čech compactification, 189
Stone–Čech theorem, 189
Stone–Weierstrass theorem
  classical, 296
  complex and real case, 293, 294
Strict inductive limit, 99
Subbase, 30
Sublattice, 11
Subset, 1
Support, 314

Tietze's extension theorem, 70
Topological groups, 126
Topological product, 83
Topological spaces, 21
Topological sum, 80
Topological vector spaces, 128, 129
Topologically complete spaces, 145
Topology, 21
  compact-open, 240

Topology (*cont'd*)
  compatible with duality, 265
  conjoining, 264
  discrete, 22
  finest locally convex, 235
  indiscrete, 22
  of joint continuity, 241
  of joint continuity on compacta, 241
  Mackey, 265
  point-open, 238
  product, 83
  quotient, 90
  relative or induced, 77
  $\mathfrak{S}$-, 248
  splitting, 264
  strong, 258
  uniform, 106
  weak, 80
  weak* or w*-, 234, 258, 279
  Whitehead, 270
Translation-invariant metric, 129
Translations, 300
Transpose of a map, 287
Two-sided ideal, 15
Tychonoff fixed-point theorem, 138
Tychonoff plank, 74
Tychonoff theorem, 86, 153

Ultrafilter, 44
Uniform continuity, 113
Uniform space, 104
Uniformity, 104
  base of, 105
  largest, 105
  smallest, 105
  subbase of, 105
  of uniform convergence, 248
Uniformizable space, 111
Union of sets, 2
Upperbound, 8
Urysohn's lemma, 69
Urysohn's metrization, 132

Vertices, 269

Weak topology
  for linear spaces, 288
  for topological sum, 80
Weierstrass theorem, classical, 421
Well-ordered set, 9
Whitehead topology, 270

Zero set, 302
Zorn's lemma, 9